Advances in Optics and Optoelectronics

The Advances in Optics and Optoelectronics series focuses on the exciting new developments in the fast emerging fields of optics and optoelectronics. The volumes cover key theories, basic implementation methods, and practical applications in but not limited to the following subject areas:

AI Photonics
Laser Science and Technology
Quantum Optics and Information
Optoelectronic Devices and Applications
Fiber-Based Technologies and Applications
Near-infrared, Mid-infrared and Far-infrared Technologies and Applications
Biophotonics and Medical Optics
Optical Materials, Characterization Methods and Techniques
Spectroscopy Science and Applications
Microscopy and Adaptive Optics
Microwave Photonics and Wireless Convergence
Optical Communications and Networks

Within the scopes of the series are monographs, edited volumes and conference proceedings. We expect that scientists, engineers, and graduate students will find the books in this series useful in their research and development, teaching and studies.

Shiyao Fu · Chunqing Gao

Optical Vortex Beams

Fundamentals and Techniques

清华大学出版社
TSINGHUA UNIVERSITY PRESS

Springer

Shiyao Fu
Beijing Institute of Technology
Beijing, China

Chunqing Gao
Beijing Institute of Technology
Beijing, China

ISSN 2731-6009 ISSN 2731-6017 (electronic)
Advances in Optics and Optoelectronics
ISBN 978-981-99-1809-6 ISBN 978-981-99-1810-2 (eBook)
https://doi.org/10.1007/978-981-99-1810-2

Jointly published with Tsinghua University Press
The print edition is not for sale in China (Mainland). Customers from China (Mainland) please order the print book from: Tsinghua University Press.

This Springer imprint is published by the registered company Springer Nature Singapore Pte Ltd.
The registered company address is: 152 Beach Road, #21-01/04 Gateway East, Singapore 189721, Singapore

Introduction

Vortex beams and their applications are one of the hot topics in optics and photonics community in recent years. Since Allen et al. demonstrated that a vortex beam with helical wavefront carries orbital angular momentum (OAM) in 1992, a bridge between microscopic photons and macroscopic laser beams was established. Originally, a vortex beam refers to phase vortex, namely, the helical phase distribution, resulting in a phase singularity at the beam center and a doughnut intensity pattern. Afterwards, vortex in lasers is expanded to polarization degree of freedom. Polarization vortex with inhomogeneous polarization distributions, also known as vector beams, is generated through tailoring both OAM and spin angular momentum (SAM). A polarization singularity is present in the beam center and also leads to a doughnut intensity pattern. Thus, phase and polarization vortex read generally vortex beams. Actually, both phase vortex and polarization vortex result from manipulating angular momenta (SAM and OAM) of photons. The amazing characteristics of vortex beams attach much attentions around the world, and lots of advanced applications are inspired. For instance, large-capacity laser communications, rotation sensing and detection, laser processing, high-resolution imaging, holographic encryption, optical computing, and also quantum scenarios.

Study on the basic principle, generation, mode recognition, and application of vortex beams becomes hotter and hotter especially in the past decades. Researches from USA, China, Great Britain, South Africa, Germany, Australia, and other countries have proposed lots of advanced and favorable works. Our team has been working on vortex beams since more than 20 years ago. Our research focuses more on vortex beam generation and OAM mode recognition. This book mainly introduces advances of our works, and also representative works demonstrated by other institutions. The structure of this book is as follows. Chapter 1 is the fundamental of laser beam propagation, which is the diffraction basis of vortex beams. Chapter 2 introduces the concepts of both vortex beam and angular momentum of photon. Chapter 3 introduces the representative approaches of generating vortex beams, including OAM mode converter, spiral phase plate, fork-shaped gratings, phase-only diffraction elements, geometric phase modulation, integrated OAM emitter, etc. The effective schemes of generating multiplexed OAM carried vortex beam are also present. Chapter 4 further extends the content of Chap. 3, introducing another degree of freedom as wavevector (diffraction order), and thus obtains vortex lattices. Chapter 5 discusses schemes of diagnosing OAM states and OAM spectra of vortex beam, which can be accomplished through interferometers and diffraction gratings. The inhomogenous media like turbulent atmosphere will introduce serious distortions for vortex beams. One must have the ability to correct distortions to meet the demands of practical scenarios. Hence, Chap. 6 introduces the distortion compensation of vortex beams. Chapter 7 further extends the above discussions from phase singularity to polarization singularity, and another vortex beam, polarization vortex or vector beam, is present. The last chapter, Chap. 8, illustrates a special vortex beam, as perfect optical vortex. The "perfect" property is reflected on the beam diameter which is independent of the topological charge. Such beam will bring more possibility for fiber scenarios.

It is worth mentioning and emphasizing that this book is the English translation of the Chinese book, *Vortex Beams*, written by us, published by Tsinghua University Press in 2019. Such book received wide acclaims in China, and recognized by National Foundation of Academic Publishment of Science and Technology of China. Hence, we would like to publish its English version for readers around the world. The past 3 years have actually been a period of rapid development in vortex beam. Many novel vortex fields or vortex packets are demonstrated, for instance, spatiotemporal vortex with transverse OAM toroidal vortices of light, OAM comb, SU (2) vortex, and so on. In addition, artificial intelligence (AI) has been well introduced in vortex beam research. The device for generating vortex beam is becoming smaller and smaller, where the vortex beam can be generated and processed even on a chip, further benefiting the applications on photonic chips. We are very sorry that the above amazing works are not involved in this book because of the restriction of English translation contract.

We would like to express our gratitude to those who have worked on vortex beams and their applications. Their hard efforts create the era of vortex beams and also the derived progress of many practical applications. Professor Horst Weber from Technical University Berlin is greatly acknowledged for his constant supports in this work and his helps in publishing this book. We acknowledge our colleagues and graduate students, Dr. Yidong Liu, Dr. Xiaoqing Qi, Dr. Jingtao Xin, Dr. Kunjian Dai, and Miss Zijun Shang, in our team, for their works on this book. We also acknowledge National Key Research and Development Program of China (2022YFB3607700), National Natural Science Foundation of China (NSFC) (11834001, 61905012, 60778002, 69908001), National Defense Basic Scientific Research Program of China (JCKY2020602C007), National Postdoctoral Program for Innovative Talents (BX20190036), and National Basic Research Program of China (973 Program)

(2014CB340002, 2014CB340004), for funding our study. And finally, we appreciate the efforts of Springer and Tsinghua University Press for publishing this book. 2022 is the 30th anniversary of Allen's pioneering work on OAM and vortex beams. We convince this book will become an effective toolkit for researchers working on vortex beams, and inspire more wonderful ideas.

December 2022

Shiyao Fu
Chunqing Gao

Contents

Chapter 1
Fundamentals of Beam Propagation

Optical vortices are a new kind of structured beams, which have unique characteristics, and have attached more and more attention in recent years. Before studying optical vortices, an important premise is that, one must be clear of the rules of beams' propagation. Hence in this chapter, we mainly discuss the fundamentals of beams propagation, including the Maxwell's equations, wave functions, polarizations and scalar diffraction theory.

1.1 Basis of Electromagnetic Theory of Light

1.1.1 Maxwell's Equations

Electromagnetic field is the ensemble of interacting and alternating electric and magnetic fields. Alternating electromagnetic fields form electromagnetic waves if they propagate according to the laws of electromagnetism. To describe the electromagnetic wave propagating in vacuum and the interaction between electromagnetic waves and media, four vectors, the electric field intensity $\vec{E}$, the magnetic induction intensity $\vec{B}$, the electric displacement $\vec{D}$, and the magnetic field intensity $\vec{H}$ are introduced.

In 1873, Maxwell summarized the fundamental laws of electromagnetism and the experimental results at that time, and put forward the electromagnetic theory of light. From such theory, light is regarded as an electromagnetic disturbance propagating in the form of a field on the basis of the laws of electromagnetism. In other words, light is a kind of electromagnetic wave. The electromagnetic theory of light can be summarized as a set of equations associated with four vectors $\vec{E}$, $\vec{B}$, $\vec{D}$, and $\vec{H}$, and named as Maxwell's equations. They describe the change of the four vectors with time and space. And most of the phenomena related to the propagation and superposition of light and beams can be explained from Maxwell's equations.

Maxwell's equations with integral form read:

S. Fu and C. Gao, *Optical Vortex Beams*, Advances in Optics and Optoelectronics,
https://doi.org/10.1007/978-981-99-1810-2_1

$$\oint_C \vec{E} \cdot \mathrm{d}l = -\iint_A \frac{\partial \vec{B}}{\partial t} \cdot \mathrm{d}s \tag{1.1.1}$$

$$\oiint_A \vec{D} \cdot \mathrm{d}s = \iiint_V \rho \mathrm{d}v \tag{1.1.2}$$

$$\oiint_A \vec{B} \cdot \mathrm{d}s = 0 \tag{1.1.3}$$

$$\oint_C \vec{H} \cdot \mathrm{d}l = \iint_A \left(\vec{J} + \frac{\partial \vec{D}}{\partial t}\right) \cdot \mathrm{d}s \tag{1.1.4}$$

Equation (1.1.1) is the integral form of Faraday's law of electromagnetic induction, the right hand of which gives the change rate of magnetic flux over time through any surface A in space. The ring integral of electric field $\vec{E}$ along the curve C in the left hand denotes the induced electromotive force. Equation (1.1.1) implies that, changing magnetic fields produce electric fields. The minus in the right hand means the induced electromotive force has the tendency of impeding the change of magnetic fields.

Equation (1.1.2) is Gauss's law of electric fields. The integral of the charge density ρ in the right hand denotes the total electric charge in volume V. And the integral of electric displacement $\vec{D}$ in the left hand denotes the electric flux through closed surface A. Equation (1.1.2) implies that, the electric flux flowing out of the volume V through the closed surface A equals to the total free charges number in the space surrounded by surface A. The negative total charge number means electric flux flowing into the volume V.

Equation (1.1.3) is Gauss's law of magnetic fields, which implies the magnetic flux flowing in the closed surface A is identical with the one flowing out.

Equation (1.1.4) is Maxwell & Ampere's law, also known as the law of total current, which is an extension of Ampere's law of steady current. In Eq. (1.1.4), $\vec{J}$ is the current density, and $\int_A \vec{J} \cdot \mathrm{d}s$ denotes the conducting current strength through surface A. The variation of electric field $\partial \vec{D}/\partial t$ is displacement current density. Hence the integral $\int_A \partial \vec{D}/\partial t \cdot \mathrm{d}s$ denotes the displacement current strength through surface A. Equation (1.1.4) implies the flowing charges create circular magnetic field.

For the sake of solving the problem of electromagnetic field vector of a fixed point, the differential form of Maxwell's equations is essential. Such differential form can be derived from Eqs. (1.1.1)–(1.1.4) through Green's theorem and Stokes' theorem as:

$$\nabla \times \vec{E} = -\frac{\partial \vec{B}}{\partial t} \tag{1.1.5}$$

$$\nabla \cdot \vec{D} = \rho \tag{1.1.6}$$

$$\nabla \cdot \vec{B} = 0 \tag{1.1.7}$$

$$\nabla \times \vec{H} = \vec{J} + \frac{\partial \vec{D}}{\partial t} \tag{1.1.8}$$

with ∇ the Hamiltonian operator.

In the differential form of Maxwell's equations, Eq. (1.1.5) implies the flux density change at a point in space creates circular electric fields. Equation (1.1.6) implies the electric displacement vector diverges outward from positive charge, or converges to negative charge. Equation (1.1.7) implies that magnetic fields are passive. And Eq. (1.1.8) implies both the conducting and displacement current contribute to circular magnetic fields.

To describe the universal law of electromagnetic fields, another set of equations associated with the medium in which the electromagnetic field resides, need to be introduced. And such equations called matter equations:

$$\vec{D} = \varepsilon \vec{E} \tag{1.1.9}$$

$$\vec{H} = \mu^{-1} \vec{B} \tag{1.1.10}$$

$$\vec{J} = \kappa \vec{E} \tag{1.1.11}$$

In Eq. (1.1.9), ε denotes permittivity. Equation (1.1.9) gives the relationship between the electric field intensity $\vec{E}$ and the electric displacement $\vec{D}$, and can be expressed as another form:

$$\vec{D} = \varepsilon_0 \vec{E} + \vec{P} \tag{1.1.12}$$

with $\varepsilon_0 = 4\pi \times 9 \times 10^{-9}$ F/m the vacuum permittivity. $\vec{P}$ is the electric polarization intensity, and corresponds to vector sum of molecular electric dipole moments in unit volume medium.

Equation (1.1.10) describes the relationship between the magnetic field intensity $\vec{H}$ and the magnetic induction intensity $\vec{B}$, where μ is a scalar constant as permeability. For non-ferromagnetic medium, μ is very close to the vacuum permeability as $\mu_0 = 4\pi \times 10^{-7}$ H/m.

Equation (1.1.11) gives the relationship between the current density $\vec{J}$ and the electric field intensity $\vec{E}$. The parameter κ is called electrical conductivity. For vacuum, $\kappa = 0$.

1.1.2 Wave Differential Equation

From Maxwell's equations, we can prove that the propagation of electromagnetic fields is fluctuating. For simplicity, the scenario of electromagnetic wave propagating in infinitely extended homogeneous, isotropic, transparent and passive media is considered. In such scenario, parameters κ, μ and ε are scalers and independent of positions. The transparent media illustrates $\kappa = 0$ and $\vec{J} = 0$, otherwise the alternating electromagnetic fields in the medium result in currents, which in turn consumes the power of electromagnetic wave. The passive media means the charge density $\rho = 0$. By now, the Maxwell's equations read:

$$\nabla \times \vec{E} = -\frac{\partial \vec{B}}{\partial t} \tag{1.1.13}$$

$$\nabla \times \vec{B} = \mu\varepsilon\frac{\partial \vec{E}}{\partial t} \tag{1.1.14}$$

$$\nabla \cdot \vec{E} = 0 \tag{1.1.15}$$

$$\nabla \cdot \vec{B} = 0 \tag{1.1.16}$$

Take the derivative with respect to time at both hands of Eq. (1.1.14), and swap the derivative order at the left hand:

$$\nabla \times \left(\frac{\partial \vec{B}}{\partial t}\right) = \mu\varepsilon\frac{\partial^2 \vec{E}}{\partial t^2} \tag{1.1.17}$$

Considering double vector product formula:

$$\nabla \times \left(\nabla \times \vec{E}\right) = \nabla\left(\nabla \cdot \vec{E}\right) - \nabla^2 \vec{E} \tag{1.1.18}$$

with ∇^2 the Laplacian operator. Equation (1.1.17) with Eqs. (1.1.13), (1.1.15) and (1.1.18) result in:

$$\nabla^2 \vec{E} = \mu\varepsilon\frac{\partial^2 \vec{E}}{\partial t^2} \tag{1.1.19}$$

Similarly, one can eliminate $\vec{E}$ and obtain:

$$\nabla^2 \vec{B} = \mu\varepsilon\frac{\partial^2 \vec{B}}{\partial t^2} \tag{1.1.20}$$

Equations (1.1.19) and (1.1.20) are called wave differential equation. Take the one-dimensional electric field $\vec{E}$ propagate along z axis as example, the general solution of Eq. (1.1.19) is:

$$\vec{E}(z,t) = \vec{E_1}\left(z - \frac{t}{\sqrt{\mu\varepsilon}}\right) + \vec{E_2}\left(z + \frac{t}{\sqrt{\mu\varepsilon}}\right) \tag{1.1.21}$$

with $\vec{E_1}$ and $\vec{E_2}$ the unary functions with independent variable $[z - t(\mu\varepsilon)^{-0.5}]$ and $[z + t(\mu\varepsilon)^{-0.5}]$, respectively.

Considering the particular solution $\vec{E_1}[z - t(\mu\varepsilon)^{-0.5}]$ firstly, supposing that $E_1(0)$ is the electric field value with $z = 0$ and $t = 0$, and then in the location $z = t(\mu\varepsilon)^{-0.5}$ such value turns to $E_1[z - t(\mu\varepsilon)^{-0.5}]$ after propagating along z axis at the velocity $v = (\mu\varepsilon)^{-0.5}$. Obviously, $E_1(0) = E_1[z - t(\mu\varepsilon)^{-0.5}]$, which is the one-dimensional wave results from the propagation of the disturbance of electric field $\vec{E}$. Hence, $\vec{E_1}[z - t(\mu\varepsilon)^{-0.5}]$ indicates that the one-dimensional electric field $\vec{E_1}$ propagate along z axis with the velocity of $v = (\mu\varepsilon)^{-0.5}$.

Similarly, $\vec{E_2}[z - t(\mu\varepsilon)^{-0.5}]$ is the one-dimensional electric field $\vec{E_2}$ propagate along the opposite direction of z axis with the velocity of $v = (\mu\varepsilon)^{-0.5}$. As for the magnetic field, one can draw the similar conclusion from Eq. (1.1.20).

The above analysis shows that alternating electric and magnetic fields travel through a medium with matter constant μ and ε in the form of waves. The traveling velocity is:

$$v = \frac{1}{\sqrt{\mu\varepsilon}} \tag{1.1.22}$$

Thus the velocity of electromagnetic wave in vacuum is:

$$c = v = \frac{1}{\sqrt{\mu\varepsilon}} = 2.99794 \times 10^8 \text{ m/s} \tag{1.1.23}$$

Scientists have already proved that light is a kind of electromagnetic wave. Obviously the light velocity in vacuum is that of electromagnetic as $c = 2.99794 \times 10^8$ m/s.

1.2 Wave Function

Optical wave is a kind of electromagnetic wave, and is the propagation of electric and magnetic fields at high frequencies in space. Therefore, optical waves can be described by the parameter $\vec{E}$ and $\vec{B}$. The optical wave function is defined as the function describing the change of the physical quantity $\vec{E}$ and $\vec{B}$ with space and time. The fluctuations of electric field $\vec{E}$ and magnetic field $\vec{B}$ are identical, meanwhile the electric field $\vec{E}$ is the main component. Thus usually electric field $\vec{E}$ is employed to represent optical wave functions. The scaler wave function E corresponds to scalar optical waves, and the vector wave function $\vec{E}$ corresponds to vector optical waves.

1.2.1 One-Dimensional Simple Harmonic Wave

The wave function E of a simple harmonic wave is cosinoidal or sinusoidal. Simple harmonics are good approximations for some practical optical waves, for instance, the waves from lasers or some other monochromaters. In addition, any complex wave can be regarded as the superposition of a series of simple harmonics from Fourier analysis. Thus simple harmonic waves is a kind of basic wave form.

The wave function of one-dimensional simple harmonic waves propagation along z axis read:

$$E(z - vt) = E_0 \cos\left[\frac{2\pi}{\lambda}(z - vt) + \varphi_0\right] \tag{1.2.1}$$

with E_0 the amplitude, λ the wavelength, v the propagation velocity, φ_0 the initial phase when $z = 0$ and time $t = 0$. The deep meaning of the above parameters can be acquired easily from other basic references as principle of optics [1] etc., and won't be discussed in detail here. Equation (1.2.1) can also be expressed as:

$$E(z, t) = E_0 \cos(kz - \omega t + \varphi_0) \tag{1.2.2}$$

where $k = 2\pi/\lambda$ denotes wave number, and $\omega = kv$ denotes time angular frequency. Figure 1.1 gives the diagram of One-dimensional simple harmonic wave described by Eq. (1.2.2).

According to Euler's formula, Eq. (1.2.2) can be regarded as the real part of a complex-exponential function:

$$E(z, t) = E_0 \cos(kz - \omega t + \phi_0) = \mathrm{Re}\{E_0 \exp[i(kz - \omega t + \phi_0)]\} \tag{1.2.3}$$

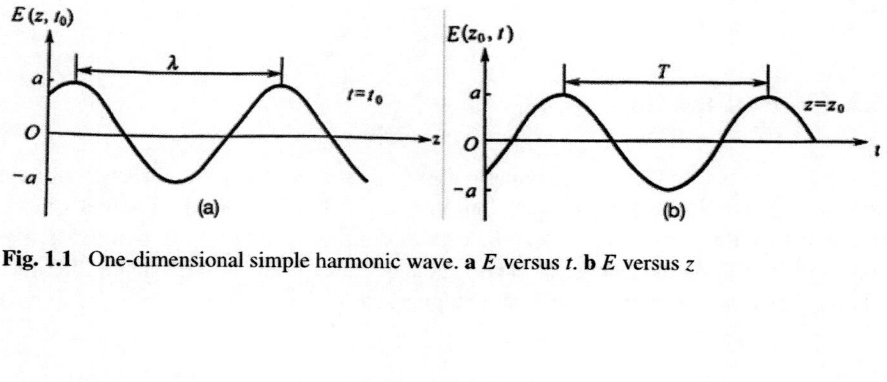

Fig. 1.1 One-dimensional simple harmonic wave. **a** E versus t. **b** E versus z

For simplicity, the symbol of real part "Re" is ignored, and the wave function is expressed directly as a complex-exponential function:

$$E(z, t) = E_0 \exp[i(kz - \omega t + \varphi_0)] \tag{1.2.4}$$

Note that only the real part of Eq. (1.2.4) represents the real fluctuation in practice.

There are many advantages to employ complex-exponentials instead of cosines. Firstly, the introduction of complex-exponential functions can separate the factors related to spatial coordinates from those related to time coordinates as:

$$E(z, t) = E(z) \exp(-i\omega t) \tag{1.2.5}$$

where

$$E(z) = E_0 \exp[i(kz + \varphi_0)] \tag{1.2.6}$$

describes how the wave changes with respect to spatial coordinates, and named as the complex amplitude. In the study of superposition and decomposition of light waves with the same initial frequency, the time factor in the wave function is not considered in the calculation since the wave has the same variations with time in different places, and the complex amplitude is used to replace the wave function directly. Secondly, the complex-exponential form can simplify the calculations, which will be reflected when dealing with the issues as multi-beam interference and diffraction analysis.

1.2.2 Three-Dimension Simple Harmonic Plane Wave

In practice, three coordinate variables are needed to determine the coordinates of the investigation points, since light waves propagate in a three-dimensional space. The most widely employed variables are three components x, y and z of the position vector $\vec{r}$ in rectangular coordinates:

$$\vec{r} = x\hat{i} + y\hat{j} + z\hat{k} \tag{1.2.7}$$

From the derivation of one-dimensional wave differential equation [Eq. (1.1.19)], one can obtain the three-dimensional scalar wave differential equation:

$$\frac{\partial^2 E}{\partial x^2} + \frac{\partial^2 E}{\partial y^2} + \frac{\partial^2 E}{\partial z^2} = \frac{1}{v^2} \cdot \frac{\partial^2 E}{\partial t^2} \tag{1.2.8}$$

Considering the Laplacian ∇, Eq. (1.2.8) reads:

$$\nabla^2 E(x, y, z, t) = \frac{1}{v^2} \cdot \frac{\partial^2 E}{\partial t^2} \tag{1.2.9}$$

and the solution of Eq. (1.2.9) has the form:

$$E(x, y, z, t) = E(k_x x + k_y y + k_z z - kvt) \tag{1.2.10}$$

with k_x, k_y and k_z the three components of the vector $\vec{k}$ in rectangular coordinates:

$$\vec{k} = k_x\hat{i} + k_y\hat{j} + k_z\hat{k} \tag{1.2.11}$$

The vector $\vec{k}$ is called wave vector, which determines the propagate direction of light waves. In addition:

$$\left|\vec{k}\right| = k = \frac{2\pi}{\lambda} = \sqrt{k_x^2 + k_y^2 + k_z^2} \tag{1.2.12}$$

$$\vec{k} = k\left(\cos\alpha\hat{i} + \cos\beta\hat{j} + \cos\gamma\hat{k}\right) \tag{1.2.13}$$

$$\begin{cases} k_x = k\cos\alpha \\ k_y = k\cos\beta \\ k_z = k\cos\gamma \end{cases} \tag{1.2.14}$$

where α, β and γ are the directional angles of the wave vector $\vec{k}$ relative to the three axes.

The introduction of wave vector $\vec{k}$ makes it easier to describe three-dimensional wave functions as:

$$E(\vec{r}, t) = E(\vec{k}\cdot\vec{r} - kvt) \tag{1.2.15}$$

Similar with one-dimensional waves, the phase of three-dimensional wave is:

$$\varphi = \vec{k}\cdot\vec{r} - kvt + \varphi_0 \tag{1.2.16}$$

with φ_0 the initial phase of three-dimensional waves in origin of coordinates. Usually for light waves, the path or collection of points with the same phase value φ at a given time is called wavefront or equiphase surface. Light waves with plane wavefronts and the fluctuations of any point on the wavefront are identical all the time are called plane waves. Obviously, if both k and v are constants, the wavefront equation:

$$\vec{k}\cdot\vec{r} - kvt = C \tag{1.2.17}$$

is a plane, where C denotes a constant, and thus Eq. (1.2.15) represents a plane wave. The wave front Π is a series plane perpendicular to the wave vector $\vec{k}$, as illustrated in Fig. 1.2.

A three-dimensional plane wave with cosine or sine wave function is

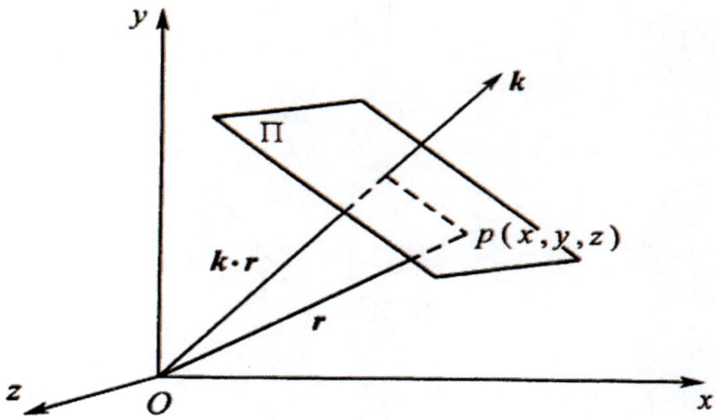

Fig. 1.2 The wavefront or the equiphase surface of a three-dimensional plane wave

$$E(\vec{r}, t) = E_0 \cos(\vec{k} \cdot \vec{r} - kvt + \varphi_0) \tag{1.2.18}$$

and such wave is called three-dimensional simple harmonic plane wave. The time parameters of such waves are identical with those of one-dimensional simple harmonic waves, while the space parameters have a little difference, reflected on the spatial period and spatial frequency.

Figure 1.3 sketches a three-dimensional simple harmonic plane wave with the wave vector $\vec{k}$, and the dashed line gives a series of equiphase surface with the phase difference 2π. The spatial period λ is the distance between two neighboring equiphase surface with the phase difference 2π along $\vec{k}$ direction, in other words, is the wavelength indeed.

Different with one-dimensional simple harmonic wave, a three-dimensional simple harmonic plane wave has various spatial period in various directions. As shown in Fig. 1.3, θ is the angle between investigation direction $\overrightarrow{Ob}$ and $\vec{k}$, then one can acquire the spatial period along $\overrightarrow{Ob}$ is:

$$T_s(\theta) = \frac{\lambda}{\cos\theta} \tag{1.2.19}$$

When $\theta < \pi/2$, $T_s(\theta) > 0$; When $\theta > \pi/2$, $T_s(\theta) < 0$. If the direction angle of $\vec{k}$ is α, β and γ, the spatial periods along the three axes are:

$$\begin{cases} T_s(x) = \frac{\lambda}{\cos\alpha} \\ T_s(y) = \frac{\lambda}{\cos\beta} \\ T_s(z) = \frac{\lambda}{\cos\gamma} \end{cases} \tag{1.2.20}$$

The spatial frequency of three-dimensional simple harmonic plane wave is defined as the reciprocal of spatial period, whose intrinsic value if $f = 1/\lambda$. Hence such spatial

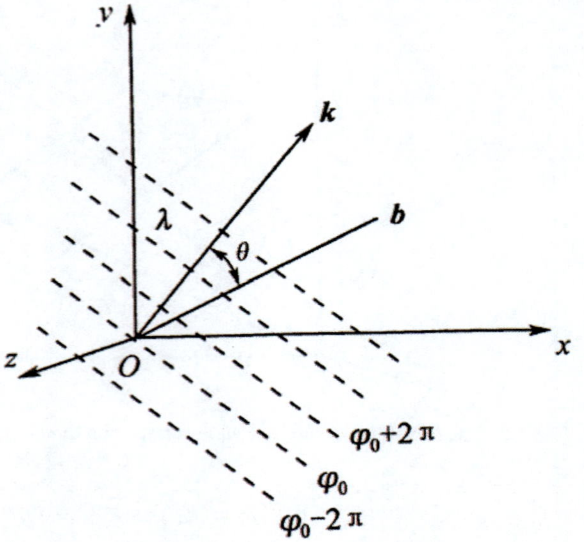

Fig. 1.3 The spatial parameter of a three-dimensional plane wave

frequency is also the function of investigation direction, and thus along $\overrightarrow{Ob}$ is:

$$f(\theta) = \frac{1}{T_s(\theta)} = \frac{\cos\theta}{\lambda} \tag{1.2.21}$$

The spatial frequencies along the three axes are:

$$\begin{cases} f_x = \frac{\cos\alpha}{\lambda} \\ f_y = \frac{\cos\beta}{\lambda} \\ f_z = \frac{\cos\gamma}{\lambda} \end{cases} \tag{1.2.22}$$

Obviously,

$$f_x^2 + f_y^2 + f_z^2 = \frac{1}{\lambda^2} = f^2 \tag{1.2.23}$$

Hence f_x, f_y and f_z are not completely independent. Once the wavelength λ is fixed, the third component can be found from other two known components. Equation (1.2.22) also illustrated that, the three components of spatial frequency gives the propagation direction of a three-dimensional simple harmonic plane wave. Therefore the wave function of three-dimensional waves can be expressed through f_x,f_y and f_z as:

$$E(\vec{r}, t) = E_0 \cos\left[2\pi(f_x x + f_y y + f_z z) - kvt\right] \tag{1.2.24}$$

If $z = 0$, the three-dimensional waves in the plane xOy is acquired as:

$$E(x, y, t) = E_0 \cos\left[2\pi(f_x x + f_y y) - kvt\right] \tag{1.2.25}$$

On the contrary, if $E(x, y, t)$ has already known, the wave function in the whole space can be obtained from Eq. (1.2.23).

Similar with one-dimensional waves, the complex exponential wave function of a three-dimensional simple harmonic plane wave reads:

$$E(\vec{r}, t) = E_0 \exp\left[i(\vec{k} \cdot \vec{r} - \omega t + \varphi_0)\right] \tag{1.2.26}$$

And the complex amplitude reads:

$$E(\vec{r}) = E_0 \exp\left[i(\vec{k} \cdot \vec{r} + \varphi_0)\right] \tag{1.2.27}$$

The wave function in the plane xOy is:

$$E(x, y, t) = E_0 \exp\left\{i\left[2\pi(f_x x + f_y y) - \omega t + \varphi_0\right]\right\} \tag{1.2.28}$$

And the complex amplitude in the plane xOy reads:

$$E(x, y) = E_0 \exp\left\{i\left[2\pi(f_x x + f_y y) + \varphi_0\right]\right\} \tag{1.2.29}$$

1.3 Polarization Basis

Polarization of light is the phenomenon that the spatial distribution of the vibration of electric displacement $\vec{D}$ of light wave loses symmetry with respect to the wave vector. Polarization is one of the most important dimensions of laser beams, and finds lots of application in the domains as optical communications, laser manufactures and so on. This section will introduce the basis of isotropic polarizations. And the complex anisotropic polarizations will be present in Chap. 7.

1.3.1 Polarizations and Jones Vector

According to the polarization features, light can be divided into natural light, partial polarizations and complete polarizations. The light from natural illuminant and general man-made illuminant directly includes all possible directions of vibration perpendicular to the wave vector of light wave, thus it does not show polarization and called nature light. Complete polarization is the light whose direction of electric displacement is fixed or regularly changed. And the partial polarization is the superposition of nature light and complete polarization light that contains various vibration directions but is more significant in one direction. Here we only focus on complete polarizations.

There are three various polarizations, the linear polarization, the elliptical polarization and the circular polarization. For linear polarization, its electric field is always vibrating in a definite linear direction that unchanged with time.

The elliptical polarizations are those, the motion trajectory of the electric field vector endpoint in the plane perpendicular to the wave vector is an ellipse. Taking the one-dimensional simple harmonic wave propagation along $+z$ axis as example, the electric displacement components on x and y axes are:

$$\begin{cases} D_x = D_{x0}\cos(kz - \omega t + \varphi_{x0}) \\ D_y = D_{y0}\cos(kz - \omega t + \varphi_{y0}) \end{cases} \tag{1.3.1}$$

with D_{x0} and D_{y0} the amplitude, φ_{x0} and φ_{y0} the initial phase of D_x and D_y respectively. Meanwhile:

$$\vec{D} = D_x\hat{i} + D_y\hat{j} \tag{1.3.2}$$

Note that in Eq. (1.3.2) the disappearance of D_z is due to $\vec{D} \perp \vec{k}$. By now, at any given time t and location z, $\vec{D}$ is always in the plane xOy, and the angle α between $\vec{D}$ and x axis is:

$$\tan\alpha(z,t) = \frac{D_y}{D_x} = \frac{D_{x0}\cos(kz - \omega t + \varphi_{x0})}{D_{y0}\cos(kz - \omega t + \varphi_{y0})} \tag{1.3.3}$$

Equation (1.3.3) can also be written as:

$$\left(\frac{D_x}{D_{x0}}\right)^2 + \left(\frac{D_y}{D_{y0}}\right)^2 - \frac{2\cos\delta}{D_{x0}}D_x D_y = \sin^2\delta \tag{1.3.4}$$

with

$$\delta = \varphi_{y0} - \varphi_{x0} \tag{1.3.5}$$

which is an elliptic equation. Equation (1.3.4) gives such elliptical motion trajectory of the endpoint of $\vec{D}$, as shown in Fig. 1.4a. Figure 1.4b sketches various elliptical trajectory with different δ. Figure 1.4 implies that, the elliptical trajectory can have various bounding rectangle whose two right-angle sides are parallel with x and y axes. And length of the two sides are $2D_{x0}$ and $2D_{y0}$ separately.

With the changing of $(kz - \omega t)$, the angle α is changing, indicating that the orientation of $\vec{D}$ is rotating. From Eq. (1.3.3), when $\sin\delta > 0$, $d\alpha/dt > 0$, the rotating direction (counter-clockwise) of $\vec{D}$ and the wave vector match the right-hand rule, and called right-handed elliptical polarization. If $\sin\delta < 0$, $d\alpha/dt < 0$, the rotating

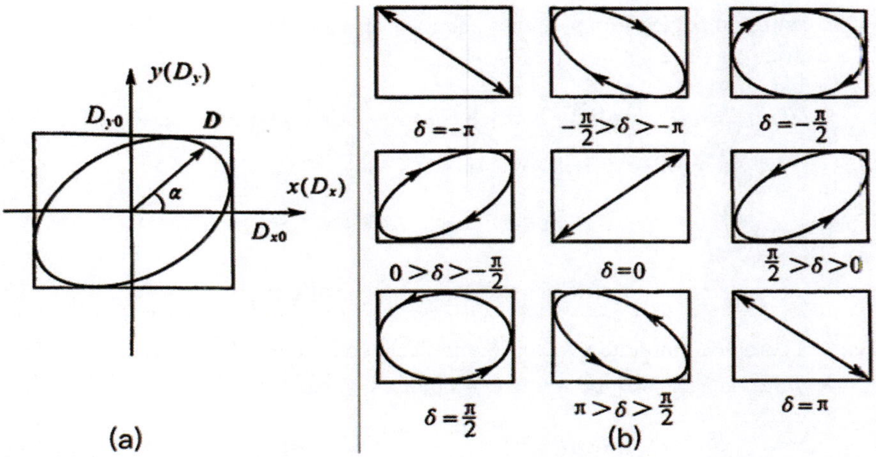

Fig. 1.4 Elliptical polarizations. **a** motion trajectory of the endpoint of $\vec{D}$; **b** elliptical polarizations with various δ

direction (clockwise) of $\vec{D}$ and the wave vector match the right-hand rule, and called left-handed elliptical polarization.

When

$$\delta = m\pi + \pi/2, \quad m \in \mathbb{Z} \tag{1.3.6}$$

$\cos\delta = 0$, $\sin^2\delta = 1$, thus Eq. (1.3.4) turn to:

$$\left(\frac{D_x}{D_{x0}}\right)^2 + \left(\frac{D_y}{D_{y0}}\right)^2 = 1 \tag{1.3.7}$$

and refers to the elliptical polarizations whose major and minor axes are parallel to x and y axes separately.

On the basis of Eq. (1.3.7), if $D_{x0} = D_{y0}$, elliptical polarizations turn to circular polarizations.

When

$$\delta = m\pi, \quad m \in \mathbb{Z} \tag{1.3.8}$$

$\cos\delta = \pm 1$, and Eq. (1.3.8) turn to equation of line:

$$\frac{D_x}{D_{x0}} = \pm\frac{D_y}{D_{y0}} \tag{1.3.9}$$

By now, one can draw the conclusion that both linear and circular polarizations are the particular cases of elliptical polarizations.

To represent polarizations, Jones vector is proposed. The electric displacement $\vec{D}$ of a c along z axis reads:

$$\begin{aligned}\vec{D}(z,t) &= D_{x0}\exp[i(kz-\omega t+\varphi_{x0})]\hat{i} + D_{y0}\exp\left[i\left(kz-\omega t+\varphi_{y0}\right)\right]\hat{j}\\ &= \overrightarrow{D_0}\exp[i(kz-\omega t)]\end{aligned} \tag{1.3.10}$$

where:

$$\overrightarrow{D_0} = D_{x0}\exp(i\varphi_{x0})\hat{i} + D_{y0}\exp\left(i\varphi_{y0}\right)\hat{j} \tag{1.3.11}$$

denotes complex amplitude vector that indicate completely the polarization state.

$\overrightarrow{D_0}$ in Eq. (1.3.11) can be written as a column vector:

$$\overrightarrow{D_0} = \begin{bmatrix} D_{x0}\exp(i\varphi_{x0}) \\ D_{y0}\exp\left(i\varphi_{y0}\right)\end{bmatrix} = \exp(i\varphi_{x0})\begin{bmatrix} D_{x0} \\ D_{y0}\exp(i\delta)\end{bmatrix} \tag{1.3.12}$$

with $\delta = \varphi_{y0} - \varphi_{x0}$ the phase difference between the components D_y and D_x.

The column vector in Eq. (1.3.12) is called Jones vector.

The intensity of the above beam is:

$$I = D_{x0}^2 + D_{y0}^2 = \overrightarrow{D_0}^{\dagger}\cdot\overrightarrow{D_0} \tag{1.3.13}$$

where † denotes Hermitian conjugate. Hence the normalized Jones vector $\widehat{\overrightarrow{D_0}}$ read:

$$\widehat{\overrightarrow{D_0}} = \frac{\overrightarrow{D_0}}{\sqrt{I}} = \exp(i\varphi_{x0})\left(D_{x0}^2 + D_{y0}^2\right)^{-0.5}\begin{bmatrix} D_{x0} \\ D_{y0}\exp(i\delta)\end{bmatrix} \tag{1.3.14}$$

Here we list a set of normalized Jones vectors of some special polarizations. For linear polarizations ($\delta = 0$ or π):

$$\widehat{\overrightarrow{D_0}} = \exp(i\varphi_{x0})\begin{bmatrix}\cos\theta \\ \sin\theta\end{bmatrix} \tag{1.3.15}$$

where

$$\begin{aligned}&\theta \in (-\pi/2, \pi/2]\\ &\cos\theta = D_{x0}\left(D_{x0}^2 + D_{y0}^2\right)^{-0.5}\\ &\sin\theta = \pm D_{y0}\left(D_{x0}^2 + D_{y0}^2\right)^{-0.5}\end{aligned} \tag{1.3.16}$$

For elliptical polarizations whose major axis is parallel to x or y axes ($\delta = \pm\pi/2$):

$$\widehat{\overrightarrow{D_0}} = \exp(i\varphi_{x0})\left(D_{x0}^2 + D_{y0}^2\right)^{-0.5}\begin{bmatrix} D_{x0} \\ \pm i D_{y0} \end{bmatrix} \tag{1.3.17}$$

For circular polarizations ($\delta = \pm\pi/2, D_{x0} = D_{y0}$):

$$\widehat{\overrightarrow{D_0}} = \frac{\sqrt{2}}{2}\exp(i\varphi_{x0})\begin{bmatrix} 1 \\ \pm i \end{bmatrix} \tag{1.3.18}$$

For two polarized beams with identical optical frequency propagating along the same direction $\overrightarrow{D_1}$ and $\overrightarrow{D_2}$, the projection of $\overrightarrow{D_1}$ onto $\overrightarrow{D_2}$ is defined as:

$$\overrightarrow{D_1}\left(\overrightarrow{D_2}\right) = \overrightarrow{D_1}^T \cdot \widehat{\overrightarrow{D_2}}^* \tag{1.3.19}$$

where superscript T and * correspond to transposition and conjugate, respectively. Similarly, the projection of $\overrightarrow{D_2}$ onto $\overrightarrow{D_1}$ is:

$$\overrightarrow{D_1}\left(\overrightarrow{D_2}\right) = \overrightarrow{D_1}^T \cdot \widehat{\overrightarrow{D_2}}^* \tag{1.3.20}$$

Under such condition, if

$$\overrightarrow{D_1}^T \cdot \widehat{\overrightarrow{D_2}}^* = \overrightarrow{D_2}^T \cdot \widehat{\overrightarrow{D_1}}^* = 0 \tag{1.3.21}$$

$\overrightarrow{D_1} \perp \overrightarrow{D_2}$, and the two polarizations are mutually orthogonal. Obviously, left circular polarizations are orthogonal with right polarizations.

1.3.2 *Wave Plates and Jones Matrix*

Wave plates are one of the most polarization modulation elements, and have played important roles in laser beam modulation systems. The basic function of a wave plate is to introduce an additional phase difference for the two known orthogonal polarization directions.

Usually, a wave plate is a uniaxial crystal sheet whose optical axis is parallel to the surface. The incident polarized beams will be decomposed into ordinary beams (linear polarizations that vibrate perpendicular to the optical axis) and extraordinary beams (linear polarizations that vibrate parallel to the optical axis) with various diffraction indices separately. The two components will suffer different phase delays, and an additional phase different is introduced to the output beams.

In detail, the optical path lengths when ordinary and extraordinary beams propagating through a wave plate are:

$$\begin{cases} L_o = dn_o \\ L_e = dn_e \end{cases} \tag{1.3.22}$$

with n_o and n_e the refractive index of ordinary and extraordinary beams separately, and d the thickness of the wave plate. Their optical path difference is:

$$\Delta L = L_e - L_o = d(n_e - n_o) \tag{1.3.23}$$

Thus the additional phase difference is:

$$\Delta\varphi = k_0 \Delta L = \frac{2\pi}{\lambda_0}(n_e - n_o) \tag{1.3.24}$$

where k_0 and λ_0 are the wave number and wavelength in vacuum, respectively.

The quarter wave plate (QWP) and the half wave plate (HWP) are the most widely used wave plates. For a QWP:

$$\Delta L = d(n_e - n_o) = (2N+1)\frac{\lambda_0}{4} \tag{1.3.25}$$

$$\Delta\varphi = (2N+1)\frac{\pi}{2} \tag{1.3.26}$$

And for a HWP:

$$\Delta L = d(n_e - n_o) = (2N+1)\frac{\lambda_0}{2} \tag{1.3.27}$$

$$\Delta\varphi = (2N+1)\pi \tag{1.3.28}$$

In Eqs. (1.3.25)–(1.3.28), N denotes an integer. Obviously, the QWP and HWP can introduce additional phase difference $\pi/2$ and π separately for the two orthogonal polarization directions. Such two orthogonal polarization directions are usually named as the fast axis and slow axis. The slow axis has the phase delay $\Delta\varphi$ compared with fast axis.

Next we will discuss how a wave plate modulate polarized beams. For a polarized beam propagating along z axis:

$$\vec{D} = \begin{bmatrix} D_x \\ D_y \end{bmatrix} \tag{1.3.29}$$

it is incident into a wave plate whose fast and slow axes u, v located in the xOy plane, as shown in Fig. 1.5. There is an angle α between uOv and xOy. Here it is stated that α is positive when the x axis turns counterclockwise to the u axis, meanwhile $\alpha \in [0, \pm\pi/2]$.

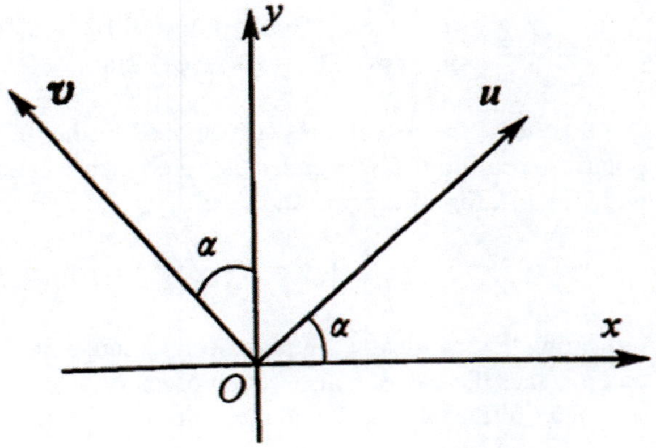

Fig. 1.5 Relationship between the coordinates *uOv* and *xOy*

For the sake of calculating the polarization $\vec{D}'$ after $\vec{D}$ passing through such wave plate, one can follow the procedures below.

① Converting the Jones vector of $\vec{D}$ to $\overrightarrow{D_{uv}}$ in *uOv* coordinates:

$$\overrightarrow{D_{uv}} = \begin{bmatrix} \cos\alpha & \sin\alpha \\ -\sin\alpha & \cos\alpha \end{bmatrix} \begin{bmatrix} D_x \\ D_y \end{bmatrix} = \begin{bmatrix} D_x\cos\alpha + D_y\sin\alpha \\ -D_x\sin\alpha + D_y\cos\alpha \end{bmatrix} = \begin{bmatrix} D_u \\ D_v \end{bmatrix} \tag{1.3.30}$$

② Let $\overrightarrow{D_{uv}}$ pass through the wave plate to obtain the Jones vector of $\vec{D}'$ in the *uOv* coordinates:

$$\overrightarrow{D'_{uv}} = \begin{bmatrix} D_u \\ D_v\exp(i\Delta\varphi) \end{bmatrix} = \begin{bmatrix} D'_u \\ D'_v \end{bmatrix} \tag{1.3.31}$$

③ Converting the Jones vector of $\overrightarrow{D'_{uv}}$ back to $\vec{D}'$ in *xOy* coordinates:

$$\vec{D}' = \begin{bmatrix} \cos\alpha & -\sin\alpha \\ \sin\alpha & \cos\alpha \end{bmatrix} \begin{bmatrix} D'_u \\ D'_v \end{bmatrix}$$
$$= \begin{bmatrix} D_x[\cos^2\alpha + \sin^2\alpha\exp(i\Delta\varphi)] & D_y[\sin\alpha\cos\alpha - \sin\alpha\cos\alpha\exp(i\Delta\varphi)] \\ D_x[\sin\alpha\cos\alpha - \sin\alpha\cos\alpha\exp(i\Delta\varphi)] & D_y[\sin^2\alpha + \cos^2\alpha\exp(i\Delta\varphi)] \end{bmatrix} \tag{1.3.32}$$

By now, one can find easily that $\vec{D}'$ can be expressed as the product of a 2×2 matrix and $\vec{D}$:

$$\begin{bmatrix} D'_x \\ D'_y \end{bmatrix} = [M] \begin{bmatrix} D_x \\ D_y \end{bmatrix} \tag{1.3.33}$$

where:

$$[M] = \begin{bmatrix} \cos^2\alpha + \sin^2\alpha \exp(i\Delta\varphi) & \sin\alpha\cos\alpha[1 - \exp(i\Delta\varphi)] \\ \sin\alpha\cos\alpha[1 - \exp(i\Delta\varphi)] & \sin^2\alpha + \cos^2\alpha \exp(i\Delta\varphi) \end{bmatrix} \tag{1.3.34}$$

Obviously, the matrix $[M]$ corresponds to the modulation of a wave plate on a polarized beam, and $[M]$ is called the Jones matrix of a wave plate. From Eqs. (1.3.33) and (1.3.34), for arbitrary α and $\Delta\varphi$:

$$|D'_x|^2 + |D'_y|^2 = |D_x|^2 + |D_y|^2 \tag{1.3.35}$$

indicating that an ideal wave plate won't change the intensity of the incident beams and matches the law of conservation of energy.

For a QWP, $\Delta\varphi = \pi/2$, and the Jones matrix is:

$$[M]_{\text{QWP}} = \begin{bmatrix} \cos^2\alpha + i\sin^2\alpha & \sin\alpha\cos\alpha(1 - i) \\ \sin\alpha\cos\alpha(1 - i) & \sin^2\alpha + i\cos^2\alpha \end{bmatrix} \tag{1.3.36}$$

For a HWP, $\Delta\varphi = \pi$, and the Jones matrix is:

$$[M]_{\text{HWP}} = \begin{bmatrix} \cos 2\alpha & \sin 2\alpha \\ \sin 2\alpha & -\cos 2\alpha \end{bmatrix} \tag{1.3.37}$$

A QWP can realize the transformation between linear polarizations and elliptical or circular polarizations. And a HWP can change the vibration direction of linear polarizations, or the rotation direction of elliptical polarizations. Such processes can be easily derived and verified through the Jones matrix given by Eqs. (1.3.36) and (1.3.37), and won't be derived in detail here.

The Jones matrix of a polarizer is given in Eq. (1.3.38), whose derivation is similar with QWPs and HWPs, and won't be introduced in detail here.

$$[M]_{\text{P}} = \begin{bmatrix} \cos^2\alpha & \sin\alpha\cos\alpha \\ \sin\alpha\cos\alpha & \sin^2\alpha \end{bmatrix} \tag{1.3.38}$$

Table 1.1 gives some of the Jones matrices of QWPs, HWPs and polarizers.

1.4 Basis of Scalar Diffraction Theory

Diffraction is a common phenomenon that beams will deviate from the linear propagation law due to the space limitation in the propagation process. The presentation of Maxwell's equations indicates that light waves are a kind of electromagnetic wave. Hence the diffraction problem can be solved strictly as the boundary problem of electromagnetic fields. Nevertheless, such scheme is too complex to get an analytic

Table 1.1 Jones matrices of some typical QWPs, HWPs and polarizers

α	0	$\pi/4$	$-\pi/4$	$\pi/2$
Polarizer	$\begin{bmatrix} 1 & 0 \\ 0 & 0 \end{bmatrix}$	$\frac{1}{2}\begin{bmatrix} 1 & 1 \\ 1 & 1 \end{bmatrix}$	$\frac{1}{2}\begin{bmatrix} 1 & -1 \\ -1 & 1 \end{bmatrix}$	$\begin{bmatrix} 0 & 0 \\ 0 & 1 \end{bmatrix}$
QWP	$\begin{bmatrix} 1 & 0 \\ 0 & i \end{bmatrix}$	$\frac{1}{2}(1+i)\begin{bmatrix} 1 & -i \\ -i & 1 \end{bmatrix}$	$\frac{1}{2}(1+i)\begin{bmatrix} 1 & i \\ i & 1 \end{bmatrix}$	$\begin{bmatrix} i & 0 \\ 0 & 1 \end{bmatrix}$
HWP	$\begin{bmatrix} 1 & 0 \\ 0 & -1 \end{bmatrix}$	$\begin{bmatrix} 0 & 1 \\ 1 & 0 \end{bmatrix}$	$\begin{bmatrix} 0 & -1 \\ -1 & 0 \end{bmatrix}$	$\begin{bmatrix} -1 & 0 \\ 0 & 1 \end{bmatrix}$

solution. In practice, the scalar diffraction theory is usually employed as an approximate solution to solve diffraction problems. Scalar diffraction theory considers only the electric field component, which is treated as a scalar, and ignores other rectangular coordinate component of the electromagnetic fields. Through the Kirchhoff formula and Rayleigh-Sommerfeld formula, the propagation process of light wave is well studied and explained. The scalar diffraction theory is an important basis of studying the generation, detection of optical vortices, and will be discussed in this section.

1.4.1 Huygens-Fresnel Principle

In 1690, Huygens proposed a hypothesis to explain the diffraction problem of beams. In such hypothesis, each surface element on the light wavefront can be regarded as a secondary perturbation center, which can produce spherical surface waves. Meanwhile, the position of the wavefront at the latter moment is the envelope surface of all these wavelet fronts [2]. Huygens' hypothesis can qualitatively discuss and explain the propagation of simple waves in homogeneous isotropic media, the diffraction of simple objects, and the law of refraction and reflection. However, it is based on the hypothesis and lacks theoretical basis and quantitative analysis.

For the sake of describing the diffraction quantitatively, Fresnel made an important supplement to Huygens principle and put forward the Huygens-Fresnel principle, where any unobstructed surface element on the wavefront can be regarded as a sub-wave source and emit spherical surface waves with the same frequency as the incident beams. The light vibration at any subsequent point results from the superposition of all the sub-waves [1]. Huygens-Fresnel principle introduces interference superposition of beams, which can be regarded as the combination of interference superposition principle and Huygens wavelet hypothesis, making it possible to describe and study the diffraction problem of beams quantitatively. Below the Huygens-Fresnel principle is described in detail by taking spherical wave diffraction via an aperture Σ as an example.

As shown in Fig. 1.6, S is a monochromatic point light source with the amplitude E_0. O is the center of the aperture. r_0 is the distance between the source and the aperture center $|SO|$. The wavefront on the point O is Ω. According to the theory of spherical wave, the complex amplitude on the wave surface Ω result from the source S is:

$$E_\Omega = \frac{E_0}{r_0}\exp(ikr_0) \tag{1.4.1}$$

The wavefront Ω' is part of the wavefront Ω that uncovered by the aperture Σ, which is divided as a set of small surface elements, and the element at any point M is set to $d\sigma$. For arbitrary point P on the imaging plane ($|MP| = r'$), from Huygens-Fresnel principle, the light wave on point P is superposed coherently by the sub-spherical wave emitted from all the small surface element on the wavefront Ω':

$$E_p = \iint_{\Omega'} E_M \mathrm{d}\sigma \tag{1.4.2}$$

where E_M denotes the complex amplitude on P emitted from M:

$$E_M = \frac{KD(\chi)E_\Omega d\sigma}{r'}\exp\left(ikr'\right) \tag{1.4.3}$$

In Eq. (1.4.3), K is a complex coefficient that gives the intensity relationship between the incident beams and sub-waves. χ denotes the angle between the outer normal direction of $d\sigma$ and the line segment MP. $D(\chi)$ is directivity factor, representing the strength of the sub-wave in different directions. The formula of Huygens-Fresnel principle can be acquired from Eqs. (1.4.1)–(1.4.3) as:

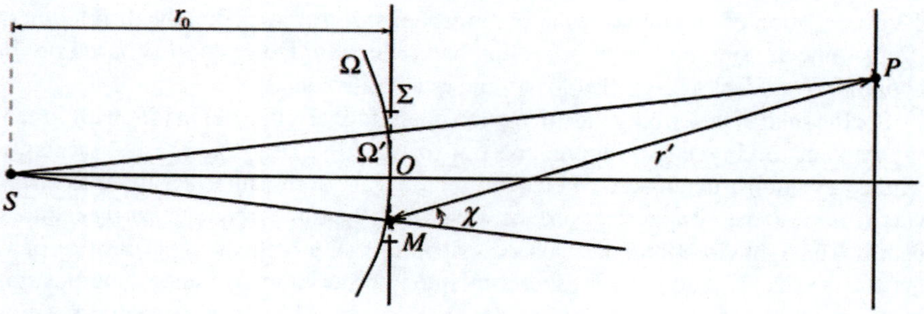

Fig. 1.6 Huygens-Fresnel principle

$$E_P = \frac{KE_0}{r_0} \exp(ikr_0) \iint_{\Omega'} D(\chi) \frac{\exp(ikr')}{r'} d\sigma \tag{1.4.4}$$

In Eq. (1.4.4), the integral surface is the uncovered wavefront Ω' when the spherical wave emitted from the source S passing through the point O. And for any sub-wave source on Ω', K is fixed. Actually, during the integration, the aperture plane Σ can also be selected as the integral surface. By now, for various sub-wave source on the plane Σ, the inconstant distances between S and the different points on Σ lead to various incident complex amplitude, thus result in different sub-wave source intensities and initial phases. Similarly, if not point source but the source that can produce more complex waves is employed, the complex amplitudes of any sub-wave sources on the plane Σ are different. If $E(\xi)$ denotes the complex amplitude of the sub-wave source on the plane Σ, where ξ represents the sources' locations on Σ, the more common form of arbitrary waves illuminating the aperture Σ can be acquired from Eq. (1.4.4) as:

$$E_P = K \iint_{\Sigma} D(\chi) E(\xi) \frac{\exp(ikr')}{r'} d\sigma \tag{1.4.5}$$

If the point source S is replaced by a parallel wave source, then the sub-wave sources on the plane Σ have isotropic complex amplitude distributions as $E(\xi) = E_C$. By now, Eq. (1.4.5) can be simplified as:

$$E_P = KE_C \iint_{\Sigma} D(\chi) \frac{\exp(ikr')}{r'} d\sigma \tag{1.4.6}$$

Therefore, Eq. (1.4.5) can be used in principle to analyze the diffraction problem in the case of any light wave irradiation. While due to the uncertainty of complex coefficient K and direction factor $D(\chi)$, there are still limitations in quantitative analysis of beams' diffraction.

1.4.2 Kirchhoff Diffraction Integral

1. Kirchhoff's law

In order to break through the limitations of Huygens-Fresnel principle, in 1882 Kirchhoff proposed an equation through Green formula that can solve diffraction problem strictly, which is known as Kirchhoff's law [3, 4].

As sketched in Fig. 1.7, for two complex functions with spatial variables $E(x, y, z)$ and $G(x, y, z)$, if E and G and their first and second partial derivatives are continuous on a closed surface S and S-enclosed space volume V, from Green formula [4] one can acquire:

$$\iiint_V (G\nabla^2 E - E\nabla^2 G)dV = \oiint_S \left(G\frac{\partial E}{\partial n} - E\frac{\partial G}{\partial n}\right)d\sigma \tag{1.4.7}$$

where the left hand is the integral with respect to volume V, and the right hand is the surface integral against the closed surface S. n is the outer normal of the closed surface S. And $d\sigma$ is the directed surface element, whose outer normal is positive, on the closed surface S.

The electromagnetic field under steady state can be described by Helmholtz equation:

$$\nabla^2\psi + k^2\psi = 0 \tag{1.4.8}$$

with $\nabla^2 = \partial_x^2 + \partial_y^2 + \partial_z^2$ the Laplacian operator, k the wave number. If $E(x, y, z)$, a complex function, is taken as the complex amplitude of the light field, and an appropriate function $G(x, y, z)$ is chosen to make both E and G meet with Eq. (1.4.8), then the complex amplitude E_P at any point P in the space can be solved by using the complex amplitude E and its normal partial derivative on any closed surface surrounding the point P.

The function G chosen by Kirchhoff is:

$$G = \frac{1}{r}\exp(ikr) \tag{1.4.9}$$

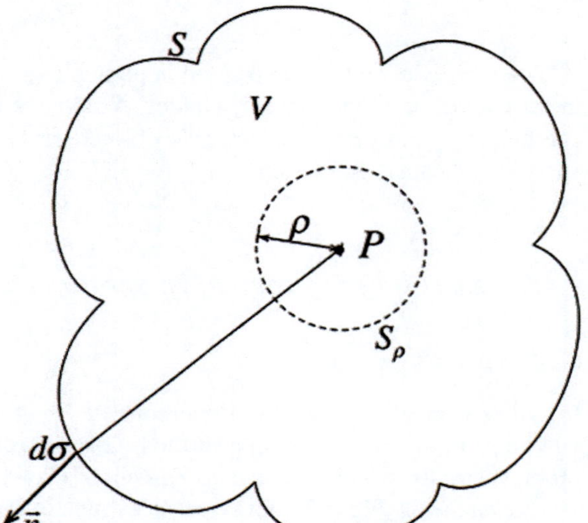

Fig. 1.7 The physical model of Kirchhoff's law

which represents the spherical sub-waves emitted by the surface element $d\sigma$ on the closed surface S. r denotes the distance between the point P and the small surface element $d\sigma$. For the point P, $r = 0$, thus $G \to \infty$. Therefore, making a spherical surface S_ρ with P the center and ρ the radius, for the volume V surrounded by the surface S and S_ρ, there must be:

$$\iiint\limits_V (G\nabla^2 E - E\nabla^2 G)dV = \oiint\limits_{S+S_P} \left(G\frac{\partial E}{\partial n} - E\frac{\partial G}{\partial n}\right)d\sigma \tag{1.4.10}$$

Due to the fact that both E and G satisfy Helmholtz Eq. (1.4.8):

$$\nabla^2 E + k^2 E = 0 \tag{1.4.11}$$

$$\nabla^2 G + k^2 G = 0 \tag{1.4.12}$$

Equation (1.4.10) with Eqs. (1.4.11) and (1.4.12) result in:

$$\oiint\limits_{S+S_P} \left(G\frac{\partial E}{\partial n} - E\frac{\partial G}{\partial n}\right)d\sigma = \iiint\limits_V (-Gk^2 E + Ek^2 G)dV = 0 \tag{1.4.13}$$

Hence:

$$\oiint\limits_{S_P} \left(G\frac{\partial E}{\partial n} - E\frac{\partial G}{\partial n}\right)d\sigma = -\oiint\limits_{S} \left(G\frac{\partial E}{\partial n} - E\frac{\partial G}{\partial n}\right)d\sigma \tag{1.4.14}$$

For the spherical surface S_ρ, the outer normal points to the center P, which is opposite with the vector from P to any point on S_ρ. Meanwhile, $r \equiv \rho$. Thus:

$$\frac{\partial G}{\partial n} = -\frac{\partial G}{\partial r} = \left(\frac{1}{r} - ik\right)\frac{\exp(ikr)}{r} = \left(\frac{1}{\rho} - ik\right)\frac{\exp(ik\rho)}{\rho} \tag{1.4.15}$$

Since the complex amplitude E and its first partial derivative are continuous at point P, they are constant for point P. When $\rho \to 0$, the left hand of Eq. (1.4.14) is:

$$\begin{aligned}\lim_{\rho\to 0}\oiint\limits_{S_P} \left(G\frac{\partial E}{\partial n} - E\frac{\partial G}{\partial n}\right)d\sigma &= \lim_{\rho\to 0}\left[4\pi\rho\exp(ik\rho)\left.\frac{\partial E}{\partial n}\right|_P - 4\pi\rho\exp(ik\rho)E_P\left(\frac{1}{\rho} - ik\rho\right)\right] \\ &= -4\pi E_P\end{aligned} \tag{1.4.16}$$

The right hand of Eq. (1.4.14) is:

$$\lim_{\rho\to 0}\left[-\oiint_S\left(G\frac{\partial E}{\partial n}-E\frac{\partial G}{\partial n}\right)d\sigma\right]=-\oiint_S\left\{\frac{\exp(ikr)}{r}\frac{\partial E}{\partial n}-E\frac{\partial}{\partial n}\left[\frac{\exp(ikr)}{r}\right]\right\}d\sigma \tag{1.4.17}$$

By now the Kirchhoff's law can be acquired from Eqs. (1.4.14), (1.4.16) and (1.4.17) as:

$$E_P=\frac{1}{4\pi}\oiint_S\left\{\frac{\exp(ikr)}{r}\frac{\partial E}{\partial n}-E\frac{\partial}{\partial n}\left[\frac{\exp(ikr)}{r}\right]\right\}d\sigma \tag{1.4.18}$$

The Kirchhoff's law indicates that, the complex amplitude E_P of any point P in space can be derived from the complex amplitude E of all points on the closed surface S that surround the point P and their partial derivatives $f_n E$ along the direction of the outer normal line n.

2. Kirchhoff diffraction integral formula

Although Eq. (1.4.18) has already given the scheme of calculating the complex amplitude of any point in space, there still have difficulties in analyzing diffraction fields. In this sub-section we will continue simplify Eq. (1.4.18) to obtain a more simple form thus to serve the calculation.

Firstly a diffraction model is built, as sketched in Fig. 1.8. A point source illuminates an infinite opaque plane Σ_1 with an aperture Σ. The size of the aperture Σ is larger than the wavelength, but smaller than the distance from the aperture Σ to point P. According to the Kirchhoff's law, selecting a closed surface, which is made up of the aperture Σ, the right side of plane Σ_1, and a spherical surface with P the center and infinite radius R. Then Eq. (1.4.18) can be written as:

$$E_P=\frac{1}{4\pi}\oiint_{\Sigma+\Sigma_1+\Sigma_2}\left(G\frac{\partial E}{\partial n}-E\frac{\partial G}{\partial n}\right)d\sigma \tag{1.4.19}$$

where G is expressed by Eq. (1.4.9). Form Kirchhoff boundary conditions:

① In the aperture Σ, the electric field E and its normal partial differential $\partial E/\partial n$ is determined by the nature of the incident wave, and independent of plane Σ_1.
② In the right hand of Σ_1, E and $\partial E/\partial n$ equal to zero, completely unaffected by hole Σ.

then one can acquire:

$$\iint_{\Sigma_1}\left(G\frac{\partial E}{\partial n}-E\frac{\partial G}{\partial n}\right)d\sigma=0 \tag{1.4.20}$$

For the spherical surface Σ_2, its center is P, and $R\to\infty$, thus the any point on Σ_2 meets:

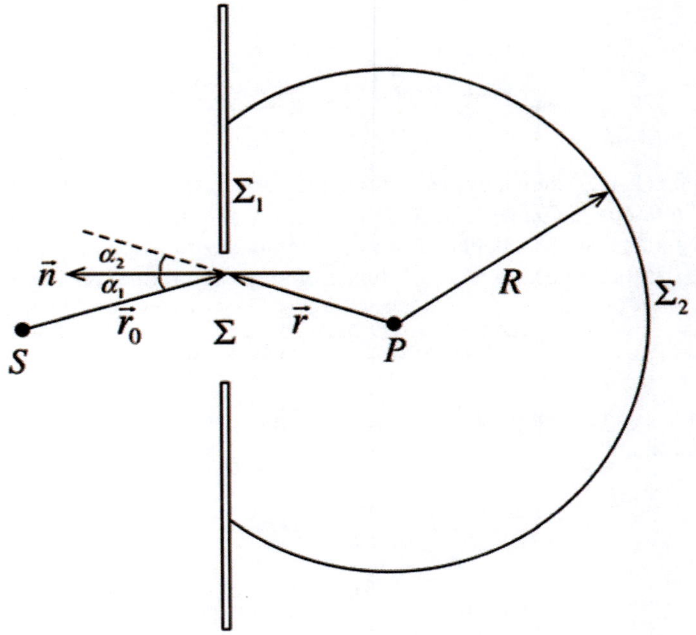

Fig. 1.8 The diffraction model that an aperture Σ is illuminated by a point source

$$G = \frac{1}{R}\exp(ikR) \tag{1.4.21}$$

$$\frac{\partial G}{\partial n} = \frac{\partial G}{\partial n}\left[\frac{\exp(ikr)}{r}\right]\Bigg|_{r=R} = \left(ik - \frac{1}{R}\right)G = ikG \tag{1.4.22}$$

Therefore, the integral of Eq. (1.4.19) on the spherical surface Σ_2 is

$$\iint\limits_{\Sigma_2}\left(G\frac{\partial E}{\partial n} - E\frac{\partial G}{\partial n}\right)d\sigma = \iint\limits_{\Omega} GR\left(\frac{\partial E}{\partial n} - ikE\right)Rd\Omega = \iint\limits_{\Omega}\exp(ikR)\left(\frac{\partial E}{\partial n} - ikE\right)Rd\Omega \tag{1.4.23}$$

where Ω is the solid angle of the spherical surface Σ_2 to the point P. When $R \to \infty$, any illumination can be regarded as the superposition of the illumination of many point sources. For point source illumination:

$$\lim_{R\to\infty} R\left(\frac{\partial E}{\partial n} - ikE\right) = 0 \tag{1.4.24}$$

which is also known as Sommerfeld radiation conditions and always true for linear optics. By now, Eq. (1.4.23) equals to zero. Equations (1.4.19), (1.4.20) and (1.4.23) result in:

$$E_P = \frac{1}{4\pi} \oiint_{\Sigma} \left(G\frac{\partial E}{\partial n} - E\frac{\partial G}{\partial n} \right) d\sigma \tag{1.4.25}$$

Equation (1.4.25) indicates that, when light waves passing through an aperture Σ on an infinite opaque plane in space, the complex amplitude E_P of arbitrary point behind the plane can be calculated from the complex amplitude on the aperture Σ. When $r \gg \lambda$, there must be $k \gg r^{-1}$, thus from Eq. (1.4.9) one can acquire:

$$\frac{\partial G}{\partial n} = \cos\langle \vec{n}, \vec{r} \rangle \frac{\partial G}{\partial r} = \cos\alpha_2 \cdot ikG \tag{1.4.26}$$

Then the complex amplitude E_Σ and its normal partial differential of the waves in aperture Σ are:

$$E_\Sigma = \frac{A}{r_0} \exp(ikr_0) \tag{1.4.27}$$

$$\frac{\partial E_\Sigma}{\partial n} = \cos\langle \vec{n}, \vec{r}_0 \rangle \frac{\partial E_\Sigma}{\partial r_0} = -\cos\alpha_1 \cdot ikE_\Sigma \tag{1.4.28}$$

with A the amplitude in the source S, r_0 the distance between S and Σ. When introducing Eqs. (1.4.9), (1.4.26), (1.4.27) and (1.4.28) into Eq. (1.4.25), one can acquire:

$$E_P = \frac{1}{i\lambda} \iint_{\Sigma} E_\Sigma \frac{\exp(ikr)}{r} \frac{\cos\alpha_1 + \cos\alpha_2}{2} d\sigma \tag{1.4.29}$$

which is the well-known Kirchhoff diffraction integral formula.

Equation (1.4.29) gives the case that spherical waves emitted from a point source and illuminate an aperture Σ. In the process of derivation, there is no restriction for the complex amplitude distribution E_Σ of the wave in the aperture Σ, therefore, one can always work out the complex amplitude behind the plane Σ_1 no matter which kind of the source dose the incident wave come from. When plane waves whose wave vector is parallel to the axis (perpendicular to the plane Σ_1), which can be regarded as spherical waves generated from a point source located in the negative infinity of the optical axis, are incident, $\cos\alpha_1 = 0$. Then one can acquire Kirchhoff diffraction integral formula for plane wave illuminating as

$$E_P = \frac{1}{i\lambda} \iint_{\Sigma} E_\Sigma \frac{\exp(ikr)}{r} \frac{1 + \cos\alpha_2}{2} d\sigma \tag{1.4.30}$$

The Kirchhoff diffraction integral formula enables us to calculate the complex amplitudes of any spatial position in the direction of beams' transmission from the complex amplitudes in a certain spatial plane. Comparing Eq. (1.4.5) with Eq. (1.4.29), obviously Kirchhoff diffraction integral formula has the same expression as the diffraction formula of Huygens-Fresnel principle. However, Kirchhoff diffraction integral formula is based on a solid mathematical basis, giving the specific form and physical meaning of each parameter, which is different from the hypothesis-based Huygens-Fresnel principle.

3. Rayleigh-Sommerfeld diffraction integral formula

Although the Kirchhoff diffraction integral formula can well calculate the diffraction field, its boundary conditions are not self-consistent obviously, which reflected on the point that both the complex amplitude E and its normal partial derivative $\partial E/\partial n$ behind the opaque plane are 0. However, for three-dimensional wave equations, if the complex amplitude and its normal partial derivative of optical fields on any infinitesimal plane are zero, the solution is also zero in the whole space, which contradicts the boundary conditions. Meanwhile, when the point P close to the aperture Σ, one cannot acquire the complex amplitude through Eq. (1.4.29). To remedy this self-inconsistency, Sommerfeld chose a new function G that different from Kirchhoff's as [1, 3]:

$$G = \frac{1}{r}\exp(ikr) - \frac{1}{r'}\exp(ikr') \tag{1.4.31}$$

where r denotes the distance between P and any other point in space, and r' is the distance between P' (the virtual point of P from an infinite mirror) and any point in space. Then the Rayleigh-Sommerfeld diffraction integral formula is acquired from Eqs. (1.4.31) and (1.4.10) as:

$$E_P = \frac{1}{i\lambda}\iint\limits_{\Sigma} E_\Sigma \frac{\exp(ikr)}{r}\cos\alpha_2 d\sigma \tag{1.4.32}$$

Compared Eq. (1.4.29) with Eq. (1.4.32), the only difference between Kirchhoff diffraction integral formula and Rayleigh-Sommerfeld diffraction integral formula reflects on the direction factor $K(\theta)$. For Kirchhoff diffraction integral formula:

$$K(\theta) = \frac{1}{2}(\cos\alpha_1 + \cos\alpha_2) \tag{1.4.33}$$

and for Rayleigh-Sommerfeld diffraction integral formula:

$$K(\theta) = \cos\alpha_2 \tag{1.4.34}$$

Usually, the diffraction integral formula is written as:

$$E_P = \frac{1}{i\lambda} \iint\limits_{\Sigma} E_\Sigma \frac{\exp(ikr)}{r} K(\theta) d\sigma \tag{1.4.35}$$

For most of situations in practice, the two boundary conditions of Kirchhoff diffraction integral formula are effective. Boundary conditions are not valid, only if the size of the aperture is of wavelength magnitude, or the observing point is close enough to the opaque plane. Therefore, both the Kirchhoff diffraction integral formula and the Rayleigh-Sommerfeld diffraction integral formula can well describe beams' diffraction.

4. The general form of the diffraction integral formula

In practice, when the size of the diffractive aperture is much smaller than the distance from the inspection plane, and the effective area of the source and the inspection hole is very small compared with the opening angle of the aperture's center, then $\cos\alpha_1 = \cos\alpha_2 = 0$. Thus $K(\theta) = 0$. From Eq. (1.4.35) the diffraction integral formula is acquired under paraxial approximation as:

$$E_P = \frac{1}{i\lambda} \iint\limits_{\Sigma} E_\Sigma \frac{\exp(ikr)}{r} d\sigma \tag{1.4.36}$$

By now, both the Kirchhoff diffraction integral formula and the Rayleigh-Sommerfeld diffraction integral formula have the same form under paraxial approximation. For a practical diffraction system, usually a unified coordinate system should be defined to make Eq. (1.4.36) more universal. As sketched in Fig. 1.9, the Cartesian coordinates for the diffractive aperture plane is (u, v), for the inspection or imaging plane is (x, y). The distance between aperture plane and the inspection plane is d. The complex amplitude that incident on the plane (u, v) is $E_0(u, v)$. And the transmittance function of the aperture is $T(u, v)$. Then:

$$r = \sqrt{(x-u)^2 + (y-v)^2 + d^2} \tag{1.4.37}$$

Considering Eq. (1.4.36), one can easily acquire the diffraction integral formula under Cartesian coordinates with arbitrary aperture being illuminated by arbitrary waves:

$$E_P(x, y) = \frac{1}{i\lambda} \iint\limits_{\infty} E_0(u, v) T(u, v) \frac{\exp\left(ik \cdot \sqrt{(x-u)^2 + (y-v)^2 + d^2}\right)}{\sqrt{(x-u)^2 + (y-v)^2 + d^2}} du dv \tag{1.4.38}$$

In summary, if the distance between the point P and the diffractive aperture is not small enough, the complex amplitude behind the diffractive plane can be calculated through Eq. (1.4.38).

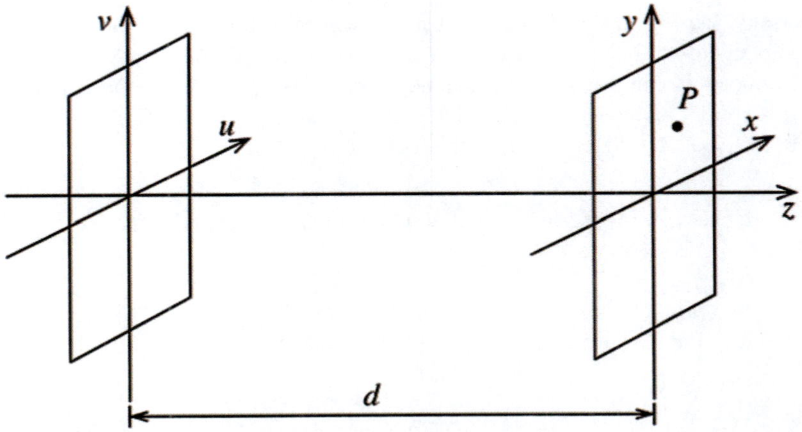

Fig. 1.9 The general model of the diffraction integral formula

1.4.3 Angular Spectrum

For diffractive situations of complex waves or beams are incident, the diffraction problem can be treated with a theoretical framework that slightly different from Kirchhoff diffraction theory, namely angular spectrum theory.

The Fourier transformation (FT) and inverse Fourier transformation (IFT) of arbitrary complex field $E(x, y, z)$ in xy plane:

$$a(f_x, f_y, z) = \iint_{\infty} E(x, y, z) \exp[-2\pi i(xf_x + yf_y)]dxdy \tag{1.4.39}$$

$$E(x, y, z) = \iint_{\infty} a(f_x, f_y, z) \exp[2\pi i(xf_x + yf_y)]df_x df_y \tag{1.4.40}$$

From Eq. (1.4.39), the spatial frequency spectrum $a(f_x, f_y, z)$ can be regarded as the complex amplitude density of plane waves with the spatial frequency (f_x, f_y) on xy plane, and the spatial frequency (f_x, f_y) determines the propagation direction of $a(f_x, f_y, z)$. If the direction cosine of the plane wave vector is $(\cos\alpha, \cos\beta, \cos\gamma)$, then:

$$(\cos\alpha, \cos\beta, \cos\gamma) \tag{1.4.41}$$

Thus the spatial frequency spectrum $a(f_x, f_y, z)$ can be expressed through the direction cosine as $a(\cos\alpha/\lambda, \cos\beta/\lambda, z)$, and named as the angular spectrum of a complex optical field $E(x, y, z)$ on xy plane.

From Eq. (1.4.40), the optical fields of $E(x, y, z)$ on xy plane can be regarded as the superposition of multiple plane waves $\exp[2\pi i(xf_x + yf_y)]$ propagating along various directions. During the propagation of plane waves in free-space, the wavefront is unchanged, and only a phase displacement associated with the propagating distance is introduced. Hence, the spatial frequency spectrum $a(f_x, f_y, d)$ of the xy plane with location $z = d$ can be derived from that of $z = 0$ [$a(f_x, f_y, 0)$]. In Eq. (1.4.40), when $z = 0$, the optical field $E(x, y, 0)$ in the location $z = 0$ is:

$$E(x, y, 0) = \iint_{\infty} a(f_x, f_y, 0) \exp[2\pi i(xf_x + yf_y)]df_x df_y \tag{1.4.42}$$

From the propagation equation of plane waves, the optical field $E(x, y, z)$ on xy plane in arbitrary location z is:

$$E(x, y, z) = \iint_{\infty} a(f_x, f_y, 0) \exp[2\pi i(xf_x + yf_y)] \exp(2\pi izf_z)df_x df_y \tag{1.4.43}$$

Compared Eq. (1.4.40) with Eq. (1.4.43), obviously:

$$a(f_x, f_y, z) = a(f_x, f_y, 0) \exp(2\pi izf_z) \tag{1.4.44}$$

From Eq. (1.4.41):

$$\lambda^2(f_x^2 + f_y^2 + f_z^2) = 1 \tag{1.4.45}$$

there must be:

$$f_z = \frac{1}{\lambda}\sqrt{1 - \lambda^2 f_x^2 - \lambda^2 f_y^2} \tag{1.4.46}$$

Combined with Eq. (1.4.44), and let $z = d$, then:

$$H_A(x, y) = \frac{a(f_x, f_y, z)}{a(f_x, f_y, 0)} = \exp\left(ikd\sqrt{1 - \lambda^2 f_x^2 - \lambda^2 f_y^2}\right) \tag{1.4.47}$$

$H_A(x, y)$ is called angular spectrum transfer function (ASTF), which is independent with the incident functions, indicating that the nature of diffraction problems is the transformation of the optical field through a linear spatially invariant system. In Eq. (1.4.47), obviously only the angular spectrum components that satisfy $1 - \lambda f_x{}^2 - \lambda f_y{}^2 > 0$ can reach the imaging plane. The component $1 - \lambda f_x{}^2 - \lambda f_y{}^2 < 0$ exists in the form of evanescent wave, whose optical fields decay exponentially with the increasing of parameter d, and only present in a very thin area near the diffraction plane. Therefore, the propagation process of optical fields in free-space is equivalent to passing through an ideal low-pass filter with a radius of $1/\lambda$.

From Eqs. (1.4.42) and (1.4.47), the essence of angular spectrum theory is the Fourier decomposition and synthesis of optical fields. Meanwhile, these derivations also give the scheme of how to calculate the complex amplitude $E_d(x, y)$ if $E_0(x, y)$ and the propagation distance d are already known. That is, the frequency spectrum of the initial field $a_0(f_x,f_y)$ is obtained through the FT of the field at the diffractive plane. And then the frequency spectrum on the receiving plane $a_d(f_x,f_y)$ is obtained by multiplying the initial spectrum by the ASTF $H_A(x, y)$. Finally, the IFT is carried out to obtain the optical field $E_d(x, y)$ after propagating by distance d. Such process reads:

$$E_d(x, y) = \mathcal{F}^{-1}\{\mathcal{F}[E_0(x, y)] \cdot H_A(f_x, f_y)\} \tag{1.4.48}$$

where $\mathcal{F}()$ and $\mathcal{F}^{-1}()$ denotes FT and IFT, respectively. Equation (1.4.48) is calculation formula for solving the diffraction problem through angular spectrum theory. It converts the cumbersome integration into FT operation, and the fast Fourier transformation (FFT) can be realized by discretizing FT. Therefore Eq. (1.4.48) is of great significance for the computer analysis of diffractions.

1.4.4 Fresnel Diffraction

1. **Fresnel approximation**

The general form of diffraction integration has been obtained in Sect. 1.4.2. However, for practical diffraction analysis, the fraction term in the integral sign of Eq. (1.4.38) is still complex, so further simplification is essential.

For the general model shown in Fig. 1.9, under paraxial approximation, it meets with:

$$\begin{cases} d \gg u \\ d \gg v \\ d \gg x \\ d \gg y \end{cases} \tag{1.4.49}$$

Thus the binomial expansion of Eq. (1.4.37) can be written as:

$$\begin{aligned} r &= \sqrt{(x-u)^2 + (y-v)^2 + d^2} \\ &= d + \frac{(x-u)^2 + (y-v)^2}{2d} - \frac{\left[(x-u)^2 + (y-v)^2\right]^2}{8d^3} + \ldots \end{aligned} \tag{1.4.50}$$

The denominator in the partial expression of Eq. (1.4.38) can be directly replaced by $r = d$, since the second and subsequent terms of Eq. (1.4.50) under paraxial approximation can be ignored. For the numerator in Eq. (1.4.38), the wave number k is of 10^6 magnitude because the wavelength λ is generally of the micron magnitude, so the second term of Eq. (1.4.50) multiplied by k is non-negligible. In Eq. (1.4.50), the terms on the right hand are decreasing successively. Thus it is stipulated that, when the phase error introduced by the third term on the right hand of Eq. (1.4.50) is less than π/2:

$$d^3 \geq \frac{1}{2\lambda}\left[(x-u)^2+(y-v)^2\right]^2 \tag{1.4.51}$$

the parameter r in the complex exponential factor can be replaced by the first two terms of Eq. (1.4.50). Such approximation is called Fresnel approximation. Diffractions that satisfying the Fresnel approximation is called Fresnel diffraction. The region satisfying Eq. (1.4.50) is called Fresnel diffraction region, where the Fresnel diffraction can be observed directly. Under Fresnel approximation, the numerator and denominator are replaced by the first two terms and the first term of Eq. (1.4.50) respectively, then Eq. (1.4.38) is simplified as:

$$E(x,y)=\frac{\exp(ikd)}{i\lambda d}\iint\limits_{\infty} E(u,v)\exp\left\{\frac{ik}{2d}\left[(x-u)^2+(y-v)^2\right]\right\}dudv \tag{1.4.52}$$

which is the Fresnel diffraction integration formula.

2. **Fresnel transfer function**

The Fresnel diffraction integration formula given in Eq. (1.4.52) is still complex for the diffraction analysis through numerical calculation. Hence it is necessary to further simplify Eq. (1.4.52) to obtain a diffraction calculation form that similar with angular spectrum theory.

The exponential term in the integral sign of Eq. (1.4.52) is expanded as:

$$\begin{aligned}E(x,y)=&\frac{\exp(ikd)}{i\lambda d}\exp\left[\frac{ik}{2d}(x^2+y^2)\right]\\&\cdot\iint\limits_{\infty} E(u,v)\exp\left[\frac{ik}{2d}(u^2+v^2)\right]\exp\left[-\frac{ik}{d}(xu+yv)\right]dudv\end{aligned} \tag{1.4.53}$$

Defining impulse response:

$$h(x,y)=\frac{\exp(ikd)}{i\lambda d}\exp\left[\frac{ik}{2d}(x^2+y^2)\right] \tag{1.4.54}$$

and let $E_0(x, y)$ the initial optical field, $E_d(x, y)$ the complex amplitude of the optical field in the plane with the distance d from the initial plane. Considering Eq. (1.4.54), Eq. (1.4.53) can be expressed as:

$$E_d(x,y)=E_0(x,y)*h(x,y) \tag{1.4.55}$$

where * corresponds to convolution. From convolution theorem:

$$\mathcal{F}[E_d(x, y)] = \mathcal{F}[E_0(x, y)] \cdot \mathcal{F}[h(x, y)] \tag{1.4.56}$$

Doing the FT of the impulse response defined in Eq. (1.4.54), then the Fresnel transfer function $H_F(f_x, f_y)$ is acquired as:

$$H_F(f_x, f_y) = \mathcal{F}[h(x, y)] = \exp\left\{ikd\left[1 - \frac{\lambda^2}{2}(f_x^2 + f_y^2)\right]\right\} \tag{1.4.57}$$

Therefore, Eq. (1.4.56) reads:

$$E_d(x, y) = \mathcal{F}^{-1}\{\mathcal{F}[E_0(x, y)] \cdot H_F(f_x, f_y)\} \tag{1.4.58}$$

By now, the diffraction calculation form is very similar with the that of angular spectrum theory [Eq. (1.4.48)]. Firstly the frequency spectrum of the complex amplitude on the initial plane is calculated through FT. Then multiplied by the Fresnel transfer function $H_F(f_x, f_y)$ to obtain the frequency spectrum of the receiving plane. And finally the complex amplitude on the receiving plane can be obtained by the IFT. The only difference between Eq. (1.4.58) and Eq. (1.4.48) is the transfer function.

The difference between the Fresnel transfer function $H_F(f_x, f_y)$ and ASTF $H_A(f_x, f_y)$ is discussed as follows. The Taylor expansion of ASTF [Eq. (1.4.47)] is

$$\begin{aligned} H_A(x, y) &= \exp\left(ikd\sqrt{1 - \lambda^2 f_x^2 - \lambda^2 f_y^2}\right) \\ &= \exp\left[ikd\left(1 - \frac{1}{2}\lambda^2(f_x^2 + f_y^2) + \frac{1}{8}\lambda^4(f_x^2 + f_y^2)^2 + \ldots\right)\right] \end{aligned} \tag{1.4.59}$$

Just keeping the first two terms:

$$H_A(f_x, f_y) \approx \exp\left\{ikd\left[1 - \frac{\lambda^2}{2}(f_x^2 + f_y^2)\right]\right\} = H_F(f_x, f_y) \tag{1.4.60}$$

indicating that the description of Fresnel diffraction in frequency domain is actually an approximation of the diffraction description of angular spectrum theory.

3. **Fresnel diffraction transformation**

The inverse operation of diffraction, where the complex amplitude on the initial plane is calculated from that of the receiving plane, is inevitable in lots of scenarios, especially the beam shaping [5], distort optical vortices compensations [6] and so on. Such inverse operation can be accomplished through Fresnel diffraction transformation.

From Eq. (1.4.58):

$$\mathcal{F}[E_d(x, y)] = \mathcal{F}[E_0(x, y)] \cdot H_F(f_x, f_y) \tag{1.4.61}$$

Thus:

$$\mathcal{F}[E_0(x,y)] = \frac{\mathcal{F}[E_d(x,y)]}{H_F(f_x, f_y)} = \mathcal{F}[E_d(x,y)] \cdot H_F^*(f_x, f_y) \quad (1.4.62)$$

Equation (1.4.62) indicates that, the frequency spectrum of the initial optical field can be obtained by the frequency spectrum of field in the receiving plane multiplied by the complex conjugate of Fresnel transfer function. Equation (1.4.57) with Eq. (1.4.62):

$$E_0(x,y) = \mathcal{F}^{-1}\left\{\mathcal{F}[E_d(x,y)] \cdot \exp\left\{-ikd\left[1 - \frac{\lambda^2}{2}\left(f_x^2 + f_y^2\right)\right]\right\}\right\} \quad (1.4.63)$$

From convolution theorem:

$$E_0(x,y) = E_d(x,y) * \left\{\frac{\exp(-ikd)}{-i\lambda d}\exp\left[-\frac{ik}{2d}\left(x^2+y^2\right)\right]\right\} \quad (1.4.64)$$

if $E(u, v)$ denotes the initial optical field $E_0(x, y)$, and $E(x, y)$ denotes the complex amplitude $E_d(x, y)$ of the field with the distance d from the initial plane, then Eq. (1.4.64) can be written as:

$$E(u,v) = \frac{\exp(-ikd)}{-i\lambda d}\iint\limits_{\infty} E(x,y)\exp\left\{-\frac{ik}{2d}\left[(u-x)^2+(v-y)^2\right]\right\}dxdy \quad (1.4.65)$$

Equation (1.4.65) is called inverse Fresnel diffraction integration formula. Obviously, Eq. (1.4.65) has a very symmetric form compared with Fresnel diffraction integration formula given by Eq. (1.4.52). The optical fields in the receiving plane can be derived from those in initial plane through Eq. (1.4.52). And the optical fields in the initial plane can be derived from those in the receiving plane through Eq. (1.4.65). The spectrum calculation form of Eq. (1.4.52) [Eq. (1.4.58)] is defined as the Fresnel diffraction transformation (FDT), denoted as $F_{(d)}()$. Similarly, the spectrum calculation form of Eq. (1.4.65) [Eq. (1.4.63)] is defined as the inverse Fresnel diffraction transformation (IFDT), denoted as $F^{-1}{}_{(d)}()$. By now FDT and IFST can be simply expressed as:

$$E_d(x,y) = F_{(d)}[E_0(x,y)] = \mathcal{F}^{-1}\{\mathcal{F}[E_0(x,y)] \cdot H_F(f_x, f_y)\} \quad (1.4.66)$$

$$E_0(x,y) = F_{(d)}^{-1}[E_d(x,y)] = \mathcal{F}^{-1}\{\mathcal{F}[E_d(x,y)] \cdot H_F^*(f_x, f_y)\} \quad (1.4.67)$$

where $E_0(x, y)$ and $E_d(x, y)$ are Fresnel diffraction pairs. The FDT and the IFDT have similar calculation methods. Their transfer functions are complex conjugates of each other. The FDT and IFDT imply that, the complex amplitudes in either the initial or receiving planes can be derived from that in receiving and initial planes, providing that the Fresnel transfer function $H_F(f_x, f_y)$ is already known.

1.4.5 Fraunhofer Diffraction

Now considering a more extreme case, on the basis of Fresnel approximation, increasing the distance d between the receiving plane and the initial aperture plane, then the diffraction patterns on the receiving plane are also increased. By now, Eq. (1.4.49) is also satisfied. Meanwhile the maximum value of the diffraction patterns' coordinate (x, y) on the receiving plane must be much larger than that of the diffraction spot coordinate (u, v) on the initial diffraction aperture plane. If:

$$\frac{k}{2d}\left(u^2 + v^2\right) \leq \frac{\pi}{2} \tag{1.4.68}$$

that is, when d exceeds a value such that the phase error introduced by the quadratic sum of the diffraction aperture plane's coordinates is less than $\pi/2$, the numerator $(u^2 + v^2)$ of the second term in Eq. (1.4.50) can be ignored. By now Eq. (1.4.50) reads:

$$r \approx d + \frac{x^2 + y^2}{2d} - \frac{xu + yv}{d} \tag{1.4.69}$$

which is known as Fraunhofer approximation. Diffractions who meet with Fraunhofer approximation is called Fraunhofer diffraction. From Eq. (1.4.68),

$$d \geq \frac{2}{\lambda}\left(u^2 + v^2\right) \tag{1.4.70}$$

The region satisfying Eq. (1.4.70) is called Fraunhofer diffraction region. Under Fraunhofer approximation, Eq. (1.4.38) can be simplified as:

$$E(x, y) = \frac{\exp(ikd)}{i\lambda d} \exp\left[\frac{ik}{2d}\left(x^2 + y^2\right)\right] \iint\limits_{\infty} E(u, v) \exp\left[-\frac{ik}{d}(xu + yv)\right] du dv \tag{1.4.71}$$

which is Fraunhofer diffraction integration formula. Equation (1.4.71) implies, the diffraction fields $E(x, y)$ can be regarded as the Fourier transformation of the initial fields $E(u, v)$ in Fraunhofer diffraction region. In practice, the observed far-field diffraction at infinity distance is Fraunhofer diffraction.

From the discussions above, only in the Fraunhofer diffraction region can the Fraunhofer diffraction be observed. For instance, if a near-infrared beam with the wavelength 1.55 μm passing through a 3 mm diameter aperture, the Fraunhofer diffraction will be observed at distance larger than 11.61 m from the aperture according to Eq. (1.4.70). In addition, it is easy to conclude from the Fraunhofer approximation that an accurate Fraunhofer diffraction can only be observed theoretically at an infinite distance from the initial diffraction plane.

For the sake of observing the exact Fraunhofer diffraction at a limited distance, a thin convex lens is introduced. According to the theory of wave optics, a thin convex lens is a phase-only transformation element, which can produce various phase delays for different annulus of the wavefront of incident beams. If a thin convex lens is placed in diffraction fields, the optical fields behind the lens can be regarded as the diffraction of the incident fields passing through an optical element. Therefore, the transmittance function $T_l(x, y)$ of thin convex lenses must be derived firstly to study the their effects on diffraction fields. Since a thin convex lens only changes the phase, but do nothing on other information, of the incident beams, the simplest incident optical field can be used to analyze and obtain the $T_l(x, y)$ that applicable to all different forms of incident fields. Here the simplest case of plane wave incident is employed.

As sketched in Fig. 1.10, when a plane wave incident in a thin convex lens along the optical axis, all the rays must converge at the image focal point. That is, a thin convex lens transforms an incident plane wave into a convergent spherical wave. From Fig. 1.10, obviously the introduced additional phases are associated with the positions on the lens. The closer to the edge, the greater the phase delay, and the closer to the center, the smaller the phase delay. From geometry, the additional phase introduced by a thin convex lens at different positions with the distance r from the center is:

$$\phi_l(r) = k\left(f - \sqrt{f^2 + r^2}\right) \tag{1.4.72}$$

where $r^2 = x^2 + y^2$, and f denotes the focal length. $f > 0$ implies convex lens, and $f < 0$ implies concave lens. Under paraxial approximation, $f \gg r$, thus binomial expanding the right hand of Eq. (1.4.72) results in:

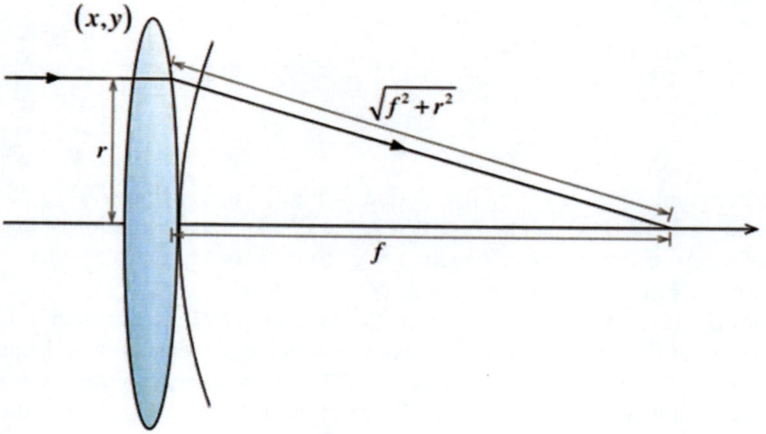

Fig. 1.10 The modulation of a thin convex lens on optical fields

$$\phi_l(r) = k\left[f - \left(f + \frac{r^2}{2f} - \frac{r^4}{8f^3} + \ldots\right)\right] \tag{1.4.73}$$

In the right hand of Eq. (1.4.73), the third and the following terms is almost 0 and can be ignored, thus:

$$\phi_l(r) = -\frac{kr^2}{2f} \tag{1.4.74}$$

Therefore, the transmittance function of a convex lens is:

$$T_l(x, y) = \exp[i\phi_l(x, y)] = \exp\left[-\frac{ik\left(x^2 + y^2\right)}{2f}\right] \tag{1.4.75}$$

Next considering a monochrome point source S, the spherical wave from which incident in an aperture with the transmittance function of $T(u, v)$ at the distance p from the source. Behind the aperture a thin convex lens with the focal length f, whose incident plane is coincident with the aperture, is placed, as shown in Fig. 1.11. The receiving plane is in the location with the distance q from the lens. Under Fresnel approximation, the complex amplitude on the aperture of the spherical wave produced from the source S is [1]:

$$U(u, v) = U_0 \exp\left[\frac{ik}{2p}\left(u^2 + v^2\right)\right] \tag{1.4.76}$$

with U_0 the complex amplitude of spherical wave at the center of an object plane. For a thin lens, its thickness is ignored, thus the complex amplitude on its exit plane is:

$$E_0(u, v) = U(u, v) \cdot T(u, v) \cdot T_l(u, v)$$

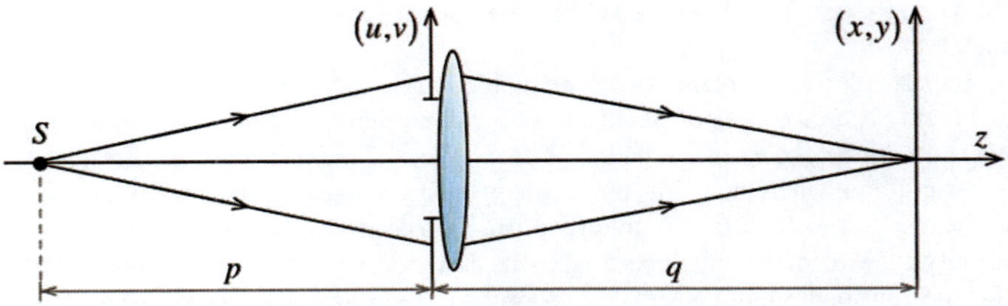

Fig. 1.11 The Fraunhofer diffraction model when light waves illuminate an aperture superposed with a thin convex lens

$$= U_0 \exp\left[\frac{ik}{2p}\left(u^2+v^2\right)\right]\exp\left[-\frac{ik\left(u^2+v^2\right)}{2f}\right]\cdot T(u,v) \tag{1.4.77}$$

Substituting Eq. (1.4.77) into the Fresnel diffraction integral formula Eq. (1.4.52), one can acquire the complex amplitude $E(x, y)$ in the receiving plane:

$$E(x,y) = \frac{U_0}{i\lambda q}\exp\left[ik\left(q+\frac{x^2+y^2}{2q}\right)\right] \cdot \iint\limits_{\infty} T(u,v)\exp\left[\frac{ik}{2}\left(\frac{1}{p}+\frac{1}{q}-\frac{1}{f}\right)\left(u^2+v^2\right)\right]\exp\left[-\frac{ik}{q}(xu+yv)\right]dudv \tag{1.4.78}$$

Once:

$$\frac{1}{p}+\frac{1}{q}=\frac{1}{f} \tag{1.4.79}$$

the source S is conjugated with the receiving plane, that is, the receiving plane is located on the conjugate image plane of the source S with respect to the thin convex lens. Hence Eq. (1.4.78) can be simplified as:

$$E(x,y) = \frac{U_0}{i\lambda q}\exp\left[ik\left(q+\frac{x^2+y^2}{2q}\right)\right]\cdot \iint\limits_{\infty} T(u,v)\exp\left[-\frac{ik}{q}(xu+yv)\right]\mathrm{d}u\mathrm{d}v \tag{1.4.80}$$

which has the same form as Fraunhofer diffraction integration formula Eq. (1.4.71). indicating that when monochromatic spherical wave from a point source illuminating a diffraction aperture with transmittance function $T(u, v)$ the Fraunhofer diffraction can be observed on the conjugate image plane of the point source by superimposing thin lens on the aperture. If the aperture is located at the exit plane of the thin lens, the same form as Eq. (1.4.71) can still be obtained according to the same derivation method.

In particular, when plane waves are incident, $p \to \infty$, then from Eq. (1.4.79) $q = f$. Therefore, in the case of plane waves incident, the Fraunhofer diffraction of the aperture can be observed on the image focal plane of the thin lens.

In addition, when the point source S happens to be placed at the object focal point of the thin convex lens, the Fraunhofer diffraction needs to be observed at the location infinity away from the lens according to the above derivation. It can be understood as that the thin convex lens converts the spherical wave produced by the point source into a plane wave, which makes the diffraction problem into a situation where the plane wave directly irradiates the diffraction aperture. Therefore, an accurate Fraunhofer diffraction can only be observed at infinity.

In summary, although it is difficult to observe Fraunhofer diffraction directly, one can introduce a thin lens, where the accurate Fraunhofer diffraction can be obtained on the conjugate image plane of the source. And such scheme is the simplest but effect way.

References

1. Born M, Wolf E. Principles of optics. Cambridge University Press;1999.
2. Li J, Xiong B. The theory and computation of information optics. Beijing: Science Press; 2009. (in Chinese).
3. Wang Y. Mathematical physics equations and special functions. Beijing: Higher Education Press; 2001. (in Chinese).
4. Xie J, Zhao D, Yan J. Physical optics. Beijing: Beijing Institute of Technology Press; 2012. (in Chinese).
5. Gerchberg RW, Saxton WO. A practical algorithm for the determination of phase from image and diffraction plane pictures. Optik. 1972;35:237–50.
6. Fu S, Zhang S, Wang T, et al. Pre-turbulence compensation of orbital angular momentum beams based on a probe and the Gerchberg-Saxton algorithm. Opt Lett. 2016;41(14):3185–8.

Chapter 2
Basic Characteristics of Vortex Beams

Start from the basic definition of vortex beams, this chapter analyzes their energy flow density, orbital angular momentum (OAM) and orthogonality between various modes. This chapter also introduces the concept and derivation of orbital angular momentum spectrum, as well as several common vortex beams.

2.1 What is Vortex Beams?

Vortex beams, also known as helical beams or twist beams, are a kind of beams with helical properties. In general, vortex beams consist of two categories, phase vortices and polarization vortices. The phase vortices have helical phases or wavefronts, and the complex amplitude in cylindrical coordinates (r, φ, z) of which comprises the helical term $\exp(il\varphi)$, with l an integer. The intensity distribution of a phase vortex has a doughnut shape, due to the phase singularity in the center. Figure 2.1 displays the intensity and wavefront distributions of some phase vortices with different l values. Obviously, the number of helical periods and helical directions are determined separately by the absolute value and sign of parameter l. When $l = 0$, phase vortices are degraded into Gaussian beams. Polarization vortices have inhomogeneous helical polarization distributions. Different from linear, elliptical and circular polarizations, polarization vortices have anisotropic polarization distributions, and each point in which has its own polarization directions, showing vector properties. Thus, polarization vortices are also called vector beams, as shown in Fig. 2.2. To avoid ambiguity, in this book phase vortices are called vortex beams, and polarization vortices are called vector beams.

In some cases, the aforementioned two kinds of vortices can exist together, namely, a beam can have both the helical wavefront and helical polarization distributions simultaneously. Such complex vortex fields are called vectorial vortex beams, which will be discussed in Chap. 7 in detail.

S. Fu and C. Gao, *Optical Vortex Beams*, Advances in Optics and Optoelectronics,
https://doi.org/10.1007/978-981-99-1810-2_2

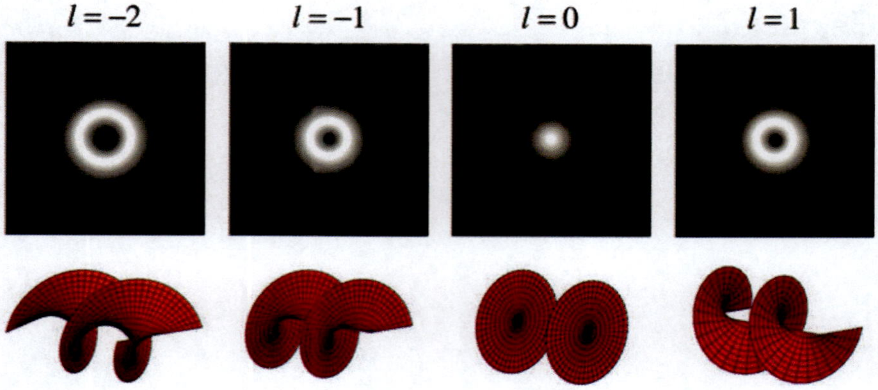

Fig. 2.1 Intensity distributions and helical phases of phase vortices with different l values

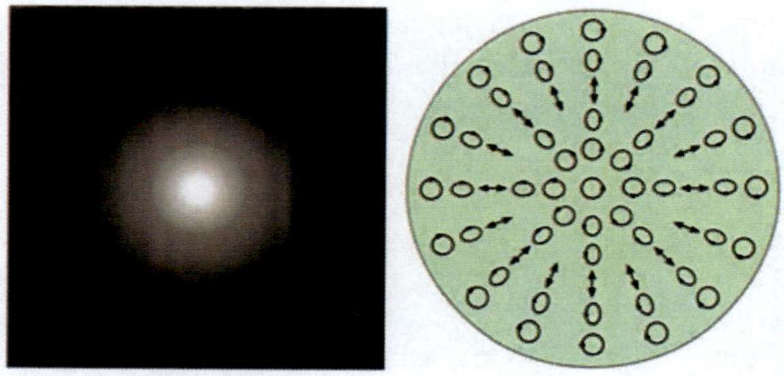

Fig. 2.2 Example of polarization vortices (vector beams) and its polarization distribution

The unique characteristics of vortex beams contribute to their widely applications. In optical communications, vortex beams can be employed as carriers, where the mode-division multiplexing is introduced, to enlarge the capacity of data transmission [1]. The rotational Doppler shift of the vortex beam makes it available to detect the rotating object [2]. When vortex beams interact with a particle, the OAM carried by the beam can be transferred to the particle to realize the particle's rotation or translation. This property has been widely used to in optical tweezers and optical spanners [3, 4]. Additionally, vortex beams can also find applications in laser material processing [5], astronomy [6], quantum techniques [7] and so on.

2.2 The Characteristics of Vortex Beams

The helical wavefront of a vortex beam brings unique properties which is different from a Gaussian beam. This section will focus on the Poynting vector, the OAM, and the orthogonality between various modes.

2.2.1 Poynting Vector

The electromagnetic field under steady state condition can be described by Helmholtz equation as

$$\nabla^2\psi + k^2\psi = 0 \tag{2.2.1}$$

with $\nabla^2 = \partial_x^2 + \partial_y^2 + \partial_z^2$ the Laplace operator, $k = 2\pi/\lambda$ the wave number, λ the wavelength. Under the paraxial approximation, the solution of Eq. (2.2.1) is

$$\psi(x, y, z) = u(x, y, z)\exp(ikz) \tag{2.2.2}$$

where $u(x, y, z)$ is the slow function along the z direction, and the phase change along the z direction is dominated by the phase term $\exp(ikz)$. Thus:

$$\begin{cases} |\partial_z u| \ll |ku| \\ \left|\partial_z^2 u\right| \ll |k\partial_z u| \ll \left|k^2 u\right| \end{cases} \tag{2.2.3}$$

Defining the horizontal two dimensional Laplace operator $\nabla_T^2 = \partial_x^2 + \partial_y^2$, and from Eqs. (2.2.1)–(2.2.3) one can obtain the paraxial wave equation

$$\nabla_T^2 u - 2ik\partial_z u = 0 \tag{2.2.4}$$

Under the paraxial approximation, the vector potential of the electromagnetic field is

$$\vec{A} = \left(\alpha\hat{x} + \beta\hat{y}\right)u(x, y, z)\exp(ikz) \tag{2.2.5}$$

where α and β are complex numbers that meet with $|\alpha|^2 + |\beta|^2 = 1$, representing the polarizations. Let ω the angular frequency, from Lorenz specification the electric field $\vec{E}$ and the magnetic field $\vec{B}$ of the electromagnetic field can be obtained separately as

$$\vec{E} = i\omega\vec{A} - \nabla\left(\frac{c^2}{i\omega}\nabla\cdot\vec{A}\right) = \left[i\omega\alpha u\hat{x} + i\omega\beta u\hat{y} - c\left(\alpha\partial_x u + \beta\partial_y u\right)\hat{z}\right]\exp(ikz) \tag{2.2.6}$$

$$\vec{B} = \nabla\times\vec{A} = \left[-\beta(\partial_z u + iku)\hat{x} + \alpha(\partial_z u + iku)\hat{y} + \left(\beta\partial_x u - \alpha\partial_y u\right)\hat{z}\right]\exp(ikz) \tag{2.2.7}$$

The Poynting vector in complex form is

$$\vec{S} = \frac{1}{2\mu_0}\vec{E}\times\vec{B}^* \equiv \vec{S} + \vec{q}i \tag{2.2.8}$$

where μ_0 denotes the vacuum permeability, $\vec{S}$ and $\vec{q}$ denote time average of power flow density and reactive power flow density, respectively. In this book, the time average value of the power flow density is abbreviated as the power flow density. From Eq. (2.2.8)

$$\vec{S} = \frac{1}{2\mu_0}\left\langle \mathrm{Re}\left(\vec{E} \times \vec{B}^*\right)\right\rangle \tag{2.2.9}$$

$$\vec{q} = \frac{1}{2\mu_0}\left\langle \mathrm{Im}\left(\vec{E} \times \vec{B}^*\right)\right\rangle \tag{2.2.10}$$

Combined with Eqs. (2.2.6) and (2.2.7), the power flow density $\vec{S}$ reads

$$\begin{aligned}\vec{S} &= \frac{1}{2\mu_0}\left\langle \mathrm{Re}\left(\vec{E} \times \vec{B}^*\right)\right\rangle \\ &= \frac{1}{4\mu_0}\left[\left(\vec{E}^* \times \vec{B}\right) + \left(\vec{E} \times \vec{B}^*\right)\right] \\ &= \frac{1}{4\mu_0}\left[i\omega\left(u\nabla u^* - u^*\nabla u\right) + 2\omega k|u|^2\hat{z} + i\omega\left(\alpha\beta^* - \alpha^*\beta\right)\left(\hat{x}\partial_y - \hat{y}\partial_x\right)|u|^2\right]\end{aligned} \tag{2.2.11}$$

Considering that

$$\left(\hat{x}\partial_y - \hat{y}\partial_x\right)|u|^2 = \nabla|u|^2 \times \hat{z} = -\frac{\partial|u|^2}{\partial r}\hat{\varphi} \tag{2.2.12}$$

Defining the polarization parameters σ:

$$\sigma \equiv i\left(\alpha\beta^* - \alpha^*\beta\right) \tag{2.2.13}$$

then Eq. (2.2.11) can be simplified as:

$$\vec{S} = \frac{1}{4\mu_0}\left[i\omega\left(u\nabla u^* - u^*\nabla u\right) + 2\omega k|u|^2\hat{z} - \omega\sigma\frac{\partial|u|^2}{\partial r}\hat{\varphi}\right] \tag{2.2.14}$$

Then, the ratio of transverse and longitudinal components of the power flow density can be obtained as

$$\vec{S}_T = \frac{i\omega}{4\mu_0}\left(u\nabla u^* - u^*\nabla u\right) - \frac{\omega\sigma}{4\mu_0}\frac{\partial|u|^2}{\partial r}\hat{\varphi} \tag{2.2.15}$$

$$\vec{S}_z = \frac{\omega k|u|^2\hat{z}}{2\mu_0} \tag{2.2.16}$$

From Eqs. (2.2.15) and (2.2.16), the power flow density $\vec{S}$ along the beam's propagation direction is independent of the polarizations, while the polarization-related term present at the transverse component. The power flow density associated with polarizations is along the azimuthal direction, whose absolute value is proportional to the radial gradient of the field intensity and polarization parameter σ. For the case of linear polarizations, the imaginary parts of parameters α and β are zero, thus $\sigma = 0$, then Eq. (2.2.15) can be simplified as

$$\vec{S}_T = \frac{i\omega}{4\mu_0}\left(u\nabla u^* - u^*\nabla u\right) \tag{2.2.17}$$

The linearly polarized vortex beams with helical wavefronts under cylindrical coordinate read:

$$u(r, \varphi, z) = u_0(r, z)\exp[i\phi(r, z)]\exp(il\varphi) \tag{2.2.18}$$

with u_0 and ϕ the amplitude and initial phase, respectively. The power flow densities of vortex beams along the three coordinates axes are acquired if we substitute Eq. (2.2.18) into Eqs. (2.2.16) and (2.2.17)

$$\vec{S}_r = \frac{\varepsilon_0\omega c^2}{2}\partial_r\phi\hat{r} \tag{2.2.19}$$

$$\vec{S}_\varphi = \frac{l\varepsilon_0\omega c^2 u_0^2}{2r}\hat{\varphi} \tag{2.2.20}$$

$$\vec{S}_z = \frac{\varepsilon_0\omega c^2 k u_0^2}{2}\hat{z} \tag{2.2.21}$$

where ε_0 is the vacuum dielectric constant, and $c = (\varepsilon_0\mu_0)^{-0.5}$ is the speed of light in vacuum. Considering the relationship between the momentum density and the power flow density

$$\vec{p} = \frac{\varepsilon_0}{2}\left\langle \mathrm{Re}\left(\vec{E} \times \vec{B}^*\right)\right\rangle = \frac{\vec{S}}{c^2} \tag{2.2.22}$$

One can obtain the momentum densities of vortex beams along three coordinate axes

$$\vec{p}_r = \frac{\varepsilon_0\omega}{2}\partial_r\phi\hat{r} \tag{2.2.23}$$

$$\vec{p}_\varphi = \frac{l\varepsilon_0\omega u_0^2}{2r}\hat{\varphi} \tag{2.2.24}$$

$$\vec{p}_z = \frac{\varepsilon_0 \omega k u_0^2}{2}\hat{z} \tag{2.2.25}$$

For vortex beams with small divergence, $\partial_r \phi = 0$, thus in Eq. (2.2.23) $\vec{p}_r = 0$. Obviously, the motion direction of photons in vortex beams with helical wavefront is unparallel to the optical axis, but has a deviation angle γ. From Eqs. (2.2.24) and (2.2.25), such angle is

$$\gamma = \frac{p_\varphi}{p_z} = \frac{l\lambda}{2\pi r} \tag{2.2.26}$$

From Eq. (2.2.26), the angle γ is associated with the parameter l, the radial coordinate r, and the wavelength λ. The deviation angle is a very small quantity. For instance, for a beam with the wavelength 1550 nm, the γ at the distance 1 mm from the beam center is only 0.247 mrad. Equation (2.2.26) also implies that, instead of traveling in a straight line, the energy of a vortex beam travels in a spiral, which is the biggest difference from the common Gaussian beams.

2.2.2 *Orbital Angular Momentum (OAM)*

With the development of physics, the wave-particle duality of light is proposed, and optics is extended from the wave theory to the quantum world. Similar with macroscopic objects, photons can also carry angular momentums. The angular momentum of a photon consists of two parts, the spin angular momentum (SAM) and the orbital angular momentum (OAM). The SAMs of photons are closely related to the polarization property of beams. While the OAMs of photons come from the complex amplitude distribution of the cross section of light fields, representing the wavefront property.

The study of the angular momentums of photons start from the SAMs. In 1909, Poynting has already predicted the circularly polarized beam carries the angular momentum. The change of beams' polarizations (e.g. from the linear polarization to the circular polarization) must be accompanied by an angular momentum exchange with the optical system [8]. In 1936, Beth et al. observed the SAMs of photons for the first time through the counterforce from the shift of the polarization state of a beam caused by a 1/2 wave plate suspended on a quartz filament, and the SAM value carried by each photon must be $+\hbar$ or $-\hbar$, with $\hbar$ the Planck constant divided by 2π [9].

The study on photons' OAM is later than that of the SAM. In 1950s, researchers found that the electromagnetic radiation from the process of electrical multipole radiation carries OAM [10]. However, in the decades that followed, the research on the OAM of photons remained at the theoretical stage without any substantial breakthrough. Until 1992, Allen et al. found the OAM is intrinsic property of the beam with a helical wavefront. And the OAM value that carried by each photon in a vortex beam is the integral multiple of $\hbar$ [11]. This discovery starts the research of vortex beams and their OAM. In this sub-section, we will introduce the OAM characteristics of vortex beams through formula derivation.

Equation (2.2.22) has already gives the definition of linear momentum density $\vec{p}$. From the relationship between the angular momentum density $\vec{m}$ and linear momentum density

$$\vec{m} = \vec{r} \times \vec{p} \tag{2.2.27}$$

and combined with Eqs. (2.2.22) and (2.2.14), one can obtain

$$\vec{m} = \frac{\varepsilon_0 \omega}{4}\left[i\left(u\partial_\varphi u^* - u^*\partial_\varphi u\right) - r\sigma\frac{\partial |u|^2}{\partial r}\right]\hat{z} - \frac{\varepsilon_0 \omega}{4}\left[ir\left(u\partial_z u^* - u^*\partial_z u\right) - 2k|u|^2 r\right]\hat{\varphi} \tag{2.2.28}$$

Obviously, the angular momentum consists of two components, one is along the beams' propagation direction $\hat{z}$, and the other is along the azimuth $\hat{\varphi}$ on the cross section. Here only the more practical component along $\hat{z}$ is considered, which reads:

$$m_z = \frac{\varepsilon_0 \omega}{4}\left[i\left(u\partial_\varphi u^* - u^*\partial_\varphi u\right) - r\sigma\frac{\partial |u|^2}{\partial r}\right] \tag{2.2.29}$$

The total angular momentum $\vec{M}$ is acquired when integrating Eq. (1.2.28) in the whole space:

$$\vec{M} = \iiint \vec{m} r dr d\varphi dz \tag{2.2.30}$$

By differentiating the total angular momentum with respect to time t, the flux of angular momentum flowing through the cross section in unit time is obtained as

$$\vec{J} = \frac{d\vec{M}}{dt} = \frac{d}{dt}\iiint \vec{m} r dr d\varphi dz = c\iint \vec{m} r dr d\varphi \tag{2.2.31}$$

Its components along the direction $\hat{z}$ is

$$J_z = c\iint \vec{m}\cdot\hat{z} r dr d\varphi = c\iint m_z r dr d\varphi \tag{2.2.32}$$

Substituting Eq. (2.2.29) into Eq. (2.2.32), one can obtain

$$J_z = \frac{c\varepsilon_0 \omega}{4}\iint\left[i\left(u\partial_\varphi u^* - u^*\partial_\varphi u\right) - r\sigma\frac{\partial |u|^2}{\partial r}\right] r dr d\varphi \tag{2.2.33}$$

which can be decomposed into:

$$J_z = J_{zl} + J_{z\sigma} \tag{2.2.34}$$

where

$$J_{zl} = \frac{c\varepsilon_0\omega}{4}\iint i\left(u\partial_\varphi u^* - u^*\partial_\varphi u\right)rdrd\varphi \tag{2.2.35}$$

$$J_{z\sigma} = -\frac{c\varepsilon_0\omega}{4}r^2\sigma\frac{\partial|u|^2}{\partial r}drd\varphi \tag{2.2.36}$$

From Eqs. (2.2.35) and (2.2.36), J_{zl} is independent of polarizations, which is defined as OAM flux. And $J_{z\sigma}$ is associate with polarizations, which is defined as SAM flux.

For the vortex beams with helical phase term given by Eq. (2.2.18), the OAM components on the beams' propagation direction can be derived from Eq. (2.2.29) as:

$$m_z = \frac{\varepsilon_0\omega l}{2}|u|^2 - \frac{\varepsilon_0\omega\sigma r}{4}\frac{\partial|u|^2}{\partial r} \tag{2.2.37}$$

If n denotes the number of photons in unit volume, ω denotes the angular frequency of photons, the energy density w of the beam is:

$$w = n\hbar\omega = \frac{p_z\omega}{k} \tag{2.2.38}$$

Equation (2.2.25) with Eq. (2.2.38) leads to:

$$w = \frac{\varepsilon_0\omega^2 k u_0^2}{2k} = \frac{\varepsilon_0\omega^2|u|^2}{2} \tag{2.2.39}$$

From Eqs. (2.2.37) and (2.2.39) the ratio of angular momentum density m_z and the energy density w along the beam propagation direction is derived as:

$$\frac{m_z}{w} = \frac{l}{\omega} - \frac{\sigma r}{2\omega|u|^2}\frac{\partial|u|^2}{\partial r} \tag{2.2.40}$$

Equation (2.2.40) indicates that, the ratio of local angular momentum density and the energy density of the vortex beam are influenced by the polarization parameter σ. When $\sigma = 0$, such ratio is fixed and equals to l/ω. If σ is nonzero, the ratio is related to the field distribution on the cross section. By integrating the numerator and denominator of Eq. (2.2.40) on the cross section of the beam simultaneously, the ratio of total angular momentum to total energy can be obtained as:

$$\frac{M_z}{W} = \frac{\iint m_z r dr d\varphi}{\iint w r dr d\varphi} = \frac{l + \sigma}{\omega} \tag{2.2.41}$$

Equation (2.2.41) implies that, the ratio of total angular momentum to total energy per unit length over the cross-section is independent of the field distribution, but is only affected by the parameters l and σ. The numerator and denominator of Eq. (2.2.41) multiplied simultaneously by $n\hbar$ result in

$$M_z = nl\hbar + n\sigma\hbar \tag{2.2.42}$$

where $nl\hbar$ is the OAM and $n\sigma\hbar$ is the SAM, indicating that under the paraxial approximation the OAM and SAM of a vortex beam can be separated. From Eq. (2.2.42) the angular momentum carried by each photon is

$$M_{zs} = l\hbar + \sigma\hbar \tag{2.2.43}$$

indicating that the OAM carried by each photon in a vortex beam with helical term $\exp(il\varphi)$ is $l\hbar$. In Eqs. (2.2.42) and (2.2.43), l is called the topological charge, also known as the orbital quantum number, which can be any integer. Meanwhile, the topological charge l is the eigenvalue of a vortex beam, which determines the helical wavefront. In other words, the absolute value of l determines the OAM value carried by each photon in a vortex beam. And the sign of l determines the rotating direction of the helical wavefront. Therefore, vortex beams with topological charge l is called l-th order vortex beams, represented by Dirac notation as $|l\rangle$.

The parameter σ is called the spin quantum number, which has two eigenvalues as -1 and $+1$. $\sigma = +1$ represents the right circular polarization and $\sigma = -1$ represents the left circular polarization. In previous discussions, σ can be any value of the closed interval $[-1, +1]$. This is actually the macroscopic result of multi-photon superposition, namely, any polarizations result from the linear superposition of many photons with opposite spin quantum numbers. Henceσ discussed previously can't reflect the spin properties of photons. In Chap. 7 the generation of arbitrary polarizations will be discussed in detail.

2.2.3 Orthogonality

Considering two vortex beams with different topological charges

$$u_1(r, \varphi, z) = f_1(r, \varphi)\exp(il_1\varphi) \tag{2.2.44}$$

$$u_2(r, \varphi, z) = f_2(r, \varphi)\exp(il_2\varphi) \tag{2.2.45}$$

Integrating their inner product in the azimuth results in

$$\int_{0}^{2\pi} u_1 \cdot u_2^* d\varphi = f_1 \cdot f_2^* \delta(l_2 - l_1) \tag{2.2.46}$$

where

$$\delta(\varsigma) = \begin{cases} 1, \varsigma = 0 \\ 0, \varsigma \neq 0 \end{cases} \tag{2.2.47}$$

Therefore, vortex beams with different topological charges are orthogonal with each other. The orthogonality of vortex beams indicates that, the coaxially propagated different vortex beams can be separated from each other without influences. Such features of vortex beams makes it possible to apply in large-capacity data transmissions.

2.2.4 *Mirror Image Property*

Plane mirror is a common device in optical system, which can be used to change the direction of light propagation. Single plane mirror imaging has the following properties:

(1) The plane mirror can realize ideal imaging in the whole space, where the object point and the image point are symmetric about the plane mirror.
(2) Objects and images have same sizes, but different shapes. The right-handed coordinate system in the object space is changed into the left-handed coordinate system in the image space, namely, the *x*-axis and *y*-axis are interchanged, as shown in Fig. 2.3. When observing the object plane and the image plane in the direction of the incoming and outgoing rays respectively, if the object plane rotates counterclockwise, the image plane will rotate clockwise, forming a "mirror image".

For a vortex beam system, a plane mirror do nothing on the intensity distribution of beams due to the ideal imaging property. Hence we only concern the variation

Fig. 2.3 The coordinate system transformation from specular reflection

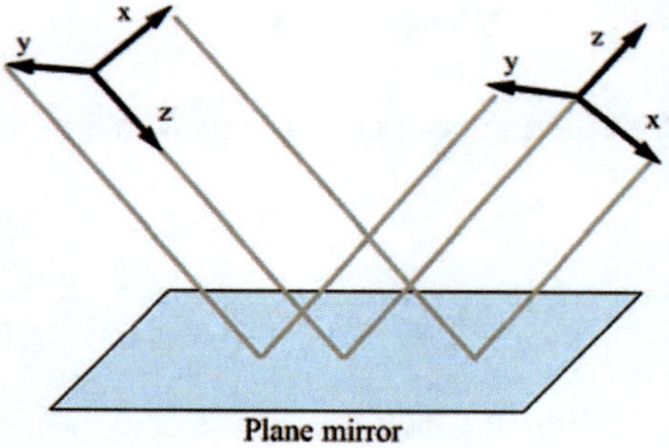

of the OAM, or the eigenvalue topological charge. As shown in Fig. 2.4, supposing there is a vortex beam in a right-handed coordinate system, the rotation direction of its wavefront is from x axis to y axis. After reflected by a plane mirror, the right-handed coordinate turns to left-handed coordinate, where the x-axis and y-axis are interchanged. However, the rotation direction of the helical wavefront is still from x axis to y axis. The rotation direction of the helical wavefront of the reflected vortex beams must be opposite compared with that of the incident vortices. Therefore, the topological charge will be opposite if a vortex beam is reflected by a plane mirror, which reads $|l\rangle \rightarrow |-l\rangle$. In other words, vortex beams with opposite topological charge are a pair of mirror image.

If a reflected optical vortex passing through another mirror, the left-handed coordinates turns to the right-hand coordinates again, and the topological charge of the twice reflected optical vortex is exactly the same as before. Obviously, when a vortex beam is reflected by the plane mirrors n times, if n is odd, the topological charge after reflections is opposite. If n is even, the topological charge won't be changed, which can be expressed as:

$$|l\rangle \xrightarrow{n \text{ times reflection}} |(-1)^n l\rangle \tag{2.2.48}$$

Therefore, effects from the specular reflection on the topological charges must be considered when designing vortex beams system. If there are an odd number of reflectors in the system, an additional reflection must be introduced in the optical path through a Dove prism or other devices to ensure that the topological charge is unchanged.

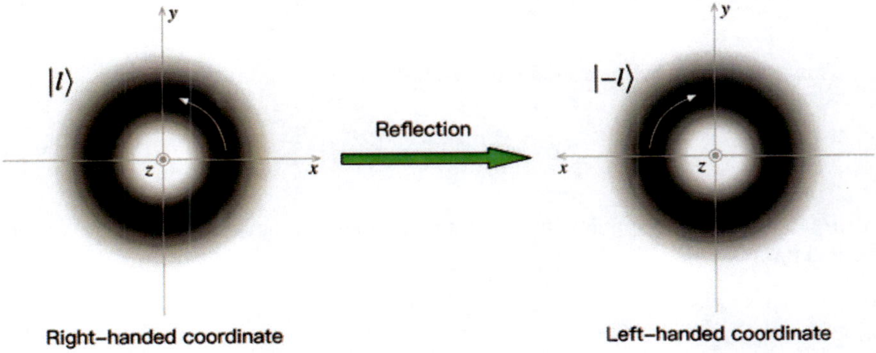

Fig. 2.4 The specular reflection of vortex beams

2.3 Orbital Angular Momentum Spectrum

Carrying OAM is one of the most important features of vortex beams. A vortex beam can carry multiple various OAM components. The proportion of intensity weight of each OAM channel determines the intensity, phase and wavefront distributions of the beam. The intensity proportions of different OAM channels in a beam is called OAM spectrum, which can reflect some of the features of a OAM-carrying beam. In this section, the OAM spectrum will be analyzed in the perspective of helical harmonic expansion and rotating operator.

2.3.1 *Helical Harmonic Expansion*

The helical harmonic $\exp(il\varphi)$ is the eigen wave function of OAM. Due to the azimuthally periodic distributions of helical harmonic, a beam can be expanded directly through the helical harmonics.

An optical field $u(x, y, z)$ can be expanded through helical harmonic $\exp(il\varphi)$ as

$$u(r,\varphi,z) = \frac{1}{\sqrt{2\pi}} \sum_{l=-\infty}^{+\infty} a_l(r,z)\exp(il\varphi) \tag{2.3.1}$$

with the coefficient

$$a_l(r,z) = \frac{1}{\sqrt{2\pi}} \int_0^{2\pi} u(r,\varphi,z)\exp(-il\varphi)d\varphi \tag{2.3.2}$$

Thus the intensity of such helical harmonic is

$$C_l = \int_0^{\infty} |a_l(r,z)|^2 r\,dr \tag{2.3.3}$$

Since the value C_l is independent of the parameter z, the relative intensity of such helical harmonic is

$$R_l = \frac{C_l}{\sum\limits_{q=-\infty}^{+\infty} C_q} \tag{2.3.4}$$

And by now the OAM spectrum is acquired.

In addition, from Eqs. (2.3.1)–(2.3.4), obviously arbitrary beams can be regarded as the linear composition of multiple helical harmonics with various orders.

2.3.2 Rotating Operator

The OAM can be understood as the orbital motion around the reference point in the spatial coordinate system. If the spatial scalar wave function Ψ rotates with the angle α around the z axis, the wave function Ψ ' after rotation is

$$\psi' = \hat{\vec{R}}_n(\alpha)\psi \tag{2.3.5}$$

where $\hat{\vec{R}}_n(\alpha)$ is the rotating operator:

$$\hat{\vec{R}}_n(\alpha) = \exp\left(-\frac{i}{\hbar}\alpha\vec{z}\cdot\hat{\vec{L}}\right) \tag{2.3.6}$$

In Eq. (2.3.6), $\hat{\vec{L}}$ is the OAM operator. From quantum mechanics,

$$\hat{\vec{L}} = -i\hbar\hat{\vec{r}} \times \nabla \tag{2.3.7}$$

with $\hat{\vec{r}}$ the radius vector operator. Therefore, OAM can be used to reflect the rotation of spatial scalar wave function, namely, the rotation of spatial coordinates.

Equation (2.3.7) with Eq. (2.3.6) results in:

$$\hat{\vec{R}}_n(\alpha) = \exp\left(-\frac{i}{\hbar}\alpha\hat{L}_z\right) \tag{2.3.8}$$

where $\hat{L}_z$ is the angular momentum operator in z direction:

$$\hat{L}_z = \hat{x}\hat{p}_y - \hat{y}\hat{p}_x = -i\hbar\partial_\varphi \tag{2.3.9}$$

Thus Eq. (2.3.8) can be simplified as:

$$\hat{\vec{R}}_n(\alpha) = \exp(-\alpha\partial_\varphi) \tag{2.3.10}$$

Considering the average function of a wave function, and let β = -α, then

$$M(\alpha) = \frac{\langle\psi|\hat{\vec{R}}_n(\alpha)|\psi\rangle}{\langle\psi|\psi\rangle} = \frac{\langle\psi|\exp(\beta\alpha_\varphi)|\psi\rangle}{\langle\psi|\psi\rangle} \tag{2.3.11}$$

Expanding the wave function shown in Eq. (2.3.11) through helical harmonic leads to

$$M(\beta) = \frac{\langle\psi|\exp(\beta\partial_\varphi)|\psi\rangle}{\langle\psi|\psi\rangle} = \sum_{m,l}\frac{\langle\psi|m\rangle\langle m|\exp(\beta\partial_\varphi)|l\rangle\langle l|\psi\rangle}{\langle\psi|\psi\rangle}$$
$$= \sum_{m,l}\frac{\langle\psi|m\rangle\langle l|\psi\rangle}{\langle\psi|\psi\rangle}\langle m|l\rangle\exp(i\beta l) = \sum_{l}\frac{\langle\psi|l\rangle\langle l|\psi\rangle}{\langle\psi|\psi\rangle}\exp(i\beta l) = \sum_{l}R_l\exp(i\beta l) \tag{2.3.12}$$

Thus R_l and $M(\beta)$ satisfy the Fourier transform relationship as

$$R_l = \frac{1}{2\pi}\int M(\beta)\exp(-i\beta l)d\beta \tag{2.3.13}$$

In Eq. (2.3.13), R_l is the relative intensity proportion of helical harmonic $\exp(il\varphi)$, namely, the OAM spectrum. Compared with the analyzing method through helical harmonic expansion, the method given in this sub-section employs the symbol representation method of quantum mechanics to describe the rotation and mean value of a vortex beam, so as to obtain the OAM spectrum conveniently.

2.4 The Common Vortex Beams

As mentioned previously, a vortex beam has a helical phase term $\exp(il\varphi)$, and each photon in which carries OAM. The common vortex beams that meet the above conditions include Laguerre–Gauss beams, Bessel beams, Bessel-Gauss beams and so on. These three vortex beams have both similarities and differences, which will be discussed in detail in this section.

2.4.1 *Laguerre–Gauss Beam*

Laguerre–Gauss (LG) beam is a kind of high-order Gaussian beams. In a stable cylindrical symmetric cavity, such as a confocal cavity with circular aperture, the higher order transverse mode is described by the product of the associating Laguerre polynomial and the Gaussian function. Therefore, a LG beam propagating along z axis reads:

$$LG_{pl}(r,\varphi,z) = \frac{C_{pl}}{\omega_0}\left(\frac{\sqrt{2}r}{\omega(z)}\right)^{|l|}L_p^{|l|}\left(\frac{2r^2}{\omega(z)^2}\right)exp\left(-\frac{r^2}{\omega(z)^2}\right)\exp(il\varphi)exp(i\Phi) \tag{2.4.1}$$

where C_{pl} is a constant, ω_0 is the waist radius of fundamental mode, l is the topological charge and p is the radial index which can be any non-negative integer. $\omega(z)$ and Φ

read

$$\omega(z) = \omega_0\sqrt{1 + \left(\frac{z}{f}\right)^2} \tag{2.4.2}$$

$$\Phi = (|l| + 2p + 1)arctan\frac{z}{f} - k\left(z + \frac{r^2}{2R}\right) \tag{2.4.3}$$

where

$$R = z + \frac{f^2}{z} \tag{2.4.4}$$

In Eqs. (2.4.2) and (2.4.3), k denotes the wave number, and f denotes the confocal parameters, also known as the Rayleigh length

$$f = \frac{\pi\omega_0^2}{\lambda} \tag{2.4.5}$$

$L^l{}_p(\varsigma)$ is the associating Laguerre polynomial:

$$L_0^{|l|}(\varsigma) = 1 \tag{2.4.6}$$

$$L_1^{|l|}(\varsigma) = 1 + |l| - \varsigma \tag{2.4.7}$$

$$L_2^{|l|}(\varsigma) = \frac{1}{2}\left[(1 + |l|)(2 + |l|) - 2(2 + |l|)\varsigma + \varsigma^2\right] \tag{2.4.8}$$

$$\vdots = \vdots$$

$$L_p^{|l|}(\varsigma) = \sum_{m=0}^{p} \frac{(p + |l|)!(-\varsigma)^m}{(|l| + m)!m!(p - m)!} \tag{2.4.9}$$

Compared with Gaussian beams, the transverse fields of LG beams are described by associating Laguerre polynomial, indicating that there are p nodal circles along the radius. Meanwhile, the complex amplitude of a LG beam comprises the helical term $\exp(il\varphi)$, implying that each photon in LG beams carries OAM.

As a high-order Gaussian beams, both the spot radius and divergence of LG beams are associated with fundamental mode Gaussian beams.. From principles of lasers, the beam radius of a LG beam reads

$$\omega_{pl}(z) = \sqrt{|l| + 2p + 1} \cdot \omega(z) \tag{2.4.10}$$

and the divergence angle θ_{pl} is

$$\theta_{pl} = \lim_{z \to \infty} \frac{\omega_{pl}(z)}{z} = \frac{2\lambda\sqrt{|l| + 2p + 1}}{\pi \omega_0} = \sqrt{|l| + 2p + 1} \cdot \theta_0 \tag{2.4.11}$$

where θ_0 corresponds to the divergence angle of fundamental modes:

$$\theta_0 = \lim_{z \to \infty} \frac{2\omega(z)}{z} = \frac{2\lambda}{\pi \omega_0} \tag{2.4.12}$$

Equations (2.4.10) and (2.4.11) indicate that, the radius and divergence angle of the LG beam will increase with the increasing of radial index p and topological charge l, and the increase with p is faster than that with l. In particular, when $p = l = 0$, from Eq. (2.4.6)$L_0^0 = 1$, thus the LG beam given in Eq. (2.4.2) is reduced to a fundamental mode Gaussian beam.

Figure 2.5 presents the intensity and phase distributions of some LG beams with various radial indices p and topological charges l when $z = 0$. Note that, for all the phase distributions and phase gratings present in this book, black denotes 0 and white denotes 2π. From the first line in Fig. 2.5, when $p = 0$, LG beams are single ring-shaped like a doughnut, else LG beams have (p-1) concentric rings. The topological charge l is reflected by the phase distribution. The sign and absolute value of l determines the direction and magnitude of spiral phase gradient respectively. The size of LG beams shown in Fig. 2.5 varies with parameters p and l, which satisfies Eq. (2.4.10).

Usually, the single-ring LG beams are the simplest vortex beams, and having higher application values. Therefore, in the following discussions in this book, vortex beams refer to single-ring LG beams without special indications.

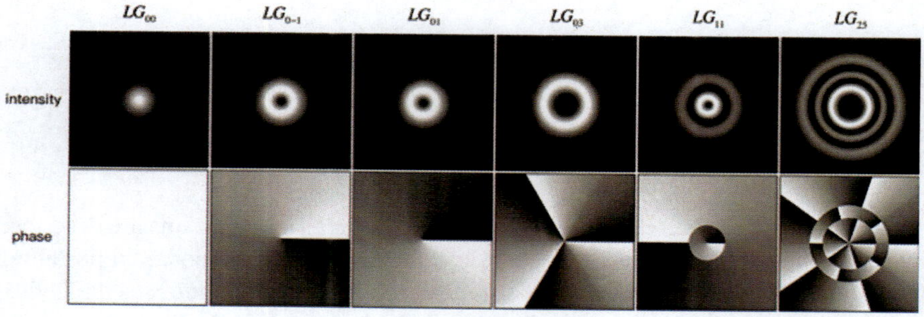

Fig. 2.5 Intensity and phase distributions of some of the LG beams with various radial indices p and topological charges l when $z = 0$

2.4.2 Bessel Beam

The Bessel beam is a kind of non-diffraction beams, which is a particular solution of the Helmholtz equation in cylindrical coordinates [12]. Similar with LG beams, the non-zeroth order Bessel beams are also vortex beams with helical wavefront and carry OAM per photon. The complex amplitude of a l-th order Bessel beam is

$$BS_l(r, \varphi, z) = A_l J_l(k_r r) \exp(ik_z z) \exp(il\varphi) \tag{2.4.13}$$

where A_l is a constant, k_r and k_z are the wave number components in radial and propagation direction separately and meet

$$k_r^2 + k_z^2 = k^2 = \frac{4\pi^2}{\lambda^2} \tag{2.4.14}$$

with k the wave number. $J_l(\varsigma)$ is l-th order Bessel function of the first kind

$$J_l(\varsigma) = \sum_{m=0}^{\infty} \frac{(-1)^m}{m!\Gamma(m+l+1)} \left(\frac{\varsigma}{2}\right)^{2m+l} \tag{2.4.15}$$

In Eq. (2.4.15), $\Gamma(\xi)$ denotes Gamma function, which can be regarded as the extension of the factorial function to non-integer variable. In real number field, Gamma function reads

$$\Gamma(\xi) = \int_0^{+\infty} t^{\xi-1} e^{-t} dt \tag{2.4.16}$$

Particularly, when ξ is a positive integer, Eq. (2.4.16) can be written as

$$\Gamma(\xi) = (\xi - 1)! \tag{2.4.17}$$

The graph of Bessel functions of the first kind with various orders derived from Eqs. (2.4.15)–(2.4.17) is present in Fig. 2.6.

The intensity distribution of Bessel beams I_{BS} can be obtained from Eq. (2.4.13) as

$$I_{BS} = |BS_l|^2 \propto J_l^2(k_r r) \tag{2.4.18}$$

Equation (2.4.18) indicates that, a Bessel beam has an infinitely extended light field distribution, that is, an infinite cross section area. Figure 2.7 presents the intensity and phase distributions of some Bessel beams. From the point of intensity distribution, there is a central bright annulus in the Bessel beam, and the central annulus of 0-th order Bessel beam is reduced to a spot, surrounded by lots of side lobes, which forming a multiple concentric rings structure. In addition, the power of each annular

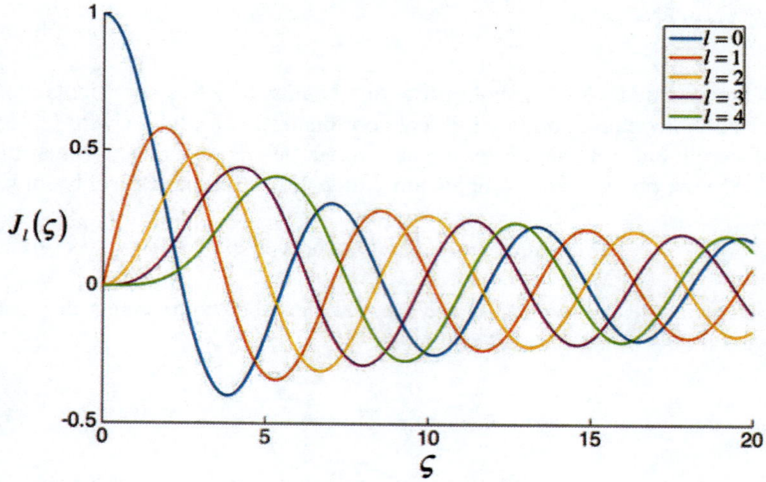

Fig. 2.6 Bessel functions of the first kind

lobe of a Bessel beam is almost identical with that of the central spot. From the point of phase, Bessel beams have similar helical phases with those of LG beams, where the sign and absolute value of the topological charge l determines the direction and magnitude of spiral phase gradient respectively. Meanwhile, the size of central bright annulus increases with the increase of $|l|$, which can be easily understood from the graph shown in Fig. 2.6.

The Bessel beam is a kind of non-diffraction beams, indicating that if there is an obstacle in its propagation path, it can recover itself after passing through the obstacle for a certain distance, that is, it has self-healing property, as shown in Fig. 2.8. This phenomenon results from that, a Bessel beam is actually a conical wave as the interference field superimposed by many plane sub-waves of equal amplitude. The angle between each of the wave vector of these plane sub-waves and the z axis is

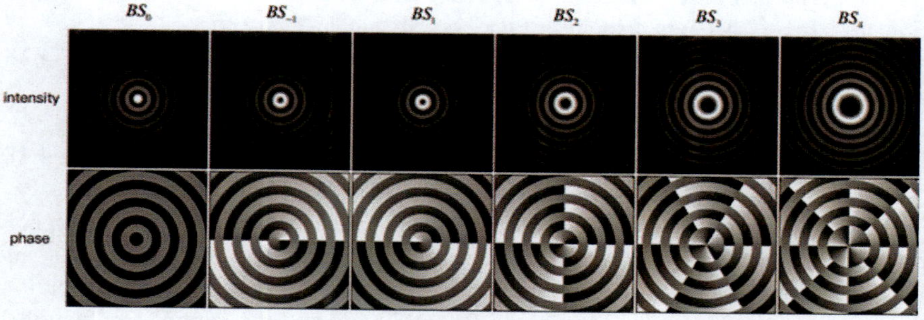

Fig. 2.7 The intensity and phase distributions of various Bessel beams when $z = 0$

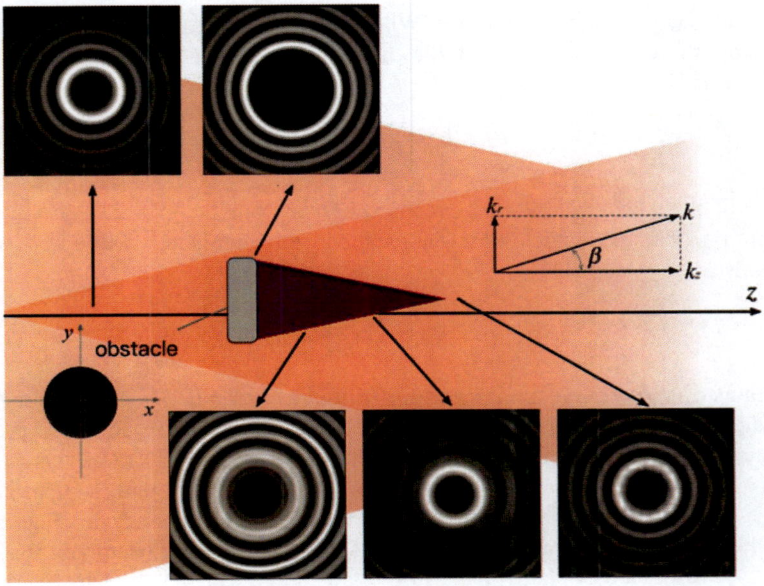

Fig. 2.8 The self-healing of Bessel beams

$$\beta = \arcsin\left(\frac{k_r}{k}\right) \tag{2.4.19}$$

with β the diffraction angle or cone angle of Bessel beams. When there is an obstacle in the propagation path of a Bessel beam, the unblocked rays will interfere behind the obstacle and form the Bessel beam again, thus realizing the self-healing.

2.4.3 Bessel-Gauss Beam

Since Bessel beams have infinitely extended field distribution, they are only ideal theoretical models, and won't exist in practice. Actually, Bessel-Gauss (BG) beams are employed as the approximation of Bessel beams [13]. BG beams have the same non-diffraction characteristics as Bessel beams in a limited propagation distance. If propagating beyond the maximum propagation distance, BG beams will no longer exist. The complex amplitude of a l-th order BG beam reads:

$$BG_l(r, \varphi, z) = A_l J_l(k_r r) \exp(ik_z z) \exp(il\varphi) \exp\left(-\frac{r^2}{\omega_0^2}\right) \tag{2.4.20}$$

with ω_0 the size of limited aperture. Compared with Eqs. (2.4.13), (2.4.20) contains an additional real term $\exp(-r^2/\omega_0^2)$, indicating that BG beams have identical phase distributions with ideal Bessel beams. When $l \neq 0$, each photon in a BG beam carries OAM with the value $l\hbar$. From (2.4.20), the intensity of a BG beam I_{BG} is:

$$I_{BG} = |BG_l|^2 \propto J_l^2(k_r r)\exp\left(-\frac{2r^2}{\omega_0^2}\right) \quad (2.4.21)$$

For a BG beam, both k_r and ω_0 are fixed. Thus from Eq. (2.4.21), the intensity distribution of a BG beam is only associated with parameters l and r. Here the function:

$$f_l(\varsigma) = J_l^2(\varsigma)\exp(-2\varsigma^2/\omega_0) \quad (2.4.22)$$

is employed to represent the intensity of different order BG beams varying with the radial coordinates, as displayed in Fig. 2.9. As a comparison, Fig. 2.10 presents the intensity of different order Bessel beams varying with the radial coordinates represented by the function $g_l(\varsigma) = J_l^2(\varsigma)$. One can acquire that, compared with $g_l(\varsigma)$, when introducing the term exp(-2ς^2),$f_l(\varsigma)$ converges much faster. Figure 2.11 gives the intensity and phase distributions of fifth order Bessel beam and BG beam when $z = 0$, obviously their phases are totally identical, but their intensities are different. The intensity of the BG beam exists only in the center area (inside the dotted box), while that of the Bessel beam is infinitely extended. Figures 2.9, 2.10 and 2.11 indicate that, BG beams are in fact the Bessel beams confined by the term $\exp(-r^2/\omega_0^2)$ and have finite cross-sectional field distributions.

BG beams exist only in a limited propagation distance, which is different from Bessel beams in full space. Such difference can be understood from the interference

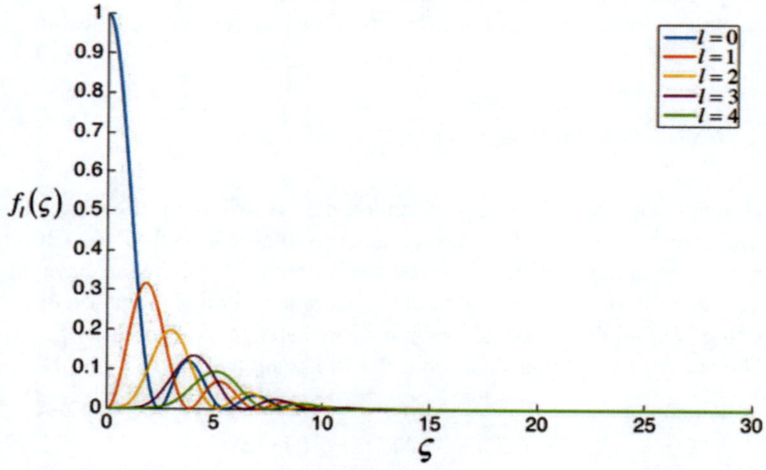

Fig. 2.9 Intensity of different order BG beams varies with the radial coordinates represented by the function $f_l(\varsigma)$

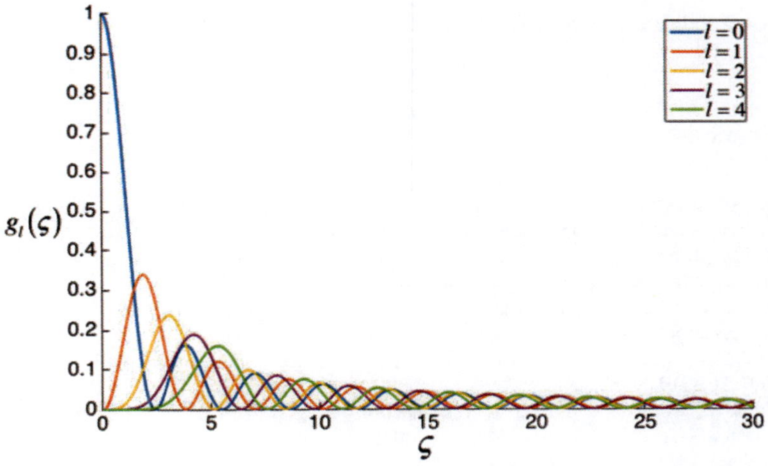

Fig. 2.10 Intensity of different order Bessel beams varies with the radial coordinates represented by the function $g_l(\varsigma)$

Fig. 2.11 Intensity and phase distributions of fifth order Bessel beam and BG beam when $z = 0$

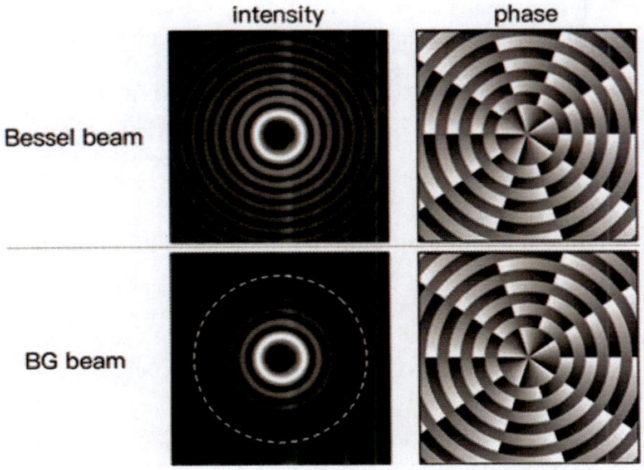

field. As mentioned in Sect. 2.4.2, the Bessel beam is superposed by a number of plane sub-waves with equal-amplitude and β angle arranged wave vector. The wavefronts of these sub-waves are infinitely extended and have no aperture limit. BG beams are also superposed by lots of plane sub-waves, but these sub-waves have aperture limits and thus have finite wavefronts, leading to that the largest propagation distance or field range $z_{\max}$ of a BG beam is associated with the aperture size ω_0 and diffraction angle β. $z_{\max}$ is obtained through a simple geometric derivation as

$$z_{\max} = \omega_0 \cot \beta \tag{2.4.23}$$

A detailed discussion of Eq. (2.4.23) will take place in Chap. 3.

References

1. Wang J, Yang JY, Fazal IM, et al. Terabit free-space data transmission employing orbital angular momentum multiplexing. Nat Photonics. 2012;6(7):488–96.
2. Lavery M, Speirits F, Barnett SM, et al. Detection of a spinning object using light's orbital angular momentum. Science. 2013;341(6145):537–40.
3. Simpson NB, Dholakia K, Allen L, et al. Mechanical equivalence of spin and orbital angular momentum of light: an optical spanner. Opt Lett. 1997;22(1):52–4.
4. Gao C, Gao M, Weber H. Generation and application of twisted hollow beams. Optik. 2004;115(3):129–32.
5. Meier M, Romano V, Feurer T. Material processing with pulsed radially and azimuthally polarized laser radiation. Appl Phys A. 2007;86(3):329–34.
6. Tamburini F, Thidé B, Molina G. Twisting of light around rotating black holes. Nat Phys. 2011;7(3):195–7.
7. Nicolas A, Veissier L, Giner L, et al. A quantum memory for orbital angular momentum photonic qubits. Nat Photonics. 2014;8(3):234–8.
8. Poynting JH. The wave motion of a revolving shaft, and a suggestion as to the angular momentum in a beam of circularly polarised light. Proc R Soc A. 1909;82(557):560–7.
9. Beth R. Mechanical detection and measurement of the angular momentum of light. Phys Rev. 1936;50:115–25.
10. Jackson JD. Classical electrodynamics, 3rd ed. Wiley; 1962.
11. Allen L, Beijersbergen MW, Spreeuw RJ, et al. Orbital angular momentum of light and the transformation of Laguerre-Gaussian laser modes. Phys Rev A. 1992;45(11):8185.
12. Durnin J. Exact solutions for nondiffracting beams. I. The scalar theory. J Opt Soc Am A, 1987; 4:651–654.
13. Gori F, Guattari G, Padovani C. Bessel-Gauss beams. Opt Commun. 1987;64(6):491–5.

Chapter 3
Generation of Vortex Beams

Abstract The generation of vortex beams carrying orbital angular momentum (OAM) is the basis of further study and practical applications. The topological charge l is the eigenvalue of vortex beams, determining the OAM value each photon carries. Therefore we will discuss the methods of generating vortex beams with arbitrary l value. They can be generated inside the laser resonator by suitable boundary conditions or outside by mode conversion. Both approaches will be discussed in this chapter.

3.1 Intra Resonator Method

3.1.1 Mode Selecting Inside the Resonator

High-order transverse mode is described by the product of associative Laguerre polynomial and Gaussian distribution function in the cylinder-symmetric stable resonator, e.g., confocal resonator with circular aperture. In such kind of resonator, Laguerre–Gauss (LG) beams, the high-order transverse mode, can be obtained by adjusting optical elements as cavity reflector.

Based on this principle, C. Tamm et al. have reported a scheme to produce LG beams directly from a laser in 1988 and 1990 [1, 2]. They introduce the loss of low-order mode inside the resonator, leading to the no fundamental mode resonance and high-order mode resonance. Then LG beams is obtained. However, such scheme is facing many problems as high loss, small topological charge output, low output power, etc.

Except for introducing low-order mode loss inside the resonator, one can also obtain high-order modes (vortex beams) through adjusting the intensity distribution of pump beams. For instance, ring-shaped pump beams along with optimized resonator parameter will result in the output of two-fold multiplexed vortices ($|+1\rangle + |-1\rangle$). If an etalon is placed at a special arranged angle inside the resonator, it will oscillate one mode but suppress the other. Then single mode vortex beams ($|+1\rangle$ or $|-1\rangle$) will be obtained [3]. The efficiency here is higher than the scheme of introducing low-order mode loss. Nevertheless, the arranged angle of the etalon

S. Fu and C. Gao, *Optical Vortex Beams*, Advances in Optics and Optoelectronics,
https://doi.org/10.1007/978-981-99-1810-2_3

must be adjusted if other-order vortex beams are desired, which means such scheme is unsuitable for producing arbitrary vortex beams flexibly.

3.1.2 *Digital Lasers*

Usually, a classical lasers can only produce one specific mode without adjusting any elements. Some of the resonator parameters must be changed when other modes are desired. If we can design a new kind of lasers, which can output arbitrary modes without changing any elements or parameters, will be of great significance for the generation of arbitrary transverse modes including vortex beams. Recently, the proposition of digital lasers makes such idea come true. It can produce lots of modes controlled only by a computer, rather than the resonator's hardware. This is also the reason why we call it digital lasers.

Digital lasers was firstly proposed by S. Ngcobo et al. [4]. The principles can be understood as: Replacing the back optical element of the cavity by a reflective phase-only liquid–crystal spatial light modulator (LC-SLM) (discussion on LC-SLM will be present in Sect. 3.3.1). By encoding different special-designed holograms on LC-SLM, various mode including vortex beams will be output. As shown in Fig. 3.1. The laser cavity is L-shaped, consists of a conventional folded resonator configuration with an Nd:YAG laser crystal as the gain medium. The laser was optically pumped by a high-power laser diode that was coupled into the cavity through a mirror coated for high transmission at the diode wavelength (808 nm) and high reflectance at the lasing wavelength (1064 nm). This folding mirror forms an L-shaped cavity so that the high-power diode beam does not interact with the LC-SLM, thus avoiding damage. Note that the intra-cavity Brewster window is to force the laser to oscillate in the desired polarization for the SLM. The light is passed out of the cavity through the output coupler (OC).

Figure 3.2 gives the various intensity distributions of the generated modes and their corresponding encoding holograms. Obviously digital lasers can produce lots of modes without adjusting any elements in the resonator. What we should do is just to change the holograms encoded on the LC-SLM.

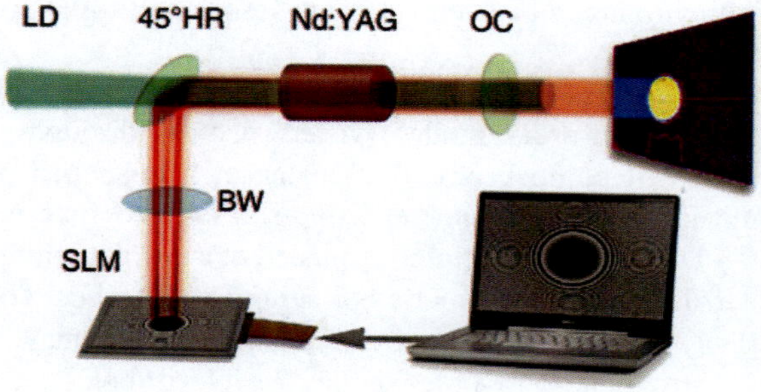

Fig. 3.1 Schematic of digital lasers. LD, laser diode; 45° HR, high reflectivity mirror at an angle of 45°; OC, output coupler; BW, Brewster window [4]

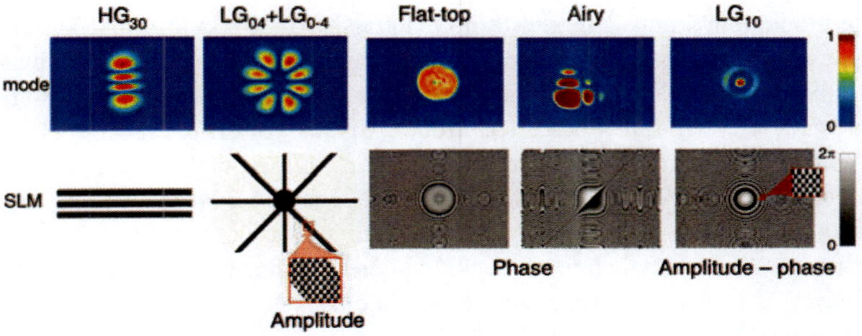

Fig. 3.2 Intensity distributions of the modes generated by the digital lasers, and their corresponding encoding holograms [4]

Although digital lasers has many advantages, limitations are still present. On one hand, the diffractive efficiency is not high, about 60% ~ 90% currently. Such low diffraction efficiency contributes to higher loss and makes it difficult to resonate. One the other hand, the damage threshold of the liquid–crystal display (LCD) of LC-SLM is low, resulting in the easy damage of the LCD. Therefore, the optical to optical conversion efficiency of digital lasers is low, making it hard to obtain high-power outputs.

In conclusion, when producing vortex beams inside the resonator one always face problems as nonadjustable topological charge, or adjustable topological charge with low generating efficiency. Currently in practice, the scheme of outside resonator is usually employed to generate vortex beams with divers topological charge.

3.2 Outside the Resonator

The so-called outside resonator method is to transform other modes as fundamental Gaussian modes into vortex beams outside the resonator. For instance, a mode converter can transform Hermite-Gauss (HG) beams into LG beams; A spiral phase plate (SPP) or fork-shaped grating can transform fundamental Gaussian beams into high-order LG beams. This section will mainly focus on three approaches of outside the resonator as mode converter, SPP and fork-shaped grating.

3.2.1 Mode Converter

Mode converters are a kind of setup that can transform high-order HG beams into high-order LG beams. Usually they consist of two or three special-arranged cylindrical lenses.

The mode converter of the first kind [5] consist of two cylindrical lenses. It has two types, π/2 converter and π converter, distinguished by the distance between the two cylindrical lenses. As sketched in Fig. 3.3, the principles of such converter are, a cylindrical lens will introduce different Gouy phase for different order HG modes. High-order HG modes (HG_{mn}) can be regarded as a superposition of multi low-order HG modes (see Fig. 3.4). Hence the passing of high-order HG modes through such mode converter will lead to the changing of relative phase of the contained low-order components. When the changes in relative phase satisfies a certain relationship, one can obtain LG beams with special topological charges.

From Figs. 3.11 and 3.12, obviously π/2 converter introduces a phase difference of π/2 for the adjacent low-order HG components of the incident HG modes HG_{mn}. And thus transform the incident HG modes HG_{mn} into LG modes LG_{pl}, meanwhile satisfying $p = \min(m,n)$, $l = m-n$. Since $m,n \geq 0$, one of the parameters m or n of the incident HG modes must be 0 if single ring LG beams ($p = 0$) are desired. π converter introduces π phase difference for the adjacent low-order HG components. Hence it can transform the incident HG_{mn} or LG_{pl} into their mirror mode HG_{nm} or $LG_{p\text{-}l}$. It should be noticed that, functions of π/2 converter and π converter are similar with quarter wave plate and half wave plate separately. Their differences are, wave plate is to modulate polarizations, while mode converter is to modulate phases.

The mode converter of the second kind [6] consists of three cylindrical lenses, as given in Fig. 3.5. The focal lengths of the three cylindrical lenses C_1, C_2 and C_3 are $f/2$, f, and $f/2$, respectively. The axes of cylindrical lenses C_1 and C_3 are parallel, while the axis of C_2 is perpendicular to that of C_1 and C_3. Both of the distance between C_1 and C_2, and between C_2 and C_3, are $f/2$. Vortex beams with helical wavefront will be output from such mode converter, when HG modes HG_{m0} whose long axis is arranged at the angle of 45° compared to the axis of C_1 are incident in.

Taking the mode converter of the second kind (three cylindrical lenses) as example, we will discuss how does it transform HG beams into vortex beams. Supposing $E(u,$

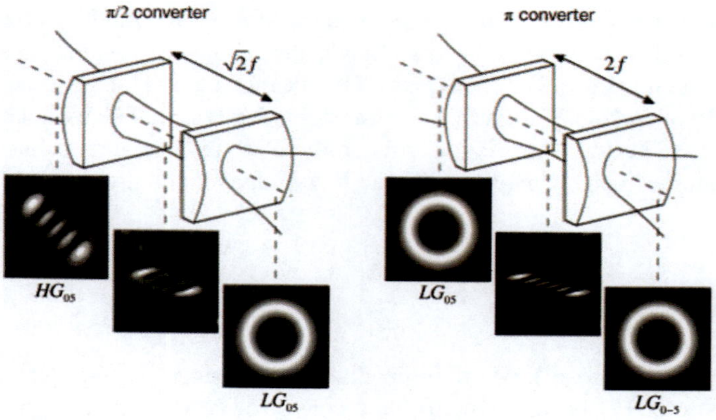

Fig. 3.3 Mode converter of the first kind: π/2 converter and π converter

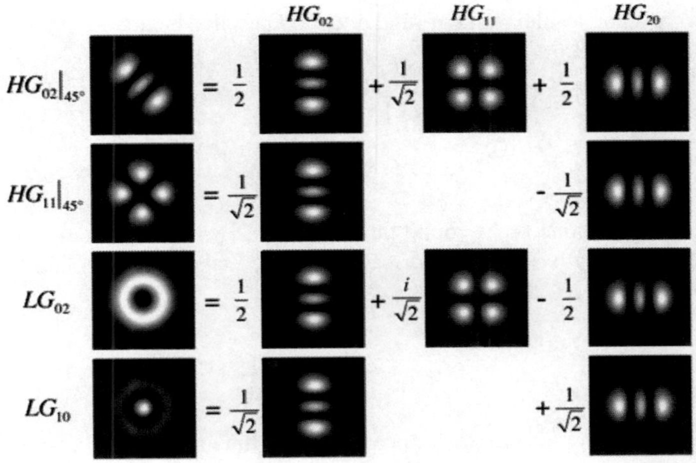

Fig. 3.4 Both of the high-order HG modes and LG modes can be regarded as a superposition of multi low-order HG modes

Fig. 3.5 Mode converter of the second kind consists of three cylindrical lenses

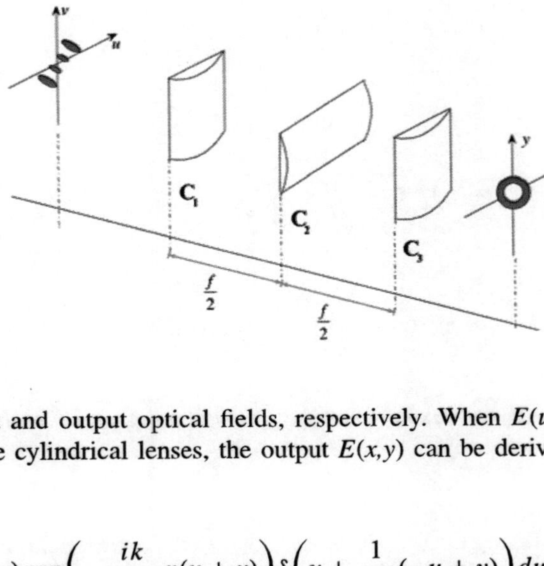

v) and $E(x, y)$ are the input and output optical fields, respectively. When $E(u,v)$ propagates through the three cylindrical lenses, the output $E(x,y)$ can be derivate from Collins integral as:

$$E(x, y) = \sqrt{\frac{-ik}{2\pi f}} \iint E(u, v) \exp\left(-\frac{ik}{\sqrt{2}f} x(u + v)\right) \delta\left(y + \frac{1}{\sqrt{2}}(-u + v)\right) dudv \tag{3.2.1}$$

Here the input $E(u,v)$ is HG beams HG_{m0}:

$$E(u, v) = \sqrt{\frac{2}{\pi}} \cdot \frac{E_0}{\sqrt{2^l l!}\omega_0} H_m\left(\frac{\sqrt{2}u}{\omega_0}\right) \exp\left(-\frac{u^2 + v^2}{\omega_0^2}\right) \tag{3.2.2}$$

with E_0 a constant, ω_0 the waist of fundamental Gaussian beams. $H_m(\varsigma)$ is mth order Hermitian polynomial:

$$H_m(\varsigma) = \sum_{k=0}^{[m/2]} \frac{(-1)^k m!}{k!(m-2k)!}(2\varsigma)^{m-2k} \tag{3.2.3}$$

In Eq. (3.2.3), [m/2] is the round number of m/2, with m a nonnegative integer. When the focal length of the cylindrical lens f is equal to the Rayleigh length (z_{Rx}, z_{Ry}) of the incident HG_{m0}:

$$f = z_{R_x} = z_{R_y} = \frac{\pi \omega_0^2}{\lambda} \tag{3.2.4}$$

considering that long axis of HG_{m0} is arranged at the angle of 45° compared to the axis of C_1, one can get:

$$x = \frac{1}{\sqrt{2}}(-u+v) \tag{3.2.5}$$

$$y = \frac{1}{\sqrt{2}}(u+v) \tag{3.2.6}$$

Equation (3.2.1) with Eqs. (3.2.4), (3.2.5) and (3.2.6):

$$E(x,y) = \frac{1}{\omega_0}\sqrt{\frac{-i}{\pi}} \iint E(u,v)\exp\left(-\frac{i\sqrt{2}}{\omega_0^2}x(u+v)\right)\delta\left(y+\frac{1}{\sqrt{2}}(-u+v)\right)dudv \tag{3.2.7}$$

According to the following integral formula [7]:

$$\int_{-\infty}^{\infty} H_m(\alpha x)\exp\{-(x-\beta)^2\}dx = \sqrt{\pi}\left(\sqrt{1-\alpha^2}\right)^m H_m\left(\frac{\alpha}{\sqrt{1-\alpha^2}}\beta\right) = \sqrt{\pi}(2\beta)^m \tag{3.2.8}$$

Taking Eq. (3.2.2) into Eq. (3.2.7), the optical fields in the focal plane of cylindrical lens C_3 are:

$$E(x,y) = \frac{E_0}{\omega_0}\sqrt{-\frac{(2)^m i}{\pi(m!)}}\left(\frac{ix-y}{\omega_0}\right)^m \exp\left(-\frac{x^2+y^2}{\omega_0^2}\right) \tag{3.2.9}$$

In polar coordinates, forcing $r^2 = x^2 + y^2$, $\varphi = \arctan(y/x)$, then Eq. (3.2.9) reads:

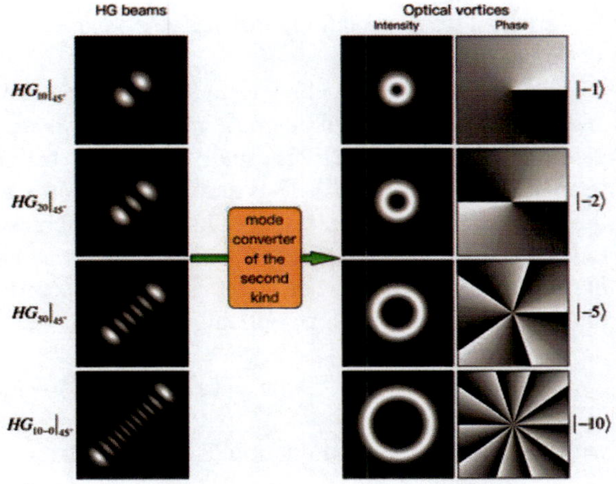

Fig. 3.6 The incident various HG beams and their corresponding vortices generated by the second kind mode converter

$$E(r,\varphi)=\frac{E_0}{\omega_0}\sqrt{\frac{-i}{\pi(m!)}}\left(\frac{\sqrt{2}ir}{\omega_0}\right)^m\exp\left(-\frac{r^2}{\omega_0^2}\right)\exp(-im\varphi) \tag{3.2.10}$$

Obviously, the optical field given in Eq. (3.2.10) comprise a helical term exp(-*imφ*), indicating that vortex beams with the topological charge –*m* are generated when HG modes HG_{m0} are incident in. In other words, HG_{m0} are transformed into vortex beams $|-m\rangle$ by the second kind mode converter. Figure 3.6 displays intensity and phase distributions of vortices generated by the second kind mode converter, when different HG beams are incident in.

In summary, mode converter can transform HG beams. Such converting has high efficiency, since all the incident power is utilized and transformed in theory. However, limitations are still present. In order to produce vortex beams, we have to generate HG beams with specific orders firstly. Additionally, mode converter has a relative complex structure, reflected by the strict demands on the positions and arranged angles of each cylindrical lens. Therefore, employing mode converter to generate vortex beams is not widely used. A much simple way is to transform Gaussian beams directly into vortex beams.

3.2.2 Spiral Phase Plate

The unique feature of vortex beams is their helical wavefront, which offers a new generation idea. If we can transform the plane wavefront of Gaussian beam into a helical wavefront, vortex beams will be generated. Based on this idea, a feasible way is to produce a diffractive optical element, which can introduce various phase delay in different azimuthal angle φ when a plane wave is incident in, and thus to bring forth helical wavefront. Such diffractive optical element is spiral phase plate (SPP), as shown in Fig. 3.7.

SPP can bring helical phase term in the complex amplitude of the incident beams, and thus has already been the most convenient way to produce vortex beams outside the resonator currently. SPP is usually made from materials transparent to the incident beams. The thickness of SPP is associated with the azimuthal angle φ, resulting in the strict control of the precision of its surface thickness when fabrication. Supposing that the refractive index of the SPP element is n, the thickness distribution function is $h(\varphi)$, and the thickness when $\varphi = 0$ is h_0, then according to the relationship between the phase difference and the optical path one can get:

$$l\varphi = (n-1)k[h(\varphi) - h_0] \tag{3.2.11}$$

with k the wavenumber. Equation (3.2.11) results in:

$$h(\varphi) = h_0 + \frac{\lambda l \varphi}{2\pi(n-1)} \tag{3.2.12}$$

Then the thickness variation along the azimuthal angle 0 ~ 2π of a SPP can be derived as:

$$\Delta h = h(2\pi) - h(0) = \frac{\lambda l}{(n-1)} \tag{3.2.13}$$

Generating vortex beams through SPP can gain high efficiency in theory. However, for the sake of outputting vortex beams with high purity, the surface of a SPP must be smooth enough, meanwhile the thickness distribution must meet with Eq. (3.2.12), providing a strict technical requirement for their actual production. In addition, one SPP is only used for generate vortex beams with specific parameters as wavelength, topological charge, etc. Different SPP have to be produced to generate vortices with

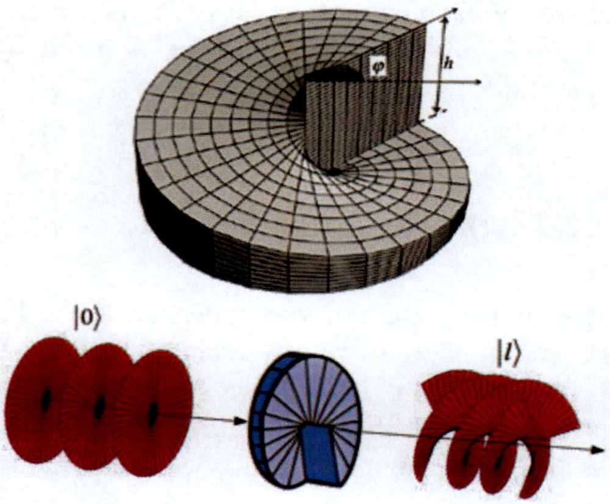

Fig. 3.7 Spiral phase plate

various parameters, increasing the cost to a certain extent and bringing in great limitations.

3.2.3 Fork-Shaped Grating

Fork-shaped grating is a kind of amplitude grating [8]. When fundamental Gaussian beams pass through a fork-shaped grating, two vortices with opposite topological charges are present at ± 1 diffraction orders separately in the far-field. Moreover, the absolute value of the topological charge is associated with the number of fork. Fork-shaped grating has become one of the mainstream approaches to generate vortex beams, for its low-cost and easily fabrication.

There are two main approaches to fabricate fork-shaped grating: ① Through the interference of plane wave and LG beams; ② Simulated by computer generated holograms. Because the principles of such two approaches are similar with holography, fork-shaped grating is also called as holographic grating. This section will focus on the first fabrication method as by means of he interference of plane wave and LG beams.

An oblique incident plane wave whose plane wavefront is not perpendicular to the z axis reads:

$$E_1 = \exp(ik_x x + ik_z z) = \exp(ikx \sin\theta + ikz \cos\theta) \quad (3.2.14)$$

where θ is the angle between the wave vector and z axis. E_2 are LG beams that propagate along z axis with the waist at $z = 0$. The complex amplitude of E_2 can be read from Eq. (1.4.1):

$$E_2 = \frac{c_{pl}}{\omega_0}\left(\frac{\sqrt{2}r}{\omega(z)}\right)^l L_p^{|l|}\left(\frac{2r^2}{\omega(z)^2}\right) exp\left(-\frac{r^2}{\omega(z)^2}\right)\exp(il\varphi)\exp(i\Phi) \quad (3.2.15)$$

When $\theta \neq 0$, the phase distribution of interference field at $z = 0$ is:

$$\phi(r, \varphi) = l\varphi + kr\cos\varphi\sin\theta \quad (3.2.16)$$

After binarization, one can get the transmission function of fork-shaped grating as:

$$a(\phi) = \begin{cases} 1, & \mathrm{mod}(\phi, 2\pi) \leq \pi \\ 0, & \text{others} \end{cases} \quad (3.2.17)$$

where mod(a,b) means taking the remainder when a divided by b. Considering that $k = 2\pi/\lambda$, then Eq. (3.2.16) can also be expressed as:

$$\phi(r, \varphi) = l\varphi + \frac{2\pi \sin\theta}{\lambda}(r\cos\varphi) \tag{3.2.18}$$

Since $x = r\cos\varphi$, one can gain the grating constant T of fork-shaped grating obviously from Eq. (3.2.18) as:

$$T = \lambda \csc\theta \tag{3.2.19}$$

Figure 3.8 displays some of the fork-shaped grating with different l value (White and black represent transparent and opaque separately for all of maps of amplitude gratings in this book). One can see clearly that the number of fork corresponds to $|l|$, while the forks' direction is associated with the sign of l. Meanwhile, fork-shaped grating will degrade into rectangle optical grating when $l = 0$. In practice, fork-shaped gratings are usually fabricated through laser ablation.

Next we will discuss how does fork-shaped grating generate vortex beams. The Fourier expansion of the transmission function $a(\phi)$ of a fork-shaped grating is:

$$a(\phi) = \sum_{b=-\infty}^{+\infty} A_b \exp(ib\phi) \tag{3.2.20}$$

with φ the phase distribution given in Eq. (3.2.16). A_b is the Fourier coefficient (Fig. 3.9):

$$A_b = \frac{\sin(b\pi/2)}{b\pi} \exp\left(-\frac{ib\pi}{2}\right) \tag{3.2.21}$$

When fundamental Gaussian beams are incident in a fork-shaped grating, the optical field output from the grating is:

$$u_0(r, \varphi) = \sqrt{\frac{2}{\pi}} \frac{1}{\omega_0} \exp\left(-\frac{r^2}{\omega_0^2}\right) a(\phi) \tag{3.2.22}$$

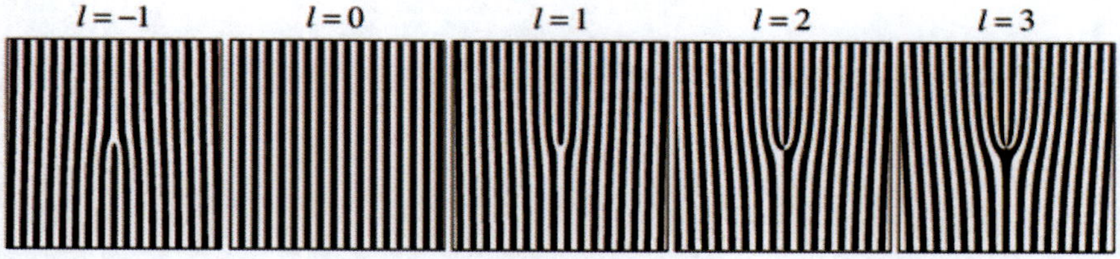

Fig. 3.8 Fork-shaped grating with different l values

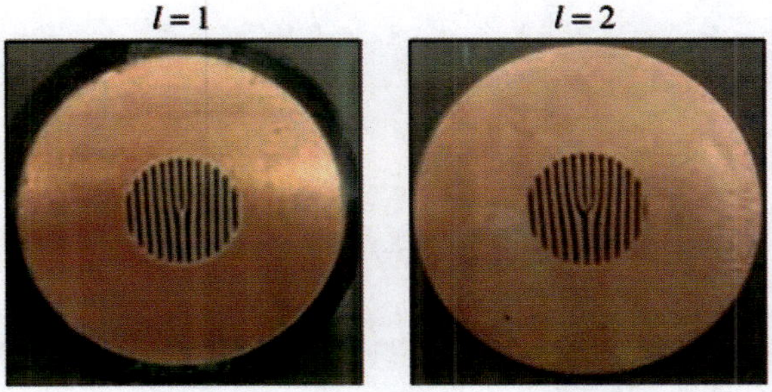

Fig. 3.9 The practical fork-shaped grating fabricated through laser ablation

According to Fraunhofer diffraction integral Eq. (2.5.4) given in Sect. 2.5.1, the far-field in Fraunhofer diffraction region can be regarded as the Fourier transformation of the original fields. Hence the far-field diffraction field when Gaussian beams passing through fork-shaped grating is:

$$u(r', \varphi') = \mathcal{F}[u_0(r, \varphi)] \tag{3.2.23}$$

with (r',φ') the spatial coordinates in the far-field. After expansion, Eq. (3.2.23) is:

$$\begin{aligned} u(r', \varphi') &= \sum_{b=-\infty}^{+\infty} A_b \mathcal{F}\left[u_0(r, \varphi) \exp(ibl\varphi) \exp(-ibkx \sin\theta)\right] \\ &= \sum_{b=-\infty}^{+\infty} A_b \mathcal{F}\left[u_0(r, \varphi) \exp(ibl\varphi)\right] * \mathcal{F}\left[\exp(-ibkx \sin\theta)\right] \end{aligned} \tag{3.2.24}$$

Since

$$\mathcal{F}\left[\exp(-ibkx \sin\theta)\right] = \delta\left(f_x + \frac{bk \sin\theta}{2\pi}\right)\delta(f_y) \tag{3.2.25}$$

where (f_x,f_y) is the Cartesian coordinates in spatial frequency domain. Considering that:

$$u_0(r, \varphi) \exp(ibl\varphi) = \sum_{p=0}^{\infty} c_p u_{p,bl}(r, \varphi) \tag{3.2.26}$$

where

$$c_p = \int_0^{2\pi} \int_0^{+\infty} u_0(r, \varphi) \exp(ibl\varphi) \cdot u^*_{p,bl}(r, \varphi) r dr d\varphi \tag{3.2.27}$$

According to the properties of Fourier transformation of LG modes, one can get:

$$\mathcal{F}\left[u_{p,bl}(r,\varphi)\right] = 2\pi i^{2p+|bl|} u_{p,bl}\left(r',\varphi'\right) \tag{3.2.28}$$

Equations (3.2.25) ~ (3.2.28) with Eq. (3.2.24):

$$u\left(r',\varphi'\right) = \sum_{b=-\infty}^{+\infty} A_b \sum_{p=0}^{+\infty} c_p i^{2p+|bl|} u_{p,bl}\left(r',\varphi'\right) \otimes \delta\left(f_x + \frac{bk\sin\theta}{2\pi}\right) \tag{3.2.29}$$

From Eq. (3.2.29), when fundamental Gaussian beams are incident in a fork-shaped grating, the far-fields are a superposition of diffractive beams in all of the diffraction orders. Specially, beams in diffraction order b are a superposition of series of vortex beams with topological charges bl. And the distance between bth and 0th order is $(bk\sin\theta)/(2\pi)$. Fork-shaped gratings $l = 1$ and $l = -2$, and their corresponding far-field diffraction patterns when Gaussian beams are incident in, are given in Fig. 3.10.

Equation (3.2.29) also implies that, the far-field diffraction patterns of fork-shaped grating contain all of the diffraction orders ($-\infty$ ~ $+\infty$). Nevertheless, only -1st ~ $+1$st diffraction orders can be observed apparently in Fig. 3.10. According to the theory of wave optics, for an amplitude rectangle grating whose ratio of slit width to grating constant is 0.5, e.g. fork-shaped grating, the diffraction efficiencies of the 0th, ±1st, ±2nd, ±3rd, ±4th and ±5th diffraction orders are 25%, 10.13%, 0, 1.13%, 0 and 0.45%, respectively. Such feature indicates that the relative power of each diffraction orders is weak except for the 0th and ±1st orders, and thus can be ignored. Hence only the ±1st diffraction orders are considered when generating vortex beams through fork-shaped grating, and the topological charges of the generated vortices are determined only by the fork value l. In real scenario an aperture stop is usually

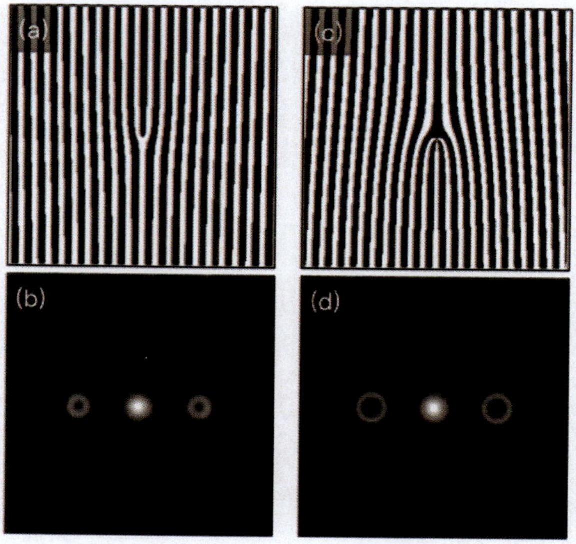

Fig. 3.10 Fork-shaped gratings and their corresponding far-field diffraction patterns when Gaussian beams are incident in. **a** Fork-shaped grating with $l = 1$; **b** Far-field diffraction patterns when Gaussian beams incident in grating (**a**); **c** Fork-shaped grating with $l = 2$; **d** Far-field diffraction patterns when Gaussian beams incident in grating (**c**)

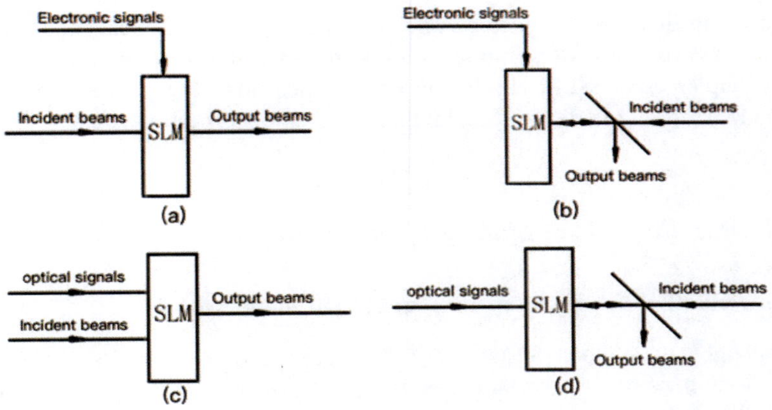

Fig. 3.11 Categories of various SLMs. **a** Transmitted electronic addressed SLM. **b** Reflective electronic addressed SLM. **c** Transmitted optical addressed SLM. (d) Reflective optical addressed SLM

Fig. 3.12 The LC-SLM produced by Holoeye in Germany

employed to filter -1st or $+$ 1st diffraction orders to obtain pure vortices, thus the efficiency is just 10.13% (such efficiency will much lower in practice). Although the generating efficiency is low, fork-shaped gratings are still widely used for their easily fabrication and low-cost.

3.3 Phase-Only Vortex Gratings

The phase-only vortex grating is a kind of optical element to produce vortex beams through modulating the phase of incident Gaussian beams, where the diffractive efficiency is greatly improved compared with amplitude gratings as fork-shaped grating, etc. However, the complex fabrication and high-cost of such elements lead to their narrow employment. The arisen of liquid–crystal spatial light modulator (LC-SLM), making it possible to imitate phase-only vortex gratings through the deflection of liquid crystal molecules controlled by electric signals, and paving the way for the using of vortex gratings. This section will start from the LC-SLM. First the principles of LC-SLM are demonstrated. Then how to generate vortex beams through phase-only vortex grating imitated by LC-SLM is discussed.

3.3.1 Liquid–crystal Spatial Light Modulator

1. **Introduction to liquid–crystal spatial light modulator**

Spatial light modulators (SLMs) are a kind devices that can encode information onto one-dimensional or two-dimensional optical data fields through employing the parallelism, inherent velocity, and interconnection capabilities of beams effectively [9]. Controlling by the time-varying optic or electric drive signals, such devices can modulate the phase, amplitude, wavelength, polarization, or even the coherence of the incident spatial optical fields.

SLM can be divided into two types as reflective one and transmitted one, depending on the direction where the beams coming out. The incident and output beams are located at two sides separately of the transmitted SLM, while at the same side of the reflective SLM. For the reflective SLM, a beam splitter is proposed to separate the output beams out from the incident beams. Additionally, SLM can also be divided as electronic and optical addressed one, according to the controlling signals. Figure 3.11 displays four different types of SLMs based on the above categories.

SLM can also be classify through their working principles, as based on the acoustic light effect, magnetic light effect, electro-optical effect and semiconductor electro-optical effect. The materials employed can be acousto-optic materials, magneto-optic materials, electro-optical crystals, liquid crystals, ferroelectric ceramics and so on. Due to the fact of the variety of SLM categorised by principle, only four common types are listed here:

① Electro-optic spatial light modulator (EO-SLM).

The light modulation of EO-SLM is based on the primary or secondary electro-optic effect of electro-optic materials. In detail, it controls the refractive index of the liquid–crystal through changing the external electric field.

② Magneto optical spatial light modulator (MO-SLM).

MO-SLM is based on the magneto-optic effect, whose write signals' modulation is realized by the induced magnetization effect of ferromagnetic materials.

③ Acousto-optic spatial light modulator (AO-SLM).

The basic principle of AO-SLM is that, the refractive index of acousto-optic materials is changed under acoustic wave field. The AO-SLM is appropriate for the modulation of one-dimensional optical signals, since usually its write signals are distributed along one dimension.

④ liquid crystal spatial light modulator (LC-SLM).

LC-SLM (Fig. 3.12) is an active optical device that can modulate incident beams through utilizing the electrically controlled birefringence (ECB) of the liquid-crystals. Usually it consists of many independent liquid crystal units, which are regularly arranged into one-dimensional or two-dimensional arrays. LC-SLM can change the wavefront of the incident beams flexibly, since each of the independent units can be controlled independently by driving signals, resulting in the alignment change of the liquid crystal molecules, and thus to effectively modulate the phase or amplitude of the incident beams. Currently, LC-SLMs are fabricated through microelectronic technology, characterized by low energy consumption, small volume, large number of pixels, easy control and low cost, and have become the mainstream of SLMs in use.

Note that all the discussion of SLM in this book is centered on LC-SLM.

2. **Principles of liquid–crystal spatial light modulator**

The key part of LC-SLM is the liquid–crystal display (LCD), which is made up of many liquid–crystal molecules arranged in a specific way. Liquid–crystal is a kind of organic compound with regular molecular arrangement between solid and liquid. Liquid-crystals not only have some of the properties of liquid and crystal (e.g. fluidity, anisotropy, etc.), but also other unique characteristics, and recently have found applications is lots of domains as display, non-destructive inspection, temperature measurement, medical diagnosis, environmental monitoring and so on.

Liquid–crystal is anisotropic, which can be categorize into three phase states according to the arrangement of molecules as nematic phase, smectic phase and cholesteric phase. It has similar optical properties with uniaxial crystals like KDP crystal, and has two kinds of refractive indices as ordinary refraction index n_o and extraordinary refractive index n_e, which implies birefringence will present if a beam is off-axis incident. If the external electric field is applied to the liquid–crystal to a certain extent, the orientation of the main axis or liquid–crystal molecules will be changed, as shown in Fig. 3.13, and such change is called Fredericks transition. Therefore, ECB will appear if a beam is incident in a liquid–crystal molecule with external electric field.

The arrangement of liquid–crystal molecules is mainly affected by three forces: intermolecular force, interfacial force and external force, where external force is

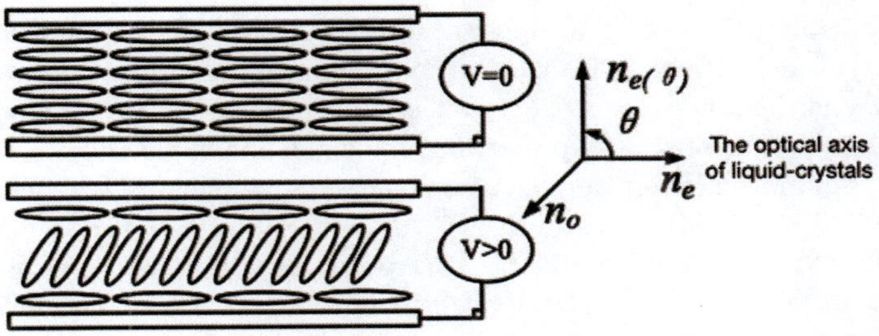

Fig. 3.13 The main axis of liquid–crystal will be changed under external electric field

usually employed to change the molecules' arrangement. According to the continuum theory of liquid crystals, where the macroscopic physical properties of an object can be described by the microscopic states of the atoms and molecules that it consists, the liquid crystal can be regarded as an elastic continuum and has elastic deformation under external forces.

The critical electric field E_c when Fredericks transition occurs is:

$$E_c = \frac{\pi}{d}\sqrt{\frac{K}{|\varepsilon|}} \tag{3.3.1}$$

where d is the thickness of liquid–crystal layer, K is the elastic constants of the elastic deformation of liquid-crystals, and ε is dielectric constant. Equation (3.3.1) implies that the critical electric field E_c is associated with the elastic constants K, the dielectric constant ε and the thickness d. Since:

$$U_{th} = E_c d \tag{3.3.2}$$

One can obtain the threshold voltage U_{th} of the Fredericks transition of liquid-crystals is:

$$U_{th} = \pi\sqrt{\frac{K}{|\varepsilon|}} \tag{3.3.3}$$

Figure 3.13 indicates that the change of the arrangement of liquid-crystals only occurs providing that the imposed voltage U is larger than the threshold voltage U_{th}. At that time, the main axis of liquid-crystals will deflect a certain angle θ along the direction of the electric field, which reads [11]:

$$\theta = \frac{\pi}{2} - 2\arctan\left[\exp\left(-\frac{U - U_{th}}{U_0}\right)\right] \tag{3.3.4}$$

with U_0 the voltage when $\theta = 49.6°$. Equation (3.3.4) means the deflection angle θ of liquid–crystal molecule is the function of, and determined only by, the imposed voltage U. Actually, the deflection angles of the liquid–crystal molecules inside the layer is not exactly the same as the those near the surface, but have a slight difference. The reason of such difference is that, the molecules near the surface are bound by the boundary, contributing to the smaller deflection angles than those of the molecules inside the layer. The relationships between the deflection angle and the position of liquid–crystal molecules are complicated under external electric field, which won't be discussed in this book.

Because of ECB, the extraordinary refractive index n_e of liquid–crystal molecule changes under external electric field. Such changed extraordinary refractive index is called effective refractive index n_{eff}, which is associated with the deflection angle θ:

$$\frac{1}{n_{eff}^2} = \frac{\cos^2\theta}{n_e^2} + \frac{\sin^2\theta}{n_o^2} \tag{3.3.5}$$

From Eq. (3.3.5) one can obtain:

$$n_{eff}(\theta) = \frac{n_o n_e}{\sqrt{n_o^2\cos^2\theta + n_e^2\sin^2\theta}} \tag{3.3.6}$$

Then the introduced optical path difference Δ under external electric field is:

$$\Delta = \int_0^d \left[n_{eff}(\theta) - n_o\right]dz \tag{3.3.7}$$

where the deflection angle θ is the function of the position z and the imposed voltage U. Then one can obtain the phase difference $\Delta\phi$:

$$\Delta\phi = k\Delta = \frac{2\pi}{\lambda}\cdot\int_0^d \left[n_{eff}(\theta) - n_o\right]dz \tag{3.3.8}$$

which indicates the phase differences $\Delta\phi$ will be introduced to beams that passing through a liquid–crystal layer with imposed voltage.

Similar with wave plate, for LC-SLM, the liquid–crystal produces different modulation characteristics for the incident beams with different polarizations. The physical properties of liquid crystal molecules are very similar with those of uniaxial crystals. From the refractive index ellipsoid of uniaxial crystal (Fig. 3.14), one can see clearly that when any linear polarizations whose wave vector is parallel to the direction of main axis are incident in a uniaxial crystal, or the orientation of the incident linear polarization is perpendicular to the main axis ($\gamma = 0$), the intersection of the plane perpendicular to the propagation direction and the surface of refractive index ellipsoid is a circle. Such circle intersection indicates the refractive indices of all the linear polarizations are identical, and no modulations are done for the incident beams. If the orientation of the incident linear polarization is non-perpendicular to the main axis ($\gamma \neq 0$), amplitude or phase modulation will be done. Figure 3.15 displays the modulation relationships between various incident linear polarizations and their corresponding amplitude and phase modulated by a LC-SLM, where β represents the angle between the polarization orientation and the liquid–crystal's main axis [10].

Obviously from Fig. 3.15, when $\beta = 0$, where the orientation of the incident linear polarization is parallel to the main axis, the liquid–crystal phase-only modulate the incident beams without any amplitude modulations, and the polarization of output beams won't be changed. When $\beta \neq 0$, not only phase but also amplitude modulations are present, and the polarization of output beams will be changed.

In summary, when using LC-SLM, if the incident beams is partially polarized, elliptically polarized, or linear polarized whose polarization direction is not parallel to the main axis, there must be polarization component non-parallel to the main axis,

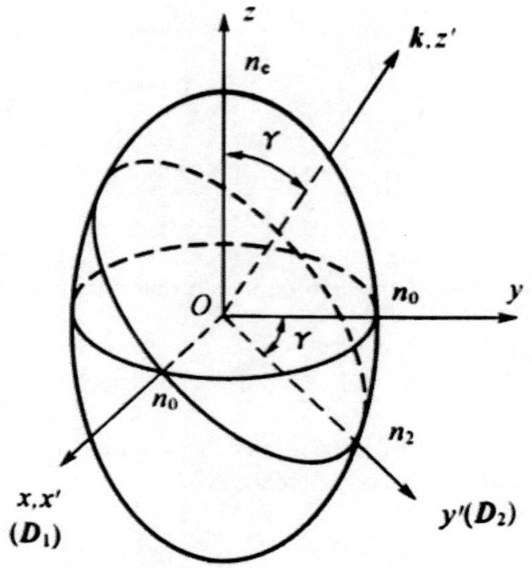

Fig. 3.14 The refractive index ellipsoid of uniaxial crystal

that is, there must be amplitude modulation. If phase-only modulation is desired, the incident beams must be linearly polarized, meanwhile the polarization orientation must be parallel to the main axis of the liquid–crystal. Usually, the main axis of commercial LC-SLM are horizontal, hence the incident beams should be horizontally linear polarization to acquire phase-only modulation.

The principles of LC-SLM, which is based on the ECB of liquid-crystals, have been introduced in detail. In practice, reflective LC-SLMs are more common, whose write signal is computer generated-hologram (CGH) with grayscale phase information. They read the gray-scale value of the CGH to control the size of the external electric field applied to the liquid–crystal molecules, and then introduce various phase modulation at different positions to imitate phase diffraction grating.

3.4 Main Parameters of Liquid–Crystal Spatial Light Modulator

Specific parameters should be used to evaluate the optical property, working mode and technical performance of LC-SLM. This sub-section lists some of the main parameters of LC-SLM, including the wavelength range, phase levels, resolution, pixel pitch, fill factor, diffraction efficiency, image frame rate, and maximum illumination.

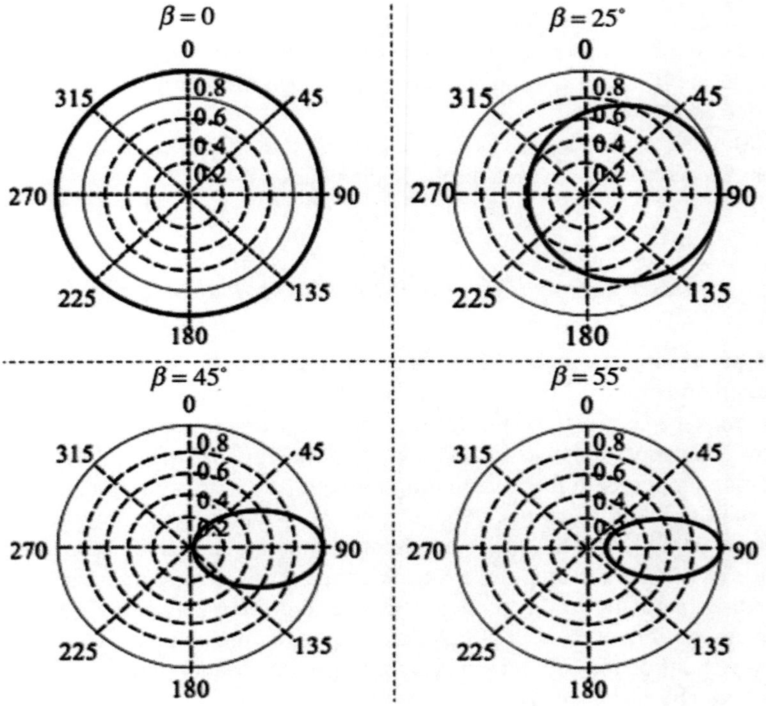

Fig. 3.15 The modulation relationships between various incident linear polarizations and their corresponding amplitude and phase modulated by a LC-SLM

① Wavelength rang: The wavelength range of the incident beams that can be modulated by the LC-SLM. Only when the incident wavelength matches the wavelength range can the favourable modulation effect be achieved.

② Phase levels: The minimum phase value that can be modulated by the LC-SLM. Generally, the current commercial LC-SLM is controlled by 8-bit (256-order) gray values, and the minimum modulated phase can be as low as minuscule $\pi/128$.

③ Resolution: The number of liquid–crystal molecules on the LCD surface. The higher resolution is, the higher the modulation accuracy of the LC-SLM and the better the modulation performance will be. Recently some of the corporations have developed LC-SLMs with 4 K resolution.

④ Pixel pitch: The size of each liquid–crystal molecule on the LCD surface. Usually the smaller pixel pitch is, the higher performance to simulate complex phase gratings will be.

⑤ Fill factor: Ratio of the total area of the liquid–crystal molecules on the LCD surface to the area of the LCD.

⑥ Diffraction efficiency: Ratio of the diffracted (output) beams to the incident beams. Diffraction efficiencies are various for LC-SLMs with different wavelength range.

⑦ Image frame rate: The switched speed of the simulated phase gratings per second. There is no specific demand on such rate for most of applications generally. But in the application of adaptive optics, high image frame rate contributes to better real-time compensation.

⑧ Maximum illumination: The maximum value of power density of incident beams. Exceed such value will results in the damage of LCD.

3.4.1 Imitating Amplitude Grating Through Phase Grating

LC-SLM can simulate various phase diffractive grating, but faces troubles to simulate amplitude grating. From Fig. 3.15, the phase-only modulation of LC-SLM can be accomplished easily through adjusting the orientation of the incident linear polarizations, but amplitude-only modulation is unavailable. However, in real scenario, simulating amplitude diffractive grating through LC-SLM is of great significance, since such simulation can test the performance of the designed amplitude grating before its actual manufacturing, and also can simulate the mixed phase-amplitude grating. To achieve the amplitude modulation, a feasible idea is to find a way that can imitate amplitude grating through phase grating. Once encoding the hologram of such phase grating onto the LC-SLM, the amplitude modulation is come true. This section will introduce two approaches to imitate amplitude gratings, including the schemes of checkerboard and blazed grating.

The checkerboard here has the phase structure that consists of alternating lattice of phase units with the phase value 0 and π. Such checkerboard phase can modulate the amplitude into zero, and thus act as the opaque part of the amplitude grating. Each phase unit in the checkboard modulates the phase only of the incident beams and do nothing with the amplitude. Hence all the phase transition on the checkerboard can be expressed by the vector in the unit circle, as shown in Fig. 3.16a. Considering two neighboring phase unit, one is 0 and the other is π. They will introduce the phase 0 and π respectively for the incident beams in neighboring positions. By means of the unit circle given in Fig. 3.16a, one can understand the reason why checkerboard can modulate the amplitude into zero very easily. If the argument (phase) of Z_1 and Z_2 is equal to 0 and π separately, then the vector sum Z_3 of Z_1 and Z_2 is null vector, implying the zero amplitude. In the case of a large number of the such arranged phase unit, their average action will modulate the amplitude of the incident beams into 0, and thus act as the opaque part of the amplitude gratings.

Actually, the principle of zero amplitude modulation of checkerboard phase can also be explained through Fourier transformation [12]. As shown in Fig. 3.16b and d, the transmittance function of the checkerboard phase $T(x,y)$ can be simply regarded as the superposition of two periodic functions in x and y directions $T(x)$ and $T(y)$ separately. In Sect. 2.5.1, we have already proven that the far-field diffraction fields are the Fourier transformation of the original fields indeed. When Gaussian beams passing through a checkerboard phase, the far-field diffraction is:

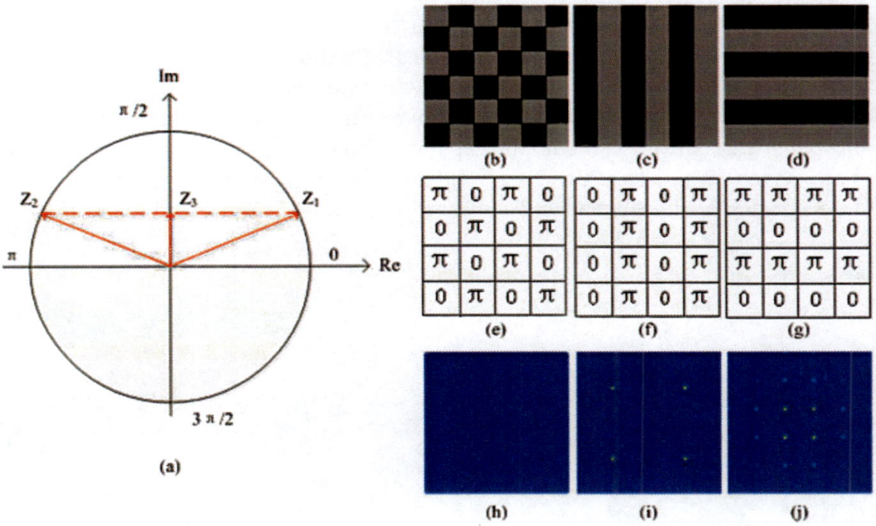

Fig. 3.16 Principles of the checkerboard phase. **a** The unit circle, where Z_1 and Z_2 represent phases of two neighboring phase units respectively. Z_3 is the vector sum of Z_1 and Z_2. **b** The structure of checkerboard phase. **c** The vertical component of a checkerboard phase. **d** The horizontal component of a checkerboard phase. **e** The corresponding phase distribution of the checkerboard phase. **f** The corresponding phase distribution of the vertical component of a checkerboard phase. **g** The corresponding phase distribution of the horizontal component of a checkerboard phase. **h–j** Simulated far-field diffraction patterns when Gaussian beams propagate through checkerboard phase with various unit sizes

$$\begin{aligned} \mathcal{F}[G(x, y) \cdot T(x, y)] &= \mathcal{F}[G(x, y)] \otimes \mathcal{F}[T(x, y)] \\ &= G\left(f_x, f_y\right) \otimes \mathcal{F}[T(x, y)] \end{aligned} \tag{3.3.9}$$

There must be δ functions in the Fourier transformation of periodic functions. When the size of each phase unit is small enough, the period of the periodic function is small enough, and after Fourier transformation the peak of the spectrum (δ functions) is very far from the origin, which means the incident beams are diffracted to other diffraction orders and the beams in 0th order are vanished and look like blocked. Figures 3.16h and j list three far-field diffraction patterns when Gaussian beams propagate through checkerboard phases whose sizes of phase units increase orderly. One can obviously see that the smaller the unit size, the smaller the period of the periodic function, the bigger the distance between each bright spot in the diffraction fields, and the better the zero amplitude modulation.

When imitate amplitude grating through checkerboard scheme, what we should do is just to design the opaque part as checkerboard phase and transparent part as 0. Figure 3.17 gives two phase gratings to imitate single-slit and fork-shaped grating through checkerboard method, and their corresponding simulated far-field diffraction patterns when Gaussian beams are incident in are shown in Fig. 3.18. The favourable diffraction patterns indicate the good performance of checkerboard method.

Different from checkerboard method, the blazed grating method is to separately diffract the transparent and opaque component of the amplitude grating along two different paths. And the two components form diffraction patterns in far-field respectively, which indirectly realizes the amplitude modulation. As given in Fig. 3.19, blazed grating is a kind of reflective phase grating, whose phase distribution function is:

$$\phi = \frac{2\pi x}{d} \tag{3.3.10}$$

Fig. 3.17 Imitating amplitude gratings through checkerboard method. **a** Single-slit. **b** Fork-shaped grating. In both **a** and **b**, The gray and black areas are the checkerboard and 0 phase, to act as opaque and transparent region separately

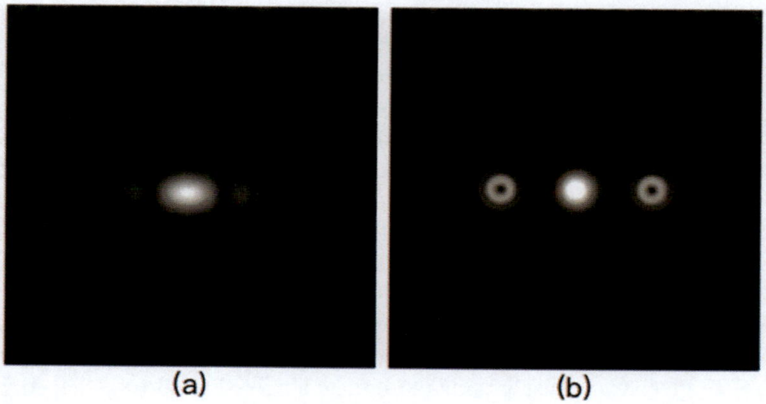

Fig. 3.18 Simulated far-field diffraction patterns when Gaussian beams propagate through the phase gratings given in Fig. 3.17

Fig. 3.19 Blazed grating

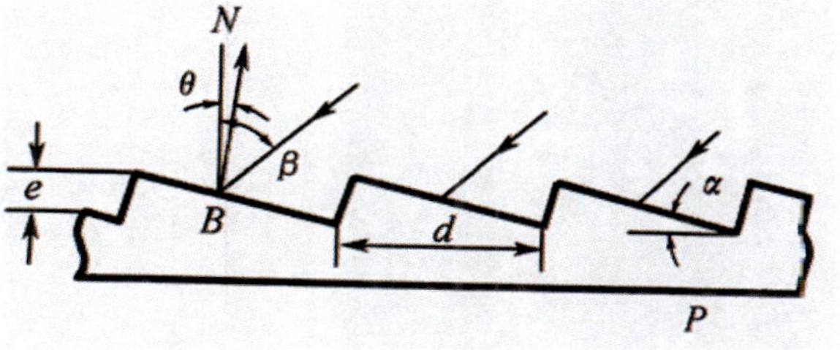

Its main structural parameters include:

Grating constant *d*: The grating period;

Blaze angle α: The angle between the sloping reflecter of blazed grating and the horizontal plane *P*;

Angle of incidence β: The angle between incident wave vector and the normal *BN* of horizontal plane *P*;

Angle of diffraction θ: The angle between diffractive wave vector and the normal *BN* of horizontal plane *P*.

The relationships between the grating constant *d*, the angle of incidence β and the angle of diffraction θ are determined by the grating equation:

$$d(\sin\beta + \sin\theta) = \lambda \tag{3.3.11}$$

For the normal incident beam, $\beta = 0$, under the paraxial approximation, from Eq. (3.3.11) one can obtain:

$$\theta = \frac{\lambda}{d} \tag{3.3.12}$$

which means after diffraction the beams intensity won't be changed. Only the angle of diffraction deviates to $\theta = \lambda/d$, reflecting on the position change of the beams in the far-field. Obviously, a blazed grating can be regarded as a slanted plane. When Gaussian beams are incident, only the first diffraction order is present and the other diffraction orders are missed. Figure 3.20 gives the diffraction results when Gaussian beam propagate through a blazed grating, where a clear position deviation is observed.

According to the feature of blazed grating discussed above, when imitating amplitude grating through blazed grating scheme, the phase of opaque and transparent area should be forced to 0 and blazed grating respectively. Thus beams diffracted by such two areas will be separated along two different directions.

Figure 3.21 shows two phase gratings to imitate single-slit and fork-shaped grating through blazed grating method. To imitate single-slit, the transparent part is blazed grating in vertical direction, and the phase of opaque part is set as 0. Hence the far-field diffraction patterns are separated along vertical direction. To imitate fork-shaped grating, the transparent part is blazed grating in horizontal direction, and the

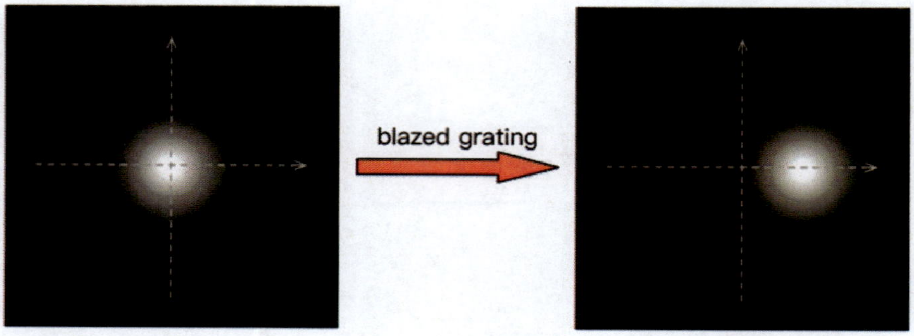

Fig. 3.20 Intensity distributions of Gaussian beams before and after passing through a blazed grating

phase of opaque part is set as 0. Hence the far-field diffraction patterns are separated along horizontal direction. Figure 3.22 shows the far-field diffraction patterns when Gaussian beams propagate through the phase gratings in Fig. 3.21, indicating that the nature of blazed grating method is to separate beams diffracted by the transparent and opaque part, thus to obtain the same diffraction performance as amplitude gratings.

Note that in Fig. 3.22b, one can clearly see that the beams in the 0th diffraction order of fork-shaped grating should be Gaussian beams, while here the petal intensity distributions are present. The reason is the introduction of horizontally blazed grating results in the superposition of the ±3rd and 0th diffraction orders of two fork-shaped gratings. And the petal patterns are interference fields. Therefore, to avoid such phenomena, the direction of a blazed grating should be perpendicular to the splitting direction of the imitated amplitude grating, as shown in Fig. 3.22a.

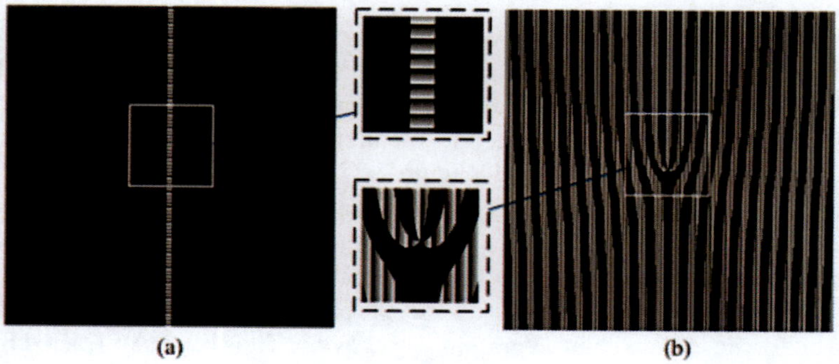

Fig. 3.21 Imitating amplitude gratings through blazed grating method. **a** Single-slit. **b** Fork-shaped grating. The black and blazed grating areas act as opaque and transparent region separately

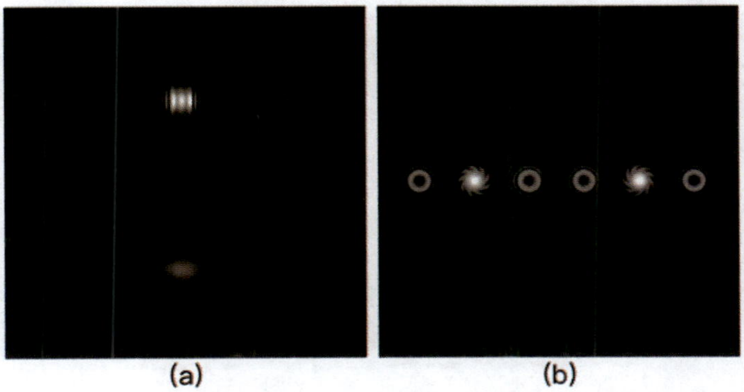

Fig. 3.22 Simulated far-field diffraction patterns when Gaussian beams propagate through the phase gratings given in Fig. 3.21

3.4.2 Generating Vortex Beams Through Phase-Only Vortex Grating

In Sect. 3.2.2, generating vortex beams through SPP has already been introduced. Such approach has simple principles and high generating efficiency. However, the complex fabrication process, high precision requirements, and fixed OAM order of the generated vortices, bring troubles for SPP's practical application. If we can design a phase grating to represent a SPP, and encode the hologram of such phase grating onto a LC-SLM, then it will be more flexible to generate vortex beams. This grating is called as phase-only vortex grating. Usually, phase-only vortex grating is calculated through computer generated hologram. Thus the hologram of vortex grating is also named as holographic SPP (HSPP).

The transmittance function of phase-only vortex grating is written under polar coordinates as:

$$T(r, \varphi) = \exp(il\varphi) \tag{3.3.13}$$

where l is the order of the vortex grating, and determine the topological charge of the generated vortices. Then the phase distribution of a vortex grating is:

$$\phi(r, \varphi) = \arg[T(r, \varphi)] = l\varphi \tag{3.3.14}$$

According to Eq. (3.3.14), vortex gratings with various orders are acquired with the help of MATLAB, as displayed in Fig. 3.23. Figure 3.23 also displays the corresponding far-field diffraction patterns and phase distributions when Gaussian beams are incident in. Obviously the phase-only vortex grating can transform Gaussian beams into vortex beams whose topological charges are identical with the grating's orders.

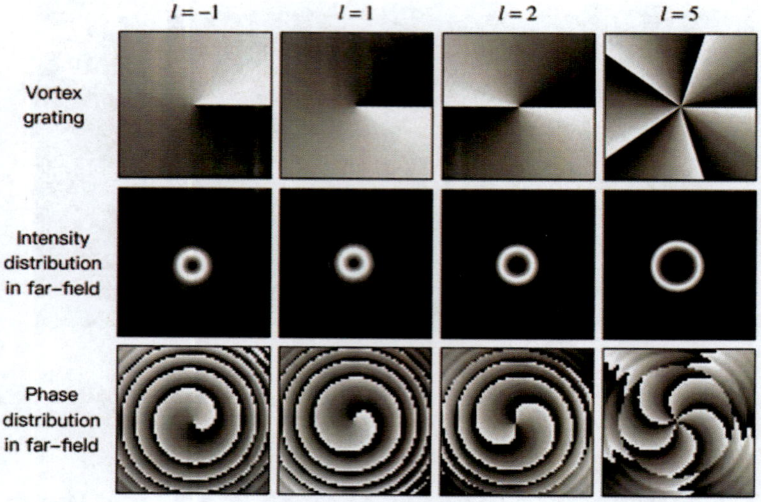

Fig. 3.23 Phase-only vortex gratings with various orders and their corresponding far-field diffractions

It must be noticed that, when encoding the HSPP on a reflective LC-SLM to generate vortex beams in practice, the topological charge of the generated vortex is opposite with the order of HSPP. According to the mirror feature of vortex beams that has been discussed in Sect. 1.2.4 as the topological charge will turn to be opposite when experience odd number reflections, a reflective LC-SLM adds one extra reflection, thus leading to the opposite topological charge.

Generating vortex beams through encoding HSPP on LC-SLM is very flexible, since the encoded HSPP can be changed timely as needed. However, when high-order vortex beams are desired, the phase leap in the center of HSPP is giant. Limited by the resolution of LC-SLM, the simulation of high-order phase vortex grating is inadequate, leading to the broaden annulus of the generated vortex beams. Figure 3.24a is a HSPP with the order 20, whose center's resolution is inadequate. Figure 3.24b shows the simulation results when Gaussian beams propagate through the HSPP. One can see clearly the broaden annulus and distorted patterns. Figure 3.24c shows the experimental results. In addition to the broaden annulus, a bright spot presents at the beam's center too. Such bright spot also results from the inadequacy of the resolution of the HSPP's center. The stronger central part of the incident Gaussian beams is almost "go through" the center of HSPP, leading to the unideal vortex beams.

Such issue can be addressed by means of optimal annulus structures [13, 14], whose basic idea is based on the correspondence before and after the Fourier transformation under polar coordinates. The broaden annulus of the generated vortices results from the inside part of HSPP. While the main and sub-main annuluses result from the outside of HSPP. In this scenario, there must be an optimal annulus, to obtain better vortex beams.

(a) (b) (c)

Fig. 3.24 **a** HSPP with the order 20. **b** and **c** are the simulated and experimental results when Gaussian beams propagate HSPP in **a**, respectively

As illustrate in Fig. 3.25, if the inner part of the HSPP is taken, and the outer part is lost by the checkerboard phase, the diffraction pattern won't be well improved. As illustrated in Fig. 3.26, if the outer part is taken and the inner part is lost by the checkerboard phase to avoid the effects result from the center part, the generated vortex beams are improved with thinner annulus. Nevertheless, sub-diffraction orders are present since the center circular disk diffraction. According to the simulated results given in Figs. 3.25 and 3.26, it is easy to understand that the poor performance of generating high-order vortex beams, mainly results from the center of HSPP.

In summary, the generating performance can be well improved through taking an optimal annulus of HSPP. The design of optimal annulus is very simple, where the detail procedure is to calculate the mean radius of the annulus on the basis of the diameters of incident beams firstly, and then compute the inner and outer radii on the basis of the desired topological charge. Here we list the above equations directly. Readers can referent to Refs. [13, 14] to acquire the detail of how to derive such equations.

Supposing that the radius of incident beams is ω, the mean radius of annulus is R_m, the desired topological charge is l, and the inner and outer radii are R_i and R_o

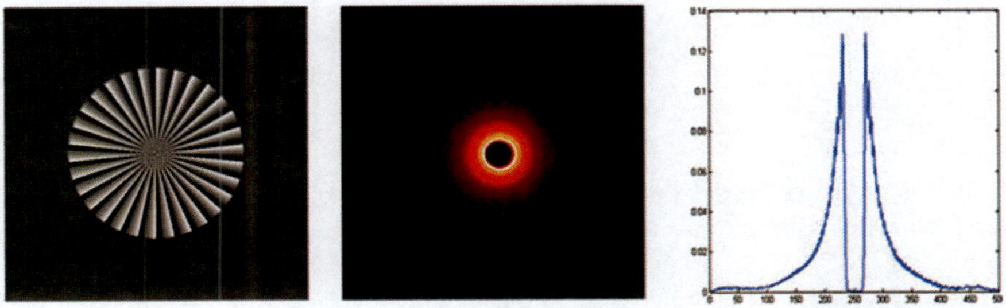

(a) (b) (c)

Fig. 3.25 **a** HSPP whose inner part is taken and outer part is lost by the checkerboard phase. **b** Simulated diffraction patterns. **c** Transverse intensity distribution

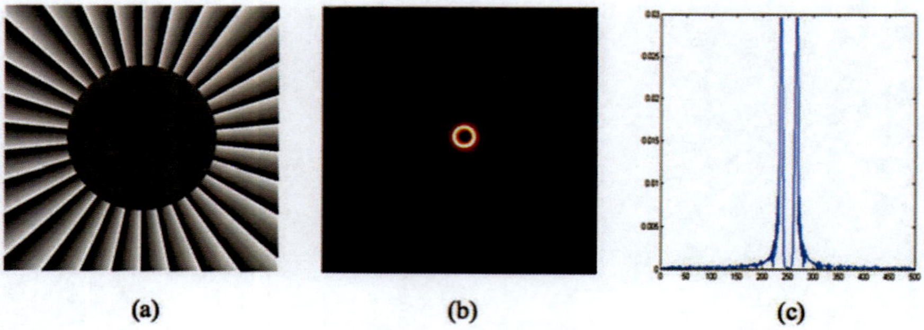

Fig. 3.26 a HSPP whose outer part is taken and inner part is lost by the checkerboard phase. b Simulated diffraction patterns. c Transverse intensity distribution

separately. They meet:

$$\omega = 1.43R_m \tag{3.3.15}$$

$$R_m = R_0 - \Delta R/2 \tag{3.3.16}$$

$$R_m = R_i + \Delta R/2 \tag{3.3.17}$$

where $\Delta R = R_o - R_i$ is the annular width.

$$\Delta R = 1.4043R_0 l^{-0.5363} \tag{3.3.18}$$

Based on Eqs. (3.3.15) ~ (3.3.18), the inner and outer radii and the width of annulus can be calculated, providing that the radius ω of incident Gaussian beams and the desired topological charge l are already known. For instance, if $\omega = 5$ mm, $l = 50$, then:

$$\begin{cases} R_m = 3.50\,\text{mm} \\ R_0 = 3.82\,\text{mm} \\ R_i = 3.18\,\text{mm} \\ \Delta R = 0.64\,\text{mm} \end{cases} \tag{3.3.19}$$

In this scenario, the designed optimal annular HSPP on the basis of Eq. (3.3.19), and the far-field diffraction patterns when Gaussian beams are incident in, are shown in Fig. 3.27.

From Fig. 3.27, the annular width of the generated vortex beams is well improved through the introducing optimal annulus for HSPP. Nevertheless, such approach has low generating efficiency, since only the beams that incident on the annulus region

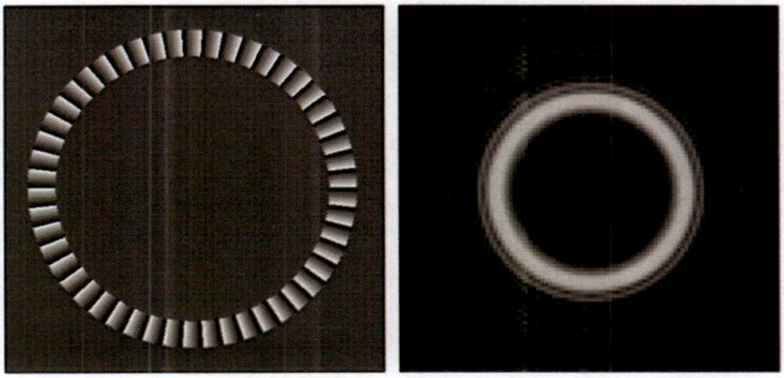

Fig. 3.27 The designed optimal annular HSPP and its corresponding far-field diffractions when Gaussian beams are incident in, providing that $\omega = 5$ mm, $l = 50$

are diffracted, and the stronger intensity part in center of the incident Gaussian beams is totally waste.

3.5 Polarization Scheme

In addition to the approaches of generating vortex beams that introduced previously, vortex beams can also be generated by means of polarization-modulation elements, which can transform Gaussian beams with specific polarizations into vortex beams. Such polarization-modulation elements include spatially variable half-wave plate (SVHWP), q-plate (QP) and so on.

3.5.1 Spatially Variable Half-Wave Plate

SVHWP is an optical element with anisotropic fast axis along azimuthal direction, as sketched in Fig. 3.28. A SVHWP consists of M sub-half wave plates (HWPs), whose fast axes are arranged at the angle of:

$$\theta(n) = \frac{2\pi m(n-1)}{M} \tag{3.4.1}$$

with n the serial number of each sub-HWP, and m the order of SVHWP.

The fabrication of SVHWP is simple:

① Calculating and sketching the geometric structure map of SVHWP on the basis of the distribution law of the fast-axes given in Eq. (3.4.1), as shown in Fig. 3.29. Then marking the fast axis and serial number of each sub-HWP, and dividing the

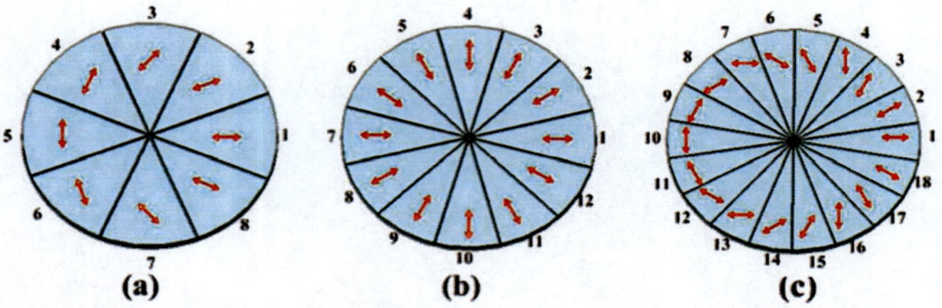

Fig. 3.28 a–c are the structures of the first, second and third order SVHWP, respectively [15]

geometric structure map into M sub-HWPs. Finally, conforming the direction of fast axes of all the sub-HWPs through their rotation.

② Arranging these sub-HWPs with consistent direction of fast axes inside a circle tightly, as shown in Fig. 3.29a and b.

③ Zooming the circles in Fig. 3.29a and b as backing, satisfying that the sizes of the circles are identical with the HWP to be cut.

④ Placing the zoomed backing on the HWP to be cut. Then cutting the HWP through a diamond cutter, and M isosceles triangle-shaped sub-HWPs are obtained.

⑤ Polishing the three edges of each sub-HWP to make it smooth. Then combining them and placing them on the quartz plate according to their serial number. Finally fixing them by bonding their outer edges with silicone rubber.

Figure 3.30 displays a picture of the second order SVHWP ($m = 2$) fabricated based on the procedures above.

A mth order SVHWP is described by Jones matrix as:

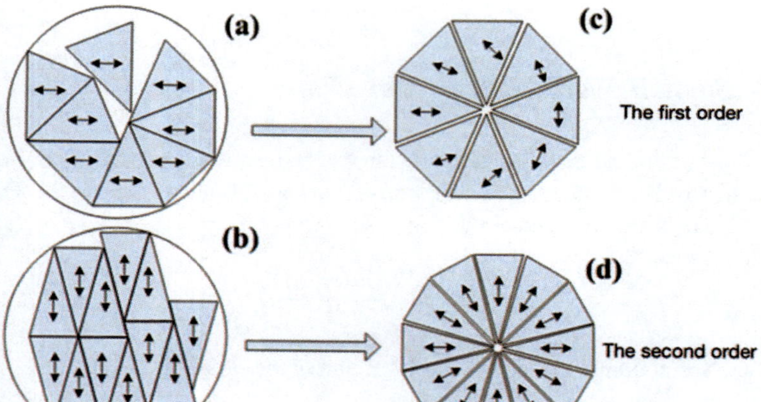

Fig. 3.29 Fabrication of SVHWP

Fig. 3.30 The fabricated second order SVHWP

$$J_{SVHWP} = \sum_{n=1}^{M} \begin{bmatrix} \cos 2\theta(n) & \sin 2\theta(n) \\ \sin 2\theta(n) & -\cos 2\theta(n) \end{bmatrix} \tag{3.4.2}$$

When $M \to \infty$, the fast axes variation along azimuthal direction of SVHWP turns to be continuous, and thus Eq. (3.4.2) is simplified as:

$$J_{SVHWP}(\varphi) = \begin{bmatrix} \cos m\varphi & \sin m\varphi \\ \sin m\varphi & -\cos m\varphi \end{bmatrix} \tag{3.4.3}$$

with φ the azimuthal angle. If two 45° arranged quarter wave plates (QWPs) are placed in front and after the SVHWP separately, the total Jones matrix is:

$$\begin{aligned} M(\varphi) &= J_{\lambda/4}(-45^\circ) \cdot J_{SVHWP}(\varphi) \cdot J_{\lambda/4}(-45^\circ) \\ &= \frac{1}{4}(1+i)^2 \begin{bmatrix} 1 & i \\ i & 1 \end{bmatrix} \begin{bmatrix} \cos m\varphi & \sin m\varphi \\ \sin m\varphi & -\cos m\varphi \end{bmatrix} \begin{bmatrix} 1 & i \\ i & 1 \end{bmatrix} \\ &= i \begin{bmatrix} \exp(im\varphi) & 0 \\ 0 & -\exp(-im\varphi) \end{bmatrix} \end{aligned} \tag{3.4.4}$$

In this scenario, obviously the matrix in Eq. (3.4.4) comprises the helical term $\exp(im\varphi)$ and $\exp(-im\varphi)$, indicating the feasibility of generating vortex beams through SVHWP.

When β angle arranged linearly polarized Gaussian beams passing through the wave plate group expressed in Fig. 3.31, the output beams are:

$$\begin{aligned} E &= M(\varphi) \begin{bmatrix} \cos\beta \\ \sin\beta \end{bmatrix} \\ &= i\exp(im\varphi)\cos\beta \begin{bmatrix} 1 \\ 0 \end{bmatrix} + i\exp(-im\varphi)\sin\beta \begin{bmatrix} 0 \\ 1 \end{bmatrix} \end{aligned} \tag{3.4.5}$$

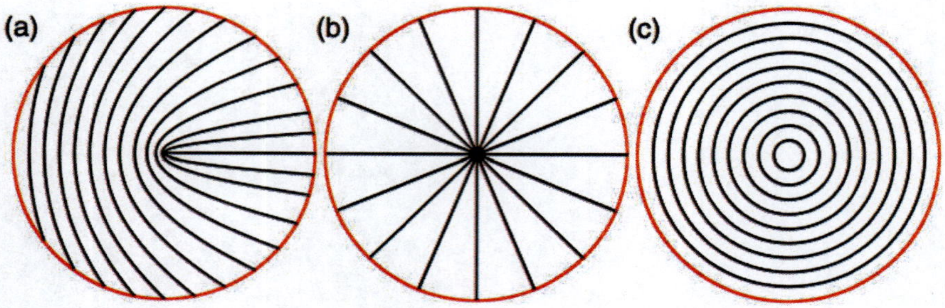

Fig. 3.31 Diagram of q-plates with various orders. **a** $q = 0.5, \alpha_0 = 0$; **b** $q = 1, \alpha_0 = 0$; **c** $q = 1, \alpha_0 = \pi/2$ [16]

which means the output beams are the coaxial combination of horizontal and vertical polarizations with vortex modes $|\text{m}\rangle$ and $|-\text{m}\rangle$, respectively. Meanwhile, the intensity proportion of such two polarizations is $(\cos\beta/\sin\beta)^2$. Therefore, a polarizer can be employed to filter the desired vortex modes or OAM states.

Particularly, when the incident beams are horizontally or vertically polarized, vertical or horizontal polarizations absent in the output beams. In this scenario, one can generate high mode purity vortex beams without a polarizer. The generation efficiency of SVHWP is associated with the number of sub-HWPs M. Previous research has shown that, the larger the M value, the higher the generation efficiency. When $M = 8$, the efficiency is 74.6%. And when $M \to \infty$, the efficiency can reach to 78.5% [15]. Therefore, when fabricate SVHWP in practice, the number of sub-HWPs should as many as possible, to obtain higher generation efficiency.

In summary, it is easy to understand from Eq. (3.4.5) that, when the horizontally or vertically polarized Gaussian beams are incident, the outputs are vortex beams with the same polarization, whose topological charges are the same or opposite with the order of SVHWP. In other words, a SVHWP along with two QWPs has the same functions with SPP, and thus has become a low-cost scheme to imitate SPP under some scenarios to generate vortex beams.

3.5.2 Q-plate

Q-plate made up of nematic liquid-crystals, is a kind of polarization modulation element, to realize the spin-to-orbital angular momentum conversion of photons [16]. It can control the non-uniform distribution of the main axis of liquid–crystal molecules, form a local half-wave plate at each point of the transverse plane, introduce helical wavefront to the modulated light, and thus give OAM to output beams. At that time, vortex beams are generated.

In polar coordinate, the main axis distributions of liquid–crystal molecule on the transverse plane of q-plate is:

$$\alpha(r, \varphi) = q\varphi + \alpha_0 \tag{3.4.6}$$

where q is the order of q-plate and should be the integer multiples of 0.5, α_0 is the direction of main axis when $\varphi = 0$. Figure 3.31 displays structures of q-plate with various orders (q values), where the solid line indicates the alignment direction of main axis of the liquid–crystal molecule at this point

The Jones matrix of a q-plate reads [16]:

$$M_q = \begin{bmatrix} \cos 2\alpha & \sin 2\alpha \\ \sin 2\alpha & -\cos 2\alpha \end{bmatrix} \tag{3.4.7}$$

where α is determined by Eq. (3.4.6). When right-circularly polarized beams are incident, the output optical fields are:

$$E = M_q \begin{bmatrix} 1 \\ i \end{bmatrix} = \exp(i2\alpha) \begin{bmatrix} 1 \\ -i \end{bmatrix} = \exp(i2q\varphi)\exp(i2\alpha_0) \begin{bmatrix} 1 \\ -i \end{bmatrix} \tag{3.4.8}$$

Equation (3.4.8) indicates that the outputs are left-circularly polarized vortex beams with helical wavefront and topological charge $2q$. Similarly, if left-circularly polarized beams are incident, the outputs are:

$$E = M_q \begin{bmatrix} 1 \\ -i \end{bmatrix} = \exp(-i2q\varphi)\exp(-i2\alpha_0) \begin{bmatrix} 1 \\ i \end{bmatrix} \tag{3.4.9}$$

which are right-circularly polarized vortex beams with the topological charge $-2q$.

If the incident beams are linear polarized, the output will be vector beams with anisotropic polarization distributions. Such phenomena will be discussed in detail in Chap. 7.

In summary, the main features of output fields when circular polarizations pass through a q-plate, are list as follows:

① The chirality of circular polarization changes, that is, from left to right and from right to left.

② The output beams have the phase delay $2\alpha_0$.

③ The outputs are vortex beams carrying OAM, whose topological charge is $2q$ or $-2q$ (determined by the incident circular polarizations).

In practice, a q-plate can be obtained, through applying the corresponding electrostatic field on the surface of the liquid–crystal layer to control the deflection of each liquid–crystal molecules. Or through polishing the upper and lower glass plates that cover the liquid–crystal layer according to the distribution of the main axes, after that filling the two axially aligned parallel glass plates with liquid–crystal molecules. Once the liquid–crystal molecules are stable in the lowest energy state, a q-plate is produced.

The SVHWP and q-plate introduced in this section, can also be employed to detect OAM states and generate vector beams, which will be discussed in the following chapters.

3.6 Integrated Vortex Emitter

The schemes of generating vortex beams introduced previously have a same characteristic: All of them are based on building a specific optical system. However, such system is not conducive to the development of miniaturization applications of vortex beams. This section will introduce an integrated vortex emitter with micrometer scale. Such emitter has simple structure, small size, easy operation, high robustness. Meanwhile, it is compatible with existing integration processes, and can generate vortex beams with selective topological charges.

3.6.1 *Whispering Gallery Mode*

Similar with mechanical wave, in the medium with column symmetry, electromagnetic wave can form the confined whispering gallery mode (WGM) through total reflection. The common column- symmetric medium including disk-shaped resonator, annular resonator (Fig. 3.32) and so on.

Actually, the WGM of electromagnetic wave is a set of solutions of Helmholtz equation. Here only the electric field component is considered, which is [17]:

$$E_r(r, \varphi, z) = E_r(r)E_r(z)\exp(\pm ip\varphi) \quad (3.5.1)$$

Fig. 3.32 The annular resonator

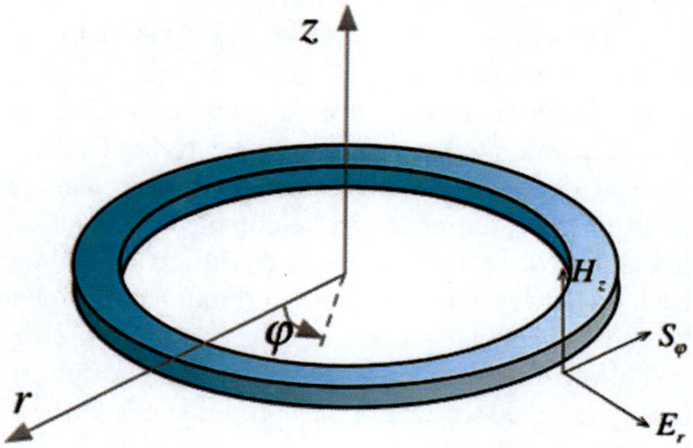

where $E_r(r)$ is the radial electric field, and usually can be expressed in the form of Bessel function or Hankel function. $E_z(r)$ is the longitudinal electric field perpendicular to the plane of the waveguide, and consistent with the longitudinal electric field component of a flat waveguide generally. p is the number of mode in the azimuthal direction of WGM. Although the WGM has a bending loss due to its radial propagation component, such loss can be ignored for the disk-shaped resonator or annular resonator with high quality factor. In the scenario, the radial or longitudinal distributions of the WGM can be regarded as standing waves. The only propagation term is $\exp(\pm ip\varphi)$, with the sign the propagation directions. Therefore, photons in WGM carry OAMs. Since the WGM needs to satisfy the phase self-consistency condition in azimuth to form a stable mode, p must be an integer, and meet with:

$$p\lambda = 2\pi n_{eff} R_{eff} \tag{3.5.2}$$

with n_{eff} and R_{eff} the effective refractive index and effective radius of the corresponding element, respectively. Based on Eq. (3.5.2), the azimuthal propagation constant of WGM is:

$$k_{\varphi}^{WGM} = \frac{2\pi n_{eff}}{\lambda} = \frac{p}{R_{eff}} \tag{3.5.3}$$

In summary, WGMs in disk-shaped or annular resonator carry OAMs. However, such modes are confined, and have zero propagation coefficient in the longitudinal propagation direction. If we can extract the OAM modes from the resonator, and give them the longitudinal momentum component to form radiation, vortex beams can be obtained. Such extraction can be accomplished through angular grating.

3.6.2 Mode Selection Through Angular Gratings

Generally, rectilinear diffraction gratings which modulate the refractive indices of media in one or more dimensions in Cartesian coordinates has been widely used. As for the annular resonator described in the previous sub-section, the angular grating, which has rotational symmetric distributions, should be employed to gain a better modulation performance. In an annular resonator, the modulation on refractive index should be in azimuth, since WGM here the has azimuthal propagation constant only. Figure 3.33 displays a simplest annularly rectangular grating [18], it diffracts WGMs in the annular resonator into longitude output mode with helical wavefront.

For the angular grating in Fig. 3.33, is grating vector is:

$$\vec{K} = K_{\varphi}\hat{\varphi} = \frac{2\pi n_{eff}}{T}\hat{\varphi} \tag{3.5.4}$$

with T the grating period. If the number of grating period is q, then T reads:

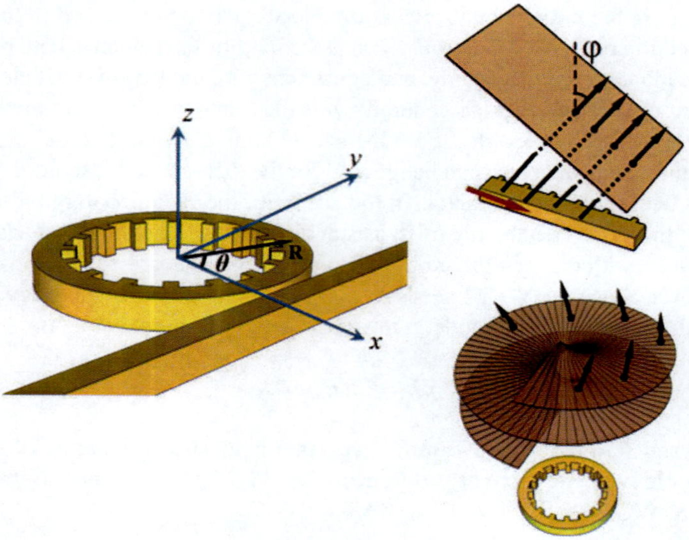

Fig. 3.33 Angular grating: The WGM in the annular resonator turns to be the longitude output mode with helical wavefront, after diffracted by an angular grating [18]

$$T = \frac{2\pi n_{eff} R_{eff}}{q} \tag{3.5.5}$$

Equation (3.5.5) with Eq. (3.5.4):

$$K_\varphi = \frac{q}{R_{eff}} \tag{3.5.6}$$

Equation (3.5.6) indicates that, the absolute value of grating vector is a discrete value determined by q, since the periodicity of an angular grating (It has q uniform periods in the section $\varphi \in [0, 2\pi)$. When an angular grating interacts with the WGM given in Eq. (3.5.1), only the modes who satisfy Bragg condition [19]:

$$k_\varphi^b = k_\varphi^{WGM} - bK_\varphi = \frac{p - bq}{R_{eff}} \tag{3.5.7}$$

can form a positive feedback, while the modes who dissatisfy will be restrained. In Eq. (3.5.7), ${k_\varphi}^b$ is the azimuthal propagation constant of the bth diffraction order. Equation (3.5.7) indicates that, when WGM with azimuthal order p interacts with an angular grating with period number q, different diffraction orders have different azimuthal propagation constant. Such phenomena are consistent with those of rectilinearly rectangular gratings in Cartesian coordinates. For the common rectilinearly rectangular gratings, different diffraction orders are reflected on various diffraction angles. While for angular gratings, different diffraction orders are reflected on different helical wavefront distributions. For the sake of radiating out the confined modes, the longitude propagation constant $k_z^{\ b}$ must be nonzero. Hence:

$$\left|k_\varphi^b\right| < k \tag{3.5.8}$$

Equation (3.5.8) with Eq. (3.5.7):

$$\left(n_{eff} - 1\right)\frac{T}{\lambda} < b < \left(n_{eff} + 1\right)\frac{T}{\lambda} \tag{3.5.9}$$

Obviously, the mode in bth diffraction order can be output from the annular resonator, as long as the appropriate choice of parameters n_{eff}, T and λ. In other words, if a certain amount of energy is injected to form a stable WGM in the annular resonator, the resonator along with an angular grating can output vortex beams carrying OAM. Such annular resonator with an angular grating is called vortex emitter. The size of such emitter can be made very small and integratable, hence it also known as integrated vortex emitter.

Next we will discuss how to determine the order or topological charge of the generated vortex beams. The uniform grating of the second order should be employed to obtain high-purity vortex beams from a vortex emitter. According to the Bragg condition of a uniform grating [19]:

$$m\lambda = 2n_{eff}T \tag{3.5.10}$$

where m is the order of the grating. For a second order grating, $m = 2$, then the grating period should be:

$$T = \frac{\lambda}{n_{eff}} \tag{3.5.11}$$

Substitute it into Eq. (3.5.9):

$$1 - \frac{1}{n_{eff}} < b < 1 + \frac{1}{n_{eff}} \tag{3.5.12}$$

Hence, b must be equal to 1, which means only the first diffraction order can be output. In this scenario:

$$k_\varphi^1 = \frac{p - q}{R_{eff}} \tag{3.5.13}$$

then the topological charge of the generated vortex beams is:

$$l = k_\varphi^1 R_{eff} = p - q \tag{3.5.14}$$

Fig. 3.34 Stereoscan photograph of an integrated vortex emitter [18]

As for a vortex emitter, q is fixed, hence the topological charge is only associated with the value of p. From Eq. (3.5.2), p is determined by the wavelength λ and effective refractive index n_{eff}, indicating that the control of topological charge can be realized by changing the input wavelength of the external coupling waveguide of the ring resonator, or by changing the temperature of the cavity to change the effective refractive index.

In 2013, X. Cai et al. demonstrated silicon-integrated vortex beam emitters, using angular gratings to extract light confined in WGMs with high OAM into free-space beams with well-controlled amounts of OAM, where the width and height of the waveguide is 500 nm and 220 nm separately, the radial of the resonator is $R = 3.9\ \mu\text{m}$, $q = 72$, as shown in Fig. 3.34 [18].

Figure 3.35 [18] shows the experimentally measured spectrum of various input wavelengths when the radial of resonator is 7.5 μm done by Cai et al. Obviously, on both side of 0th order ($l = 0$), OAM modes with topological charges ± 1, ± 2, ± 3 and so on are obtained at every interval of the free spectrum range. When the input wavelength is 1525 nm, 0th order mode ($l = 0$) together with mode splitting is obtained. Such splitting results from the second order condition ($p = q$) of the grating is exactly satisfied near the wavelength of 1525 nm, where the reflection effect of the grating, the second order diffraction, is the strongest, so that two degenerate WGMs with the same azimuthal propagation constant are allowed in the resonator. However, these two degenerate WGMs has different effective refractive indices inside the resonator, leading to two output 0th order mode with various wavelengths.

In summary, introducing an angular grating for the annular resonator, will output the confined WGM, and vortex beams are obtained in the first diffraction order. The topological charges of the generated vortex beams are determined by the wavelength and effective refractive index.

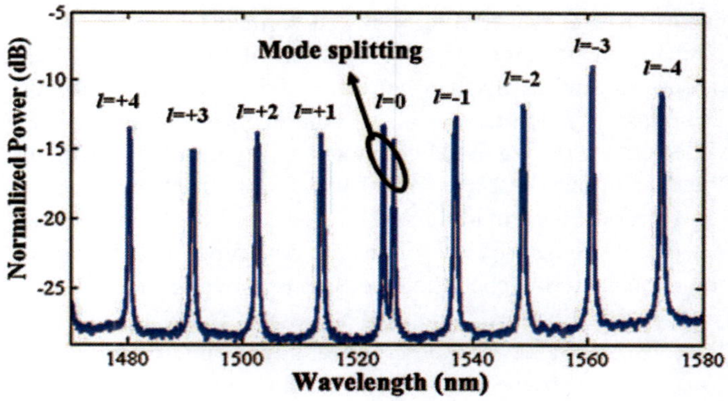

Fig. 3.35 Experimentally measured output spectrum of vortex emitter with the resonator radial $R = 7.5\ \mu\text{m}$ [18]

3.7 Generation of Multiplexed Vortex Beams

The multiplexed vortex beams consist of many different OAM components. Their intensity distributions look like petals, resulting from the interference of each OAM mode. In this section, we will focus on how to generate multiplexed vortex beams carrying multiple various OAMs.

3.7.1 *Optical Field Features of Multiplexed Vortex Beams*

Vortex beams with helical wavefront have doughnut intensity distributions due to their phase singularity. If vortex beams have more than one OAM components, the intensity distributions will consist of many petals since the interference of various OAM components. Their phase or wavefront will also be different from those of one OAM modes. Such vortices are known as multiplexed vortex beams.

Considering the simplest case, the two-fold multiplexed vortices with equal intensity proportion. The two contained topological charges are l_1 and l_2. Then complex amplitude of such two-fold multiplexed vortices ($|l_1\rangle + |l_2\rangle$) is:

$$E \propto \exp(il_1\varphi) + \exp(il_2\varphi) \tag{3.6.1}$$

Hence the intensity distributions meet with:

$$\begin{aligned} |E|^2 &\propto |\exp(il_1\varphi) + \exp(il_2\varphi)|^2 \\ &= 2 + 2(\cos l_1\varphi \cos l_2\varphi + \sin l_1\varphi \sin l_2\varphi) \\ &= 2\{1 + \cos[(l_1 - l_2)\varphi]\} \end{aligned} \tag{3.6.2}$$

which indicates that, for two-fold multiplexed vortices with equal intensity proportion, the intensity distributions are the function of the azimuthal angle φ. From Eq. (3.6.2), one can find easily that the intensity distribution has multi-petal structure, and the number of petals is $|l_1 - l_2|$. In addition, the phase profile is also changed, due to the interference of two OAM components. Figure 3.36 lists the intensity and phase distribution in the waist of some of two-fold multiplexed vortex beams with equal intensity proportion providing that $l_1 \neq l_2$.

Note that effects caused by the diameter of vortices with various topological charges are unconsidered. Hence the discussions above are merely for the smaller value of $|l_1 - l_2|$. Actually, the diameter of a vortex is associated with its topological charge l. If the absolute values of l_1 and l_2 differ greatly, interference between various OAM components will absent. In this scenario, the intensity structure of multiplexed vortices is "two concentric doughnuts", as shown in Fig. 3.37, where the intensity and phase distributions of $(|-30\rangle + |3\rangle)$ are present. From Fig. 3.37, the two OAM components don't interfere, nor do their phases affect with each other.

Multiplexed vortex beams with lots of OAM components in non-equal intensity proportions are:

$$E \propto \sum_{l=-\infty}^{+\infty} a_l \exp(il\varphi) \tag{3.6.3}$$

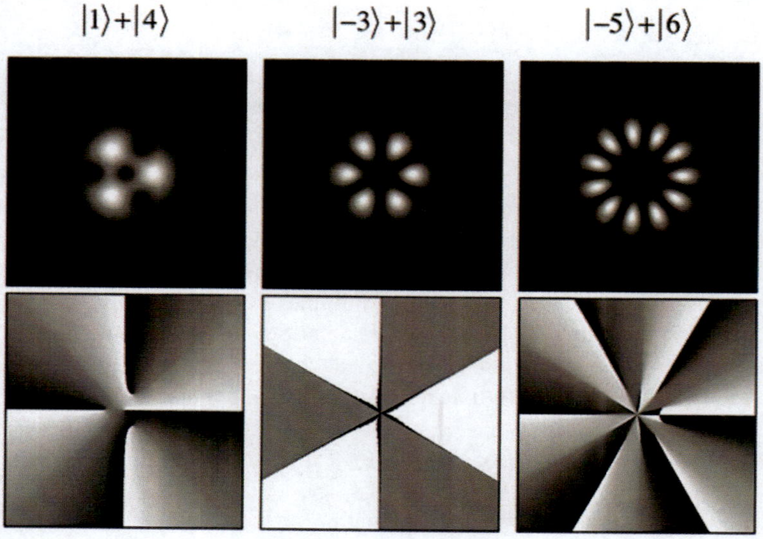

Fig. 3.36 The intensity and phase distributions in the waist of two-fold multiplexed optical vortice with equal proportion

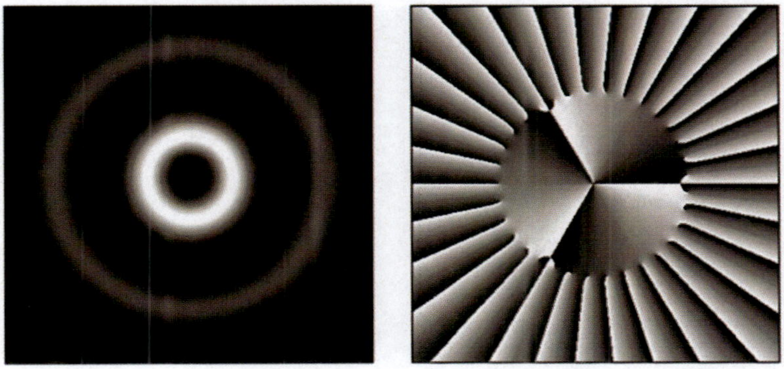

Fig. 3.37 The intensity and phase distributions in the waist of two-fold multiplexed vortex beams ($|-30\rangle + |3\rangle$)

where a_l is the complex coefficient, and $|a_l|^2$ is the intensity of mode $|l\rangle$. Actually, | Actually, || Actually, $|a_l|^2$ is the OAM spectrum introduced in Sect. 1.3, which represent the intensity proportions of various OAM modes. Figure 3.38 shows the intensity and phase distributions in the waist, and the OAM spectrum, of eight-fold multiplexed vortex beams consist of $|-14\rangle$, $|-11\rangle$, $|-8\rangle$, $|-5\rangle$, $|5\rangle$, $|8\rangle$, $|11\rangle$, and $|14\rangle$. Obviously a more complex field is present since the interference of eight OAM components. In this scenario, the OAM information can't be read directly from the point of intensity or phase distributions. And OAM spectra must be introduced to represent multiplexed vortex beams.

3.7.2 Beams Combination Method

The beams combination method is generating single mode vortex beams with various topological charges firstly, then combining them to obtain multiplexed vortices by means of specific optical systems.

Non-polarized beam splitter (Fig. 3.39) is a common beam combination element. When a beam is incident along the axis, it will be divided into two beams with equal intensity, one is transmitted, and the other is reflected whose direction is perpendicular to the incident beam. Conversely, if two vortex beams are incident along the transmission and reflection axes separately, two-fold multiplexed vortex beams will be obtained in the other two non-incident axes.

If more than two folds multiplexed are desired, scheme of cascading multi non-polarized beam splitters is available, as sketched in Fig. 3.40. Nevertheless, such generation scheme has low efficiency, since there is 50% loss once passing through a beam splitter. In addition, lots of sub-systems are needed to generate single mode vortices, leading to the huge, complex and unstable combination system.

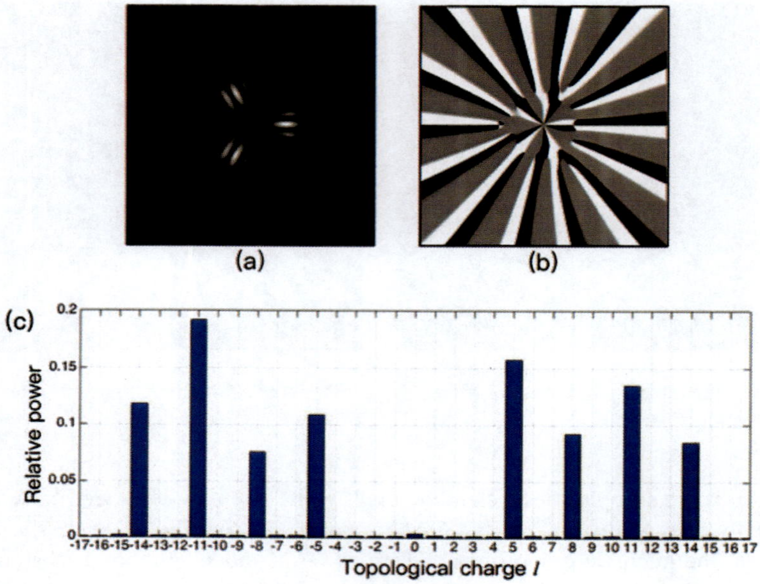

Fig. 3.38 The intensity distribution (**a**), phase distribution (**b**) in the waist, and the OAM spectrum (**c**) of eight-fold multiplexed vortex beams

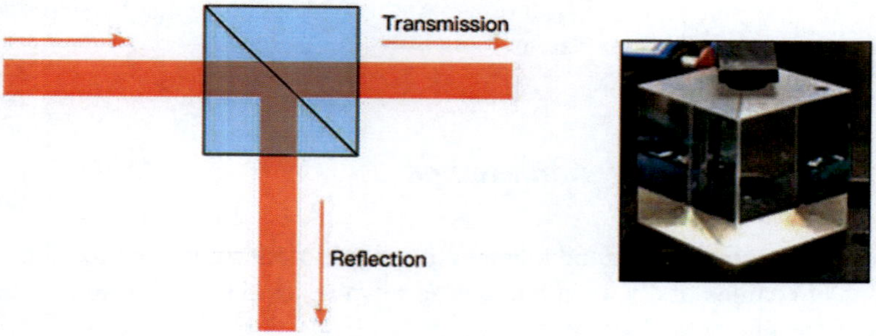

Fig. 3.39 Non-polarized beam splitter

3.7.3 Interferometer Method

The basic idea of interferometer method is identical with that of beam combination method. Their differences reflect on, the input of interferometer method is vortex beams array (which will be introduced in detail in Chap. 4), and then combining vortices in various diffraction orders by means of rotating the input array, as sketched in Fig. 3.41. In other words, the input of interferometer method is a stable vortices array, rather than multi-input of the beam combination method, which improve the robustness of the system to some extent. The common interferometers that can be used

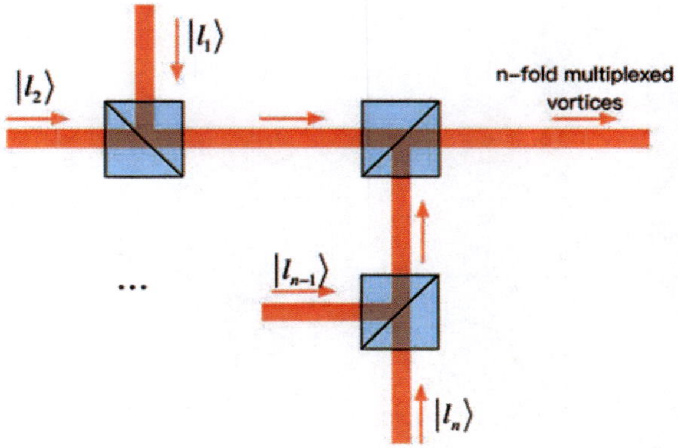

Fig. 3.40 Generating multiplexed vortex beams through cascading non-polarized beam splitters

to generate multiplexed vortices including Sagnac interferometer [20], Michelson interferometer [21], and so on.

In a Sagnac interferometer, the incident beam is divided equally into two beams. Such two beams propagate along the same path but opposite direction, and then meet at the splitting point to interfere. Figure 3.42 gives the setup to generate multiplexed vortex beams based on the Sagnac interferometer [20]. The key element here is the polarized beam splitter (PBS) and the two Dove prisms D1 and D2. Its principle is to rotate the two vortices arrays generated by the PBS together with the diffraction grating, and then superpose vortices in different diffraction orders of the array. For example, if the fork-shaped grating introduced in Sect. 3.2.3, whose diffractions contain two different vortex beams in ±1 diffraction orders, is employed here to be the diffraction grating, then after passing through the setup in Fig. 3.42, the vortices in ±1 diffraction orders will be superposed and thus two-fold multiplexed vortex beams are generated.

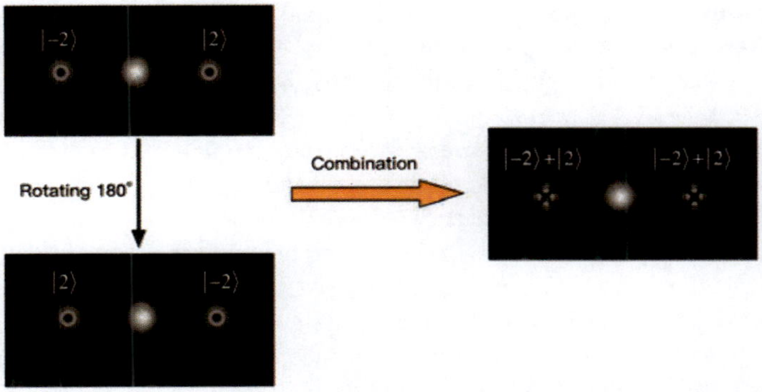

Fig. 3.41 Concept of generating multiplexed vortex beams through interferometer

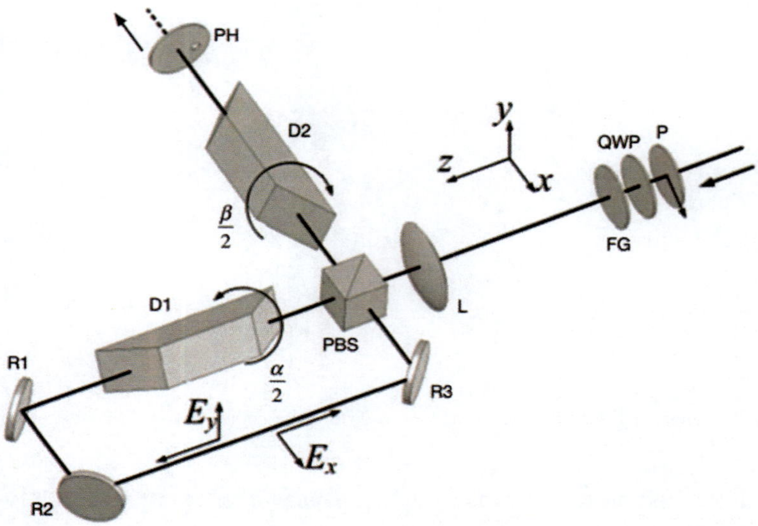

Fig. 3.42 The Sagnac interferometer for the generating of multiplexed vortex beams. P, polarizer; QWP, quarter wave plate; FG, fork-shaped grating; L, plano-convex lens; PBS, polarized beam splitter; D1 & D2, Dove prisms; R1 ~ R2, reflectors; PH, pin-hole [20]

In Fig. 3.42, circularly polarized Gaussian beams produced by a polarized together with a quarter wave plate (QWP) are diffracted by a fork-shaped grating. The reason of why circular polarization is needed here is to realize the equal-intensity splitting when pass through the PBS. Of course, a 45° linear polarization can also achieve the same effects. According to Fourier optics [22], a plano-convex lens is used to do the Fourier transformation of the fields to obtain far-field diffractions. Note that the fork-shaped grating must be located at the object focal plane of the lens, and then the far-field diffraction is obtained in the image focal plane. Between the lens and its image focal plane, a Sagnac loop is placed. The beams after passing through the lens are divided into two beams E_x and E_y with equal intensities and orthogonal linear polarizations. The beam E_x propagate along R1 → R2 → R3, while E_y propagate along R3 → R2 → R1. After back to the PBS, E_x and E_y are combined coaxially and output.

Due to that fact that E_x and E_y are propagate along opposite directions separately, after passing through the first Dove prism D1, E_x and E_y will be rotated clockwise and counterclockwise separately at the angle of α. The second Dove prism D2 outside the Sagnac loop is to clockwise rotate the combined beams at the angle of β. Therefore, after passing through D2, E_x rotates $(\beta + \alpha)$ clockwise, and E_y rotates $(\beta - \alpha)$ clockwise. If a pin-hole is placed at the image focal plane of the lens, the desired multiplexed vortex beams will be filtered. Hence, multiplexed vortices superposed by different diffraction orders are obtained through rotating the two Dove prisms.

Particularly, as for the fork-shaped grating introduced in Sect. 3.2.3, if we want to superpose beams in ±1st diffraction orders, the difference of the rotation angles of

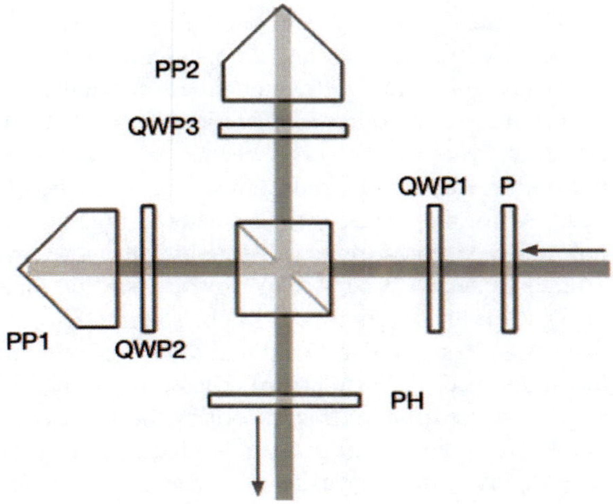

Fig. 3.43 The Michelson interferometer for the generating of multiplexed vortex beams. P, polarizer; QWP1 ~ QWP3, quarter wave plates; PBS, polarized beam splitter; PP1 & PP2, Porro prisms; PH, pin-hole [21]

E_x and E_y must be π. Thus $|(\beta + \alpha)\text{-}(\beta - \alpha)| = \pi$, and $\alpha = \pm\pi/2$. In this scenario, Dove prism D1 should be arranged at $\alpha/2 = \pm\pi/4$, and β can take arbitrary value. In fact, the Dove prism D1 is used to rotate E_x and E_y in opposite directions to accomplish the coaxial beam combination. While Dove prism D2 is used to rotate the combined beams so that the generated two-fold multiplexed vortices can be shot from the pin-hole. E_x and E_y experience identical optical path, hence this system is not sensitive to the external environment, but has very high alignment requirements.

Michelson interferometer can also be employed to generate multiplexed vortex beams, as shown in Fig. 3.43 [21]. Different from the Sagnac interferometer, two Porro prisms are employer to replace two reflectors. The rotation of two Porro prisms leads to the rotation of incident vortices array, thus to superpose various diffraction orders of the array, to generate multiplexed vortex beams.

In Fig. 3.43, the input vortex beams array (e.g., the diffraction fields of a fork-shaped grating) is transformed into circular polarization after passing through a polarizer and a QWP. The PBS divides the incident vortices array into two arrays with equal intensity and orthogonal linear polarizations, and output them to the two arms of Michelson interferometer. QWPs arranged at the angle of 45° and 135° are placed separately in the two arms, to transform linear polarizations into circular polarizations. At the end of both the two arms, a Porro prism is placed, to rotate the vortices array. And θ° rotation of the Porro prism contributes to the 2θ degree rotation of the vortices array. When the circularly polarized array reflected from the Porro prism pass through the QWP again, it will be transformed into linear polarization but perpendicular to the original linear polarization. Hence the initial reflection components will be transmitted, and the initial transmission components will be reflected, when back to the PBS. In such scenario, vortices arrays in the two arms are rotated separately and then beams of various diffraction orders are combined coaxially by the PBS. Finally two-fold multiplexed vortex beams are generated.

In summary, the nature of interferometer method is that, an improved interferometer is employed to rotate the incident vortex beams array and then superpose the beams in different diffraction orders coaxially. However, in such system, vortex beams must undergo lots of reflections. In Sect. 1.2.4, we have already studied the reflection properties of vortex beams, and find that the topological charge will be opposite for odd times reflections, and won't be changed for even times reflections. As for the Sagnac interferometer in Fig. 3.42, the split two vortex beams are reflected in 5 times (reflected once by two Dove prisms and three reflectors) and 7 times (reflected once by two Dove prisms and three reflectors, and twice by the PBS), respectively. As for the Michelson interferometer in Fig. 3.43, both the split two vortex beams are reflected in 3 times (reflected once by the PBS, and twice by the roof surface of Porro prism). Therefore, in addition to the coaxial superposition of vortices in various diffraction orders, the above two interferometers will invert the topological charge of the incident vortices. For the vortices array produced by a fork-shaped grating, such opposite is offset since the topological charges in ± 1 diffraction orders are opposite. If the incident array is more complex, reflection effects on the topological charge must be considered.

The interferometer method is a very stable way to generate multiplexed vortex beams. However, it has complex setup and require high adjusting accuracy. Additionally, an interferometer can only produce two-fold multiplexed vortices, but powerless to generated much more than two-fold multiplexing.

3.7.4 Phase Grating Method

The nature of the previously introduced two multiplexed vortices generating schemes are superposing two single-mode vortex beams coaxially through optical systems. These schemes have simple principles, but require high adjusting accuracy. Phase grating method can perfectly overcome such issues, which can generate multiplexed vortex beams directly when fundamental Gaussian beams are incident in.

Let's review the blazed grating method to imitate amplitude grating through phase-only grating discussed in Sect. 3.3.2 firstly. In such approach, the transparent and opaque parts are diffracted along different directions to form two diffraction patterns, and thus realize the amplitude modulation indirectly. When imitating fork-shaped grating through blazed-grating method, there must be an appropriate blazed grating period that lead to the superposition of $+1$st and -1st diffraction orders in the two diffraction patterns, where the superposed beams are the two-fold multiplexed vortex beams exactly, as shown in Fig. 3.44.

Figure 3.44 indicates that, under paraxial approximation, the superposition of $+1$st and -1st diffraction orders of the two diffraction patterns need the condition $\alpha_1 + \alpha_2 = \beta$, where α_1, α_2 and β are the diffraction angles of fork-shaped grating in opaque part, fork-shaped grating in transparent part and blazed grating, respectively. Usually, the ratio of slit width to grating constant of fork-shaped grating is 0.5, thus $\alpha_1 = \alpha_2$, and $\beta = 2\alpha_1$. Supposing that the grating constants of fork-shaped grating

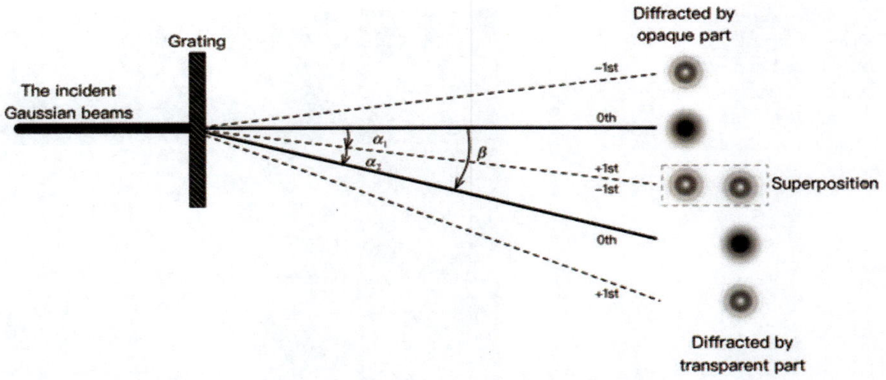

Fig. 3.44 Concept of generating two-fold multiplexed vortex beams through blaze-superposed phase-only fork-shaped grating

and blazed grating are d and Λ separately, according to grating equation, one can obtain:

$$\begin{cases} \lambda = d \sin \alpha_1 = d\alpha_1 \\ \lambda = \Lambda \sin \beta = \Lambda\beta \end{cases} \tag{3.6.4}$$

thus:

$$d = 2\Lambda \tag{3.6.5}$$

which means the +1st order of opaque part and the -1st order of transparent part will be superposed when Eq. (3.6.5) is satisfied. One can obtain the pure two-fold multiplexed vortices after filtering by an aperture. Figure 3.45 displays various blaze-superposed phase-only fork-shaped gratings under the limit of Eq. (3.6.5), and their corresponding far-field diffraction pattern when Gaussian beams are incident in. Vortex beams in ± 1 diffraction orders of fork-shaped grating have opposite topological charges but identical intensities. Hence only two-fold multiplexed vortices consist with opposite topological charge in equal intensity proportion can be generated. In addition, only two of the diffraction orders in diffraction patterns are employed, and other diffraction orders are filtered out, contributing to the low generating efficiency.

In Sects. 3.2.2 and 3.3.3, we have discussed principles and methods of generating single mode vortex beams through SPP or phase-only vortex grating. Similarly, is it possible to design a phase-only diffraction grating, when Gaussian beams are incident in, multiplexed vortex beams with arbitrary intensity proportion and OAM states distributions can be generated directly?

Supposing that the incident beam is a plane wave with the amplitude $E_0(r)$. The target multiplexed vortex beams consist of n various OAM states, their topological

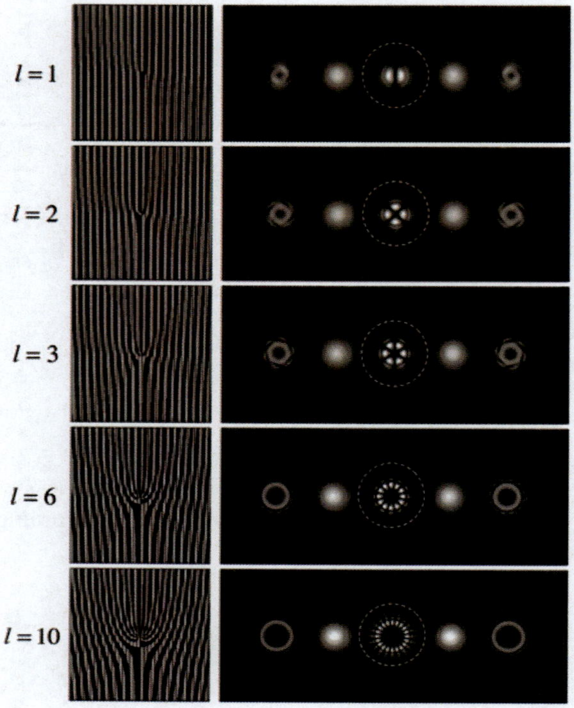

Fig. 3.45 Various blaze-superposed phase-only fork-shaped gratings and their corresponding far-field diffraction pattern when Gaussian beams are incident in

charges are $\{l_1, l_2, \ldots, l_n\}$, and normalized intensities are $\{p_1, p_2, \ldots, p_n\}$. Limiting that the phase-only grating has azimuth variable and radial invariant, and its transmission function is $T(\varphi)$. Then when the incident plane wave just passes through such grating, the fields should contain n OAM modes with different intensity distributions:

$$E = E(r)T(\varphi) = E(r)\sum_{m=1}^{n} A_m \exp(il_m\varphi) \tag{3.6.6}$$

Therefore:

$$T(\varphi) = \sum_{m=1}^{n} A_m \exp(il_m\varphi) \tag{3.6.7}$$

In Eqs.(3.6.6) and (3.6.7), $|A_m|^2 = p_m$. Equation (3.6.7) can be Fourier expanded with finite non-zero terms:

$$T(\varphi) = \sum_{m=-\infty}^{+\infty} A_m \exp(im\varphi) \tag{3.6.8}$$

where the Fourier coefficient $\{A_m\}$ is a complex array, satisfying the non-zero $\left|A_{l_a}\right|^2 = p_a$ $(a = 1, 2, \ldots n)$.

Usually, the grating given by Eq. (3.6.8) isn't phase-only. A phase-only grating $T(\varphi) = C\exp[i\varphi(\varphi)]$ must meet with:

$$T(\varphi)T^*(\varphi) = |C|^2 \tag{3.6.9}$$

As for Eq. (3.6.8), and considering Eq. (3.6.9), then:

$$\sum_{m=-\infty}^{+\infty} \sum_{m'=-\infty}^{+\infty} A_m A_{m'}^* \exp\left[i\left(m - m'\right)\varphi\right] = |C|^2 \tag{3.6.10}$$

Obviously, Eq. (3.6.10) is true providing that only one OAM mode is present. It indicates a phase-only vortex grating can only be employed to produce single mode vortex beams. When producing multiplexed vortices, such gratings must contain the amplitude term, and the phase-only modulation is inadequate. However, modulating incident beams with both amplitude and phase in a same point is impossible, leading to the undesired multiplexed vortices. Figure 3.46 displays (a) the intensity distributions and OAM spectra of three-fold multiplexed vortices consist of $|-2\rangle$, $|1\rangle$ and $|6\rangle$ with intensity proportion 1:2:1, (b) the generated phase-only grating through Eq. (3.6.8), and (c) the intensity distributions and OAM spectra of far-field diffractions when Gaussian beams are incident in the grating in (b). The OAM spectra in (c) are very different from that in (a), which is consistent with prediction.

Generating arbitrary ideal multiplexed vortex beams through a phase-only grating is infeasible. However, an approximate form of such grating can be generated through iteration method, to obtain an infinite approximation to the desired ideal multiplexed vortices.

3.7.5 Iteration Method

The idea of iteration method is: by analyzing the OAM spectra of the far-field diffraction fields of the phase-only grating continually, the parameters in the grating are modified appropriately according to the difference between the measured and desired OAM spectra, so that the mode and intensity distribution of the generated multiplexed vortices are getting closer to the desire. There are two forms of iteration method, one is done inside a computer, where a phase-only grating is produced to transform Gaussian beams directly into the desired multiplexed vortices [23]; The other is done in the optical system, where the grating encoded on the LC-SLM is switched according to the analyzed results of OAM spectra in the far-field until the desired multiplexed vortices are obtained [24]. The idea of such two forms is exactly the same. Their differences are reflected on the detection method of OAM spectra during the iteration, one is calculated while the other is actually measured.

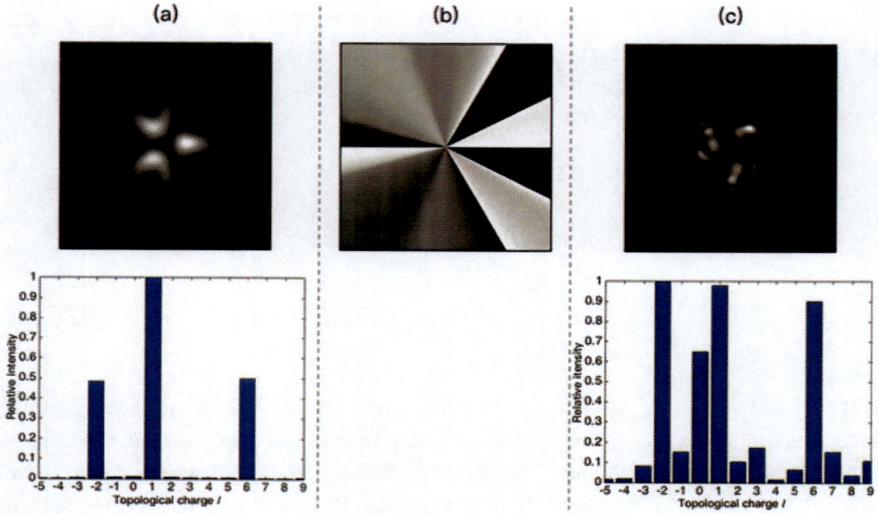

Fig. 3.46 **a** The intensity distributions and OAM spectra of three-fold multiplexed vortices with intensity proportion of 1:2:1. **b** The phase-only grating produced according to Eq. (3.6.8). **c** The intensity distribution and OAM spectra of far-field diffractions when Gaussian beams are incident in **b**

The transmittance function of a phase-only diffractive grating is:

$$g(\varphi) = \exp[i(\psi(\varphi))] \tag{3.6.11}$$

where $\psi(\varphi)$ is [23]:

$$\psi(\varphi) = \mathrm{Re}\left\{-i\ln\left[\sum_{m=-\infty}^{+\infty} C_m \exp(im\varphi)\right]\right\} \tag{3.6.12}$$

In Eq. (3.6.12), Re{ } means taking the real component, the initial value of $\{C_m\}$ is $\{C^0{}_m\}$, and the missed imaginary part is the amplitude modulation. Based on the Fourier expansion:

$$g(\varphi) = \sum_{m=-\infty}^{+\infty} B_m \exp(im\varphi) \tag{3.6.13}$$

with B_m:

$$B_m = \frac{1}{2\pi}\int_0^{2\pi} g(\varphi)\exp(-im\varphi)d\varphi \tag{3.6.14}$$

Next we will discuss the procedure of iteration method refer to Figs. 3.47 and 3.48. When Gaussian beams are incident into the phase-only grating $g(\varphi)$, one can obtain the OAM spectra of diffraction fields (Fig. 3.48b). Compared with the OAM spectra of target multiplexed vortices (Fig. 3.48a), if $|B_{l_a}|^2 < |A_{l_a}|^2$, replacing C_{l_a} in Eq. (3.6.12) with C'_{l_a}, meanwhile meet $|C'_{l_a}| > |C_{l_a}|$, so as to reduce the gap between $|A_{l_a}|$ and $|B_{l_a}|$ through increasing the absolute value of C_{l_a}. Otherwise, if $|B_{l_a}|^2 > |A_{l_a}|^2$, one must meet $|C'_{l_a}| < |C_{l_a}|$, to reduce the gap between $|A_{l_a}|$ and $|B_{l_a}|$ through decreasing the absolute value of C_{l_a}. Iterating the above process until no better results could be gotten. At that time, a series of optimal parameters $\{C_m\}$ must be acquired to obtain the multiplexed vortex beams that are closest to the target one. At the beginning of iteration, original parameters $\{C^0{}_m\}$ must be set. The choice of $\{C^0{}_m\}$ in various occasions are different, in order to achieve the fastest iterative convergence speed. In most of scenarios, the setting of $\{C_m^0\} = \{|A_m|\}$ will meet the above requirements [23].

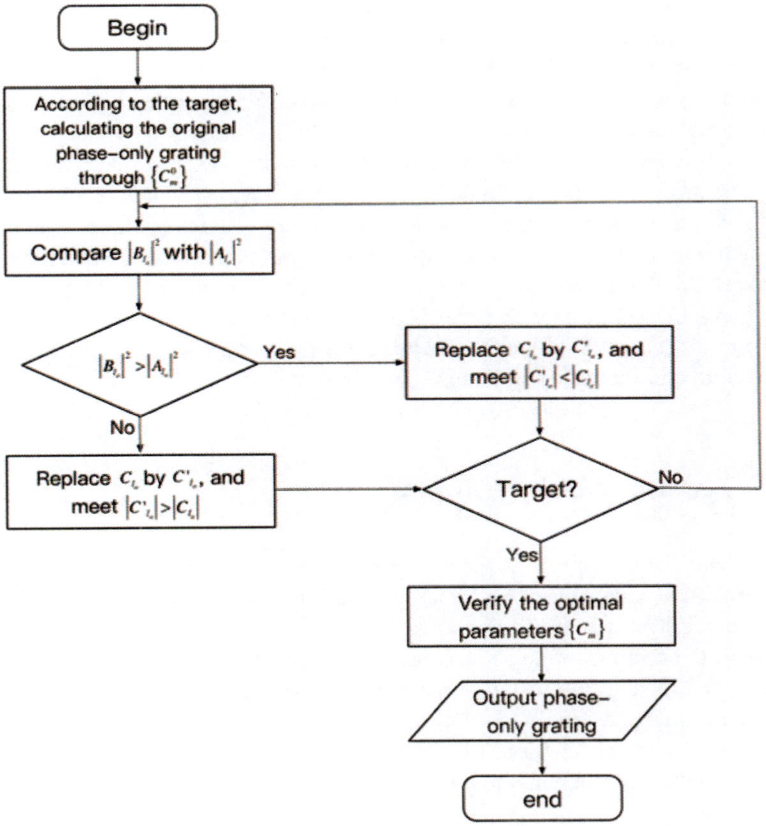

Fig. 3.47 The flow chart of the iteration method to generate multiplexed vortex beams

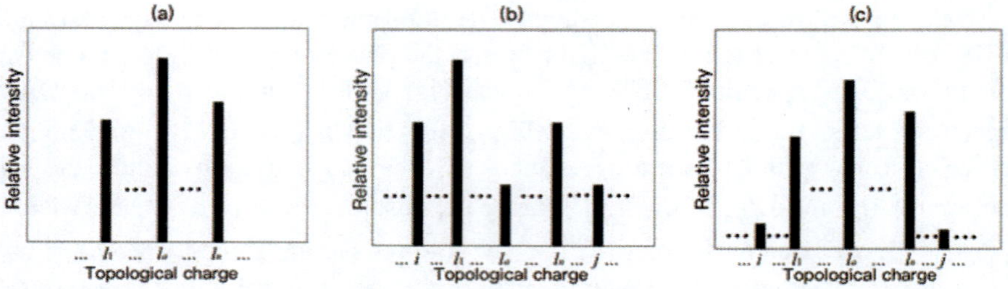

Fig. 3.48 **a** The target OAM spectra; **b** OAM spectra when Gaussian beams are incident into the phase-only grating $g(\varphi)$; **c** OAM spectra when Gaussian beams are incident into the phase-only grating optimized by iteration method

Based on the above iteration procedure, the OAM spectra of far-field diffraction fields of the optimized phase-only grating are given in Fig. 3.48c, which is very similar with that of the target (Fig. 3.48a). Their only difference is that, some of power are leaked to undesired OAM channels. Such phenomenon is inevitable, since it is impossible to generate multiplexed vortices through a phase-only diffractive grating according to the discussions in Sect. 3.6.4.

It seems very complex to generate multiplexed vortices through a phase-only grating produced by iteration method, additionally accompanied by undesired OAM channels. However, most of the incident power are concentrated on the target OAM mode, resulting in the more than 80% generation efficiency. Moreover, the produced phase-only grating can generate multiplexed vortices providing that only one Gaussian beam is incident, thus it has a simple optical system. However, when the amount of desired OAM modes is large, the choice of original parameter $\{C_m^0\} = \{|A_m|\}$ will contribute to slow convergence speed, low iteration efficiency and large difference from the target OAM spectra. In this scenario, pattern search algorithm can be employed to continuous optimizing.

3.7.6 Pattern-Search Algorithm

Pattern search algorithm (PSA) is an algorithm that used to solve an optimization problem with N independent variables. Its main idea can be understood as, continuously searching for a series of points $x_1, x_2, x_3, \ldots, x_N$, and making these points as close to the optimal value as possible. When the search termination condition is reached, the last set of points will be taken as the solution.

When optimal phase-only gratings through PSA to generate arbitrary multiplexed vortex beams, a target function, whose value is the optimal value point, must be set. Here the relative root-mean-square error (R-RMSE) R is chose as target function [25]:

$$R = \sqrt{\frac{\sum_{m=1}^{n} \left(|C_{l_m}|^2 - |A_{l_m}|^2\right)^2}{n \sum_{m=1}^{n} |C_{l_m}|^2}} \tag{3.6.15}$$

The smaller the R value is, the closer the phase-only grating is to the ideal value. Since the target mode distribution sequence $\{A_{lm}\}$ is already known, what we should do is just to optimize parameters $\{C_{lm}\}$. At this point, the issue of optimizing diffraction grating, is transformed into that of optimizing the sequence $\{C_{lm}\}$ to find the minimum R value.

Figure 3.49 gives the flow chart of PSA to find minimum R value and thus to optimize phase-only gratings. Before the pattern search, a search step size Δ_0 should be set and meet $\Delta_0 > 0$; The original iteration sequence $\{b_{lm}^0\} = \{A_{lm}\}$; The iteration counter $k = 0$. Replacing the sequence $\{C_{lm}\}$ in Eq. (3.6.12) with $\{b_{lm}^0\}$, analyzing the new OAM spectra, and then calculating the R value. Hence R can be regarded as a function of the iteration sequence $\{b_{lm}^k\}$. Next comes the pattern search. Adding $+ +\Delta_k$ or $-\Delta_k$ for each element in sequence $\{b_{lm}^k\}$, and we obtain $R(b_{lm}^{k+1})$. If $R(b_{lm}^{k+1}) < R(b_{lm}^k)$, update the search step size and meet $\Delta_{k+1} > \Delta_k$; If $R(b_{lm}^{k+1}) \geq R(b_{lm}^k)$, we set $\{b_{lm}^{k+1}\} = \{b_{lm}^k\}$, then update the search step size and meet $\Delta_{k+1} < \Delta_k$. Finally, the iteration counter $k = k + 1$, and go back to the iterative calculation, until the iteration step Δk tends to be stable, where the smallest R value is gotten. Outputting the sequence $\{b_{lm}^k\}$, setting $\{C_{lm}\} = \{b_{lm}^k\}$. After substituting such parameters into Eqs. (3.6.11) and (3.6.12), the optimized phase-only diffractive grating is obtained.

Figure 3.50 shows the OAM spectra of multiplexed vortices generated through iteration method and PSA. Wherein, Fig. 3.50a is the OAM spectrum of target multiplexed vortices consists of seven various OAM states; Fig. 3.50b and c is the OAM spectra from iteration method and PSA, respectively. Obviously, when there are many OAM modes, the iteration method can not fully converge to the target, while the result of PSA is basically the same as desire. From Eq. (3.6.15), for Fig. 3.50b the value of R-RMSE is $R = 0.0833$, and for Fig. 3.50c $R = 0.0042$, indicating the OAM spectrum obtained from PSA is much closer to that of target.

Compared with other methods introduced previously in this section, generating multiplexed vortex beams through the scheme of PSA along with iteration method has made great progress, in terms of both conversion efficiency and optical path complexity. In addition, with the development of computer technology, it is easier to calculate phase-only gratings iteratively. Therefore, the iteration calculation of phase-only gratings has become one of the most important schemes to generate multiplexed vortex beams.

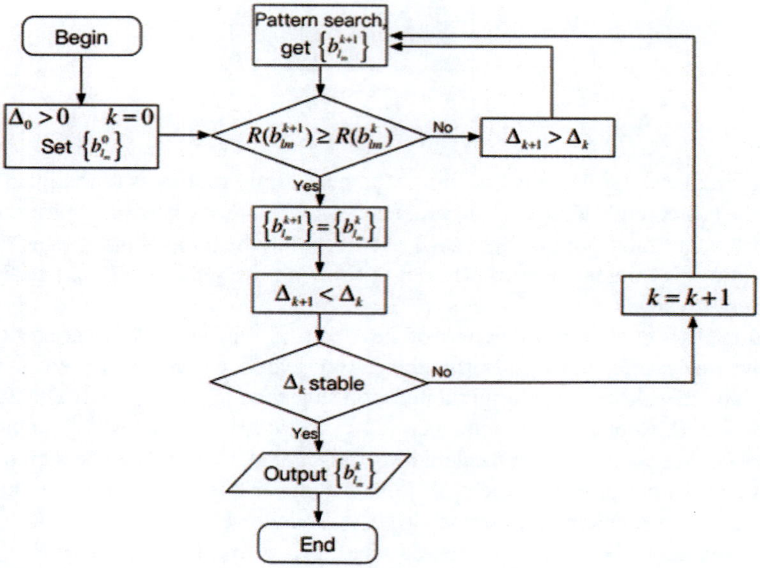

Fig. 3.49 The flow chart of PSA to generate multiplexed vortex beams

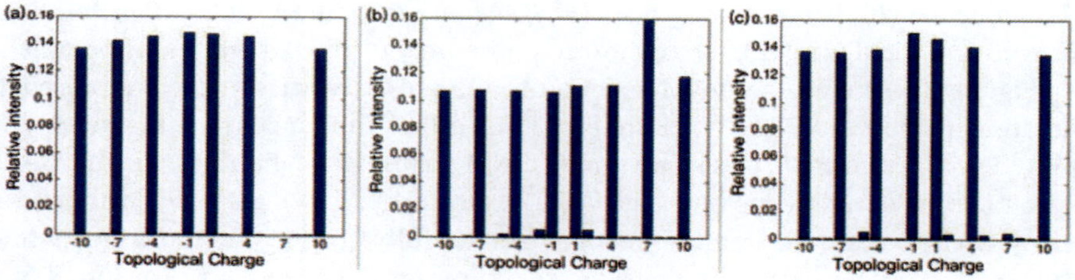

Fig. 3.50 **a** The target OAM spectrum; **b** OAM spectrum of multiplexed vortices generated from iteration method; **c** OAM spectrum of multiplexed vortices generated from PSA

3.8 Generation of Bessel-Gauss Beams

Different from the common Laguerre–Gauss (LG) beams, Bessel-Gauss (BG) beam is the actual approximation of Bessel beam. Similar with Bessel beams, BG beams are also non-diffractive, and can propagate bypass obstacles within a certain distance [26]. BG beam has a multi-concentric ring structure, which is more complex than LG beams. With the deepening of research, a variety of methods and techniques to generate BG beams have been proposed, among which the most widely and simply used ones are schemes of axion [27] and annular slit [28]. Such two schemes will be introduced in this section.

3.8.1 Axicon Method

Axicon is a phase diffractive element with a conical surface and a plane, as sketched in Fig. 3.51. It has no focusing property. When a plane wave is incident, it will be transformed into a conical wave and can be transmitted to infinity, appearing as an annular beam whose diameter is proportional to the distance and the ring width remains unchanged.

The transmittance function $T_a(r, \varphi)$ of an axicon is:

$$T_a(r,\varphi) = \exp\left(-\frac{2\pi r}{d}\right) \tag{3.7.1}$$

where d is the axicon period, represents the distance of 2π phase change along radial directions. According to Eq. (3.7.1), an axicon can be sketched as the form in Fig. 3.52, whose profile can be regarded as a sawtooth similar with the blazed grating

As discussed previously, Bessel beams and BG beams can be regarded as interference fields, which are superposed by lots of sub-plane wave with equal amplitudes and identical angles between optical axis (z-axis) and wave vector. From Fig. 3.51, when a plane wave with finite width passing through an axicon, in the diffraction fields all the incident rays will bend toward the optical axis, where the direction of wave vector in each point is in the same angle with the optical axis, and such

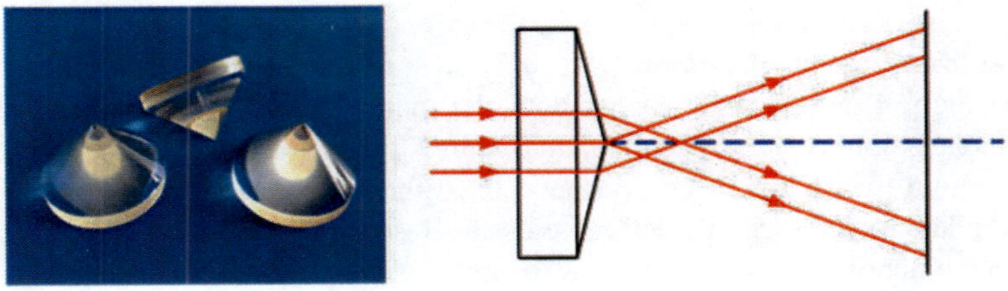

Fig. 3.51 Axicon

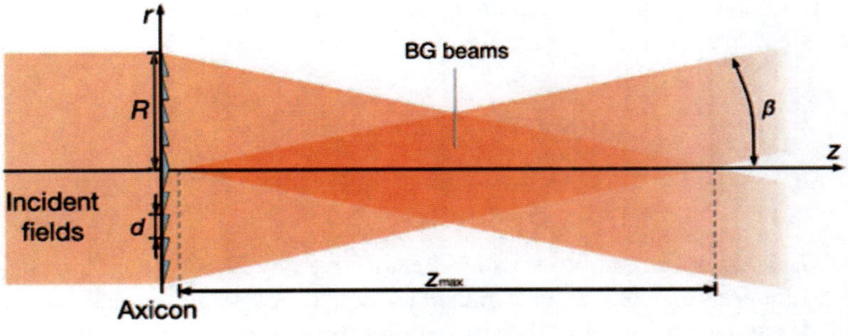

Fig. 3.52 BG beams are obtained in the overlap region behind an axicon

angle is associated with the cone apex angle of the axicon. In the region behind but near the axicon ($z < z_{max}$), the bended rays are overlapped and interfered. And the substance of such interference is the superposition of lots of sub-plane wave with equal amplitudes and identical angles between optical axis and wave vector, as sketched in Fig. 3.52. Hence obviously, when Gaussian beams are incident in, in the overlap region ($z < z_{max}$) BG beams are present, as shown in Fig. 3.7.3a. Since an axicon cannot introduce azimuthal modulation, similarly, when high-order modes, LG beams, are incident, high-order BG beams will be generated. The detail of such conversion will be discussed later.

Similar with a blazed grating, an axicon only has the first diffraction orders, hence its grating equation is:

$$\lambda = d \sin \beta \tag{3.7.2}$$

with λ the wavelength and d the axicon period. Under paraxial approximation:

$$z_{max} = \omega \cot \beta = \frac{\omega}{\sin \beta} = \frac{d\omega}{\lambda} \tag{3.7.3}$$

where ω is the waist radius of incident beams. Equation (3.7.3) indicates the overlap length z_{max} is associated with the incident waist radius ω, axicon period d and wavelength λ. When the incident waist size is larger than the axicon's diameter, parameter ω should be replaced by the radius of undersurface of the axicon. Usually, on the condition that taking the intersection of the axicon and the optical axis as the origin of coordinates, the overlap region $z \in [0, z_{max}]$ is called Bessel region.

Considering an extreme case, if both the incident plane wave and axicon are infinite extension, as $\omega \to \infty$, then we obtain $Z_{max} \to \infty$, indicating that the light fields behind the axicon at any distance from the axicon are the superposition of slanted plane wave. Such diffraction fields are Bessel beams, as shown in Fig. 3.53b. This phenomenon also implies that, BG beams are actually Bessel beams with a finite aperture constraint.

Next we will discuss how does axicon transform LG beams into BG beam in detail by means of stationary phase method [29]. According to Fresnel diffraction integral, when a LG beam passing through an axicon, the diffraction fields at the distance z can be written from Eq. (2.4.4) as:

$$E(x, y) = \frac{\exp(ikz)}{i\lambda z} \iint_{\infty} LG_{pl}(u, v) \cdot T_a(u, v) \exp\left\{ \frac{ik}{2z} \left[(x-u)^2 + (y-v)^2 \right] \right\} du dv \tag{3.7.4}$$

where LG_{pl} is complex amplitude of LG beams given by Eq. (1.4.1), T_a is the transmittance function of an axicon determined by Eq. (3.7.1). In cylindrical coordinates, Eq. (3.7.4) can be expressed as the separation form of angular and radial

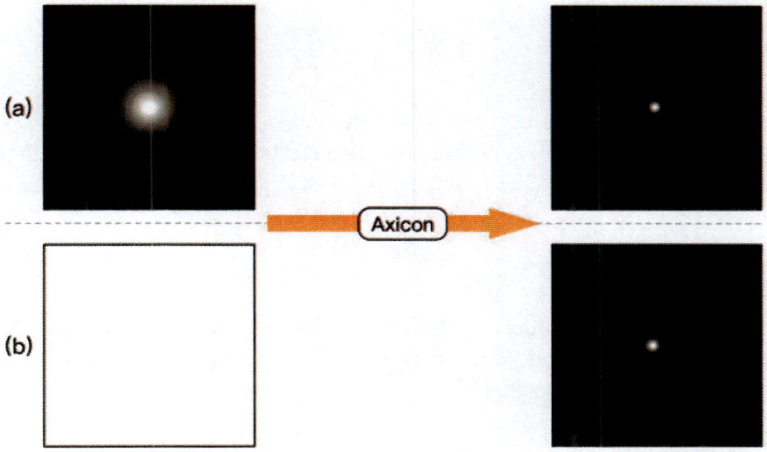

Fig. 3.53 Diffraction patterns when **a** Gaussian beams and **b** infinitely extended plane wave are incident in an axicon

$$E(r,\varphi,z)=\frac{1}{i\lambda z}\exp\left[ik\left(z+\frac{r^2}{2z}\right)\right]\int_0^R r'\left[A\left(\frac{\sqrt{2}r'}{\omega_0}\right)^l\exp\left(-\frac{r'^2}{\omega_0^2}\right)T_a(r',\varphi')\right]$$
$$\cdot\exp\left(\frac{ikr'^2}{2z}\right)\int_0^{2\pi}\exp(il\varphi)\exp\left[\frac{-ikr'r\cos(\varphi-\varphi')}{z}\right]d\varphi' dr' \tag{3.7.5}$$

where (r', φ') is coordinates of original fields, R is the limited diameter, and A is a complex constant:

$$A=\frac{C_{pl}}{\omega_0}\exp(i\Phi) \tag{3.7.6}$$

The detailed parameters in Eq. (3.7.6) have already given in Sect. 1.4.1. Firstly integrating Eq. (3.7.5) in the angular direction:

$$E(r,\varphi,z)=\frac{1}{i\lambda z}\exp\left[ik\left(z+\frac{r^2}{2z}\right)\right]\exp(il\varphi)\cdot\int_0^R f_l(r')\exp[-ik\mu(r')]dr' \tag{3.7.7}$$

where:

$$f_l(r')=2\pi(-i)^l\left[A\left(\frac{\sqrt{2}r'}{\omega_0}\right)^l\exp\left(-\frac{r'^2}{\omega_0^2}\right)r' J_l\left(\frac{krr'}{z}\right)\right] \tag{3.7.8}$$

$$\mu(r') = \frac{r'^2}{2z} - \frac{k_r r'}{k} \quad (3.7.9)$$

where, $J_l()$ is lth order Bessel function of first kind, $k_r = 2\pi/d$ is radial wave number or radial propagation constant. Taking the first derivative of Eq. (3.7.9) with respect to r', and force it to 0, then the inflection point of Eq. (3.7.6) is obtained as:

$$r'_c = \frac{k_r z}{k} \quad (3.7.10)$$

indicating that Eq. (3.7.9) has one inflection point only. According to the principles of stationary phase method, if the change in the stable phase region is small, the following approximation exists:

$$\int_0^R f_l(r') \exp[-ik\mu(r')]dr' \propto \frac{f_l(r'_c) \exp[ik\mu(r'_c)]}{\sqrt{kz^{-1}}} \quad (3.7.11)$$

By ignoring the position independent terms, the intensity distribution after the axicon reads:

$$I(r, \varphi, z) \propto z^{2l+1} \exp\left(-\frac{2z^2}{z_{max}^2}\right) J_l^2(k_r r) \quad (3.7.12)$$

where z_{max} is given as Eq. (3.7.3). Equation (3.7.12) means that the diffraction fields after passing through an axicon are mainly determined by lth order Bessel function of first kind. Meanwhile, Eq. (3.7.7) has already contain the helical term $\exp(il\varphi)$. Therefore, one can draw the conclusion that lth order BG beams can be transformed from lth order LG beams through an axicon, and the generated BG beams are only present at the Bessel region, as shown in Fig. 3.54.

If we want to generate BG beams directly from Gaussian beams, scheme of cascading a SPP and an axicon can be employed, where Gaussian beams are transformed into lth order vortex beams (LG beams) firstly, and then transformed into lth

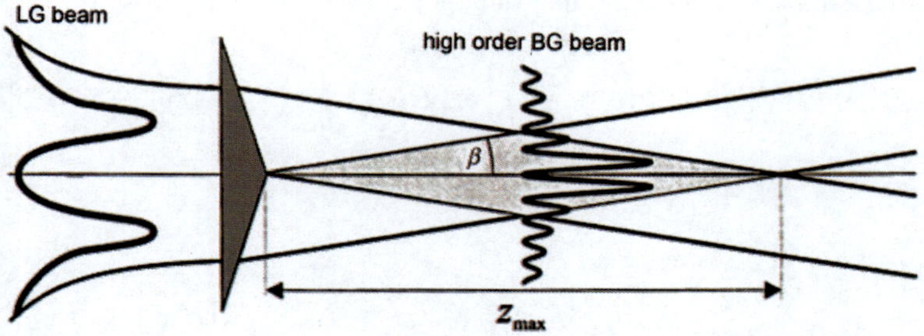

Fig. 3.54 lth order BG beams can be generated when lth order LG beams passing through an axicon

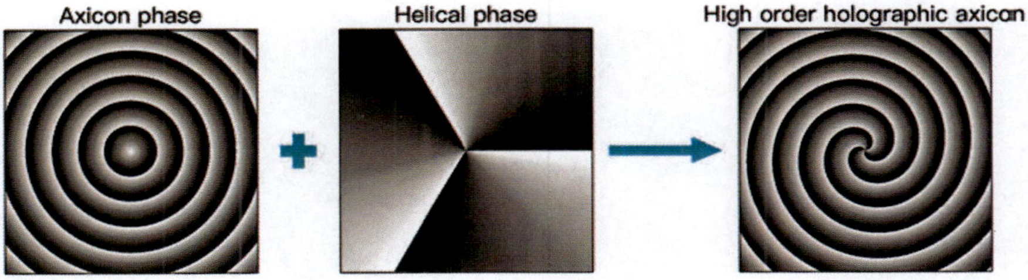

Fig. 3.55 Production of high order holographic axicon

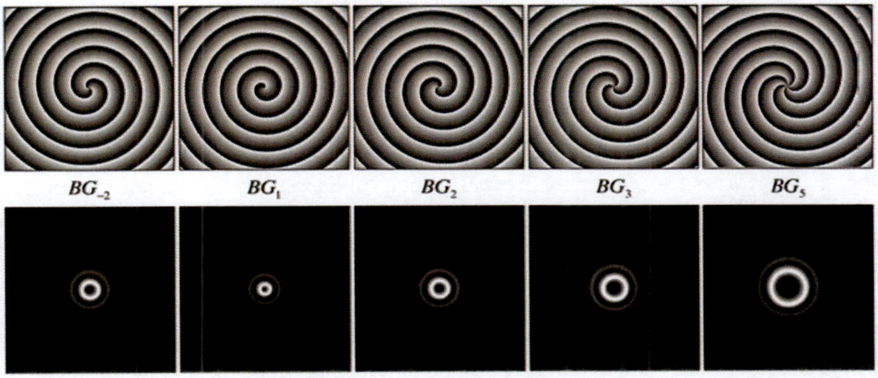

Fig. 3.56 Various high-order holographic axicon and their corresponding generated BG beams when Gaussian beams are incident

order BG beams. Another way is to design a new phase-only diffractive grating called high order holographic axicon, which consists of the helical phase and axicon phase. Then a SLM encoded by such high order holographic axicon can transform Gaussian beams directly into high order BG beams. Due to the flexibility of SLM, such scheme can generate BG beams of selective orders without adjusting any elements, and has been one of the most popular way to produce BG beams.

The production of high order holographic axicon is very easy. What we should do is just to superpose phase distributions of an axicon given by Eq. (3.7.1) with vortex grating shown in Fig. 3.23, as sketched in Fig. 3.55. Figure 3.56 gives some examples of various high order holographic axicon and their corresponding generated BG beams when Gaussian beams are incident.

3.8.2 Annular Slit Method

As displayed in Fig. 3.57, annular slit is a kind of amplitude diffractive grating, whose transmittance function is:

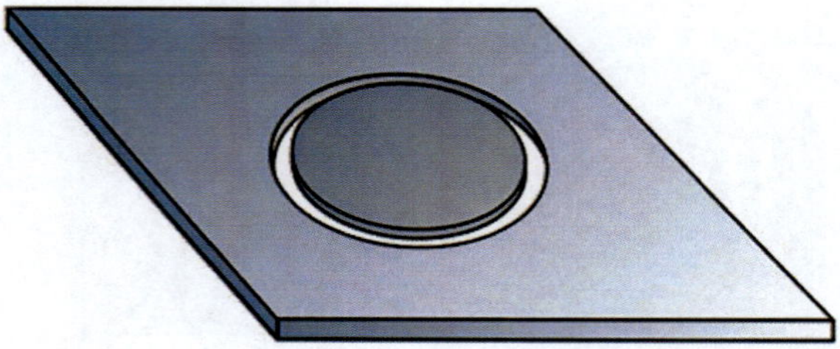

Fig. 3.57 Annular slit

$$T_r(r,\varphi) = \begin{cases} 1, |r - r_0| \le \dfrac{\Lambda}{2} \\ 0, |r - r_0| > \dfrac{\Lambda}{2} \end{cases} \tag{3.7.13}$$

with r_0 the center radius, and Λ the annular width. When Gaussian beams passing through an annular slit, an annular-shaped optical field will present at the surface of the slit. According to Huygens Fresnel principle, each point in the annular optical field can be regarded as a sub-light source, beams emitted by which are focused by a lens located at focal distance from the annular slit, and then interfered with each other in the overlapping area behind the lens to form BG beams, as sketched in Fig. 3.58. According to principles of Fourier optics, the field in the image focal plane is the Fourier transformation of that in object focal plane. While in Fraunhofer diffraction integral, the diffraction field is the Fourier transformation of initial field. Hence it is easy to understand that BG beams are the Fraunhofer diffraction of annular optical fields.

Similar with axicon method, when lth order LG beams passing through an annular slit, their Fraunhofer diffractions are lth order BG beams. Therefore, when generating high order BG beams from Gaussian beam directly through an annular slit, one can

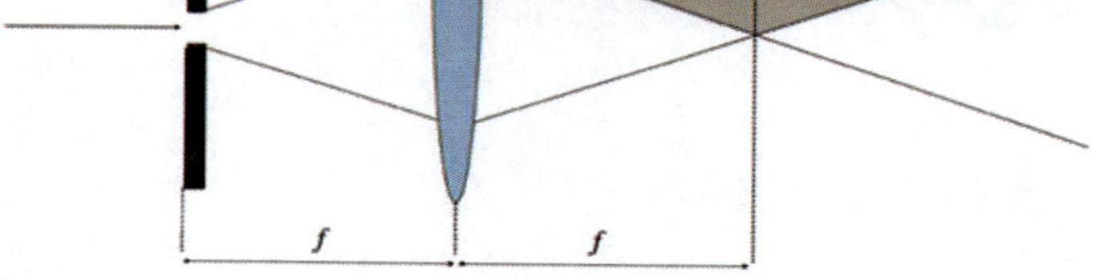

Fig. 3.58 Generating BG beams through annular slit

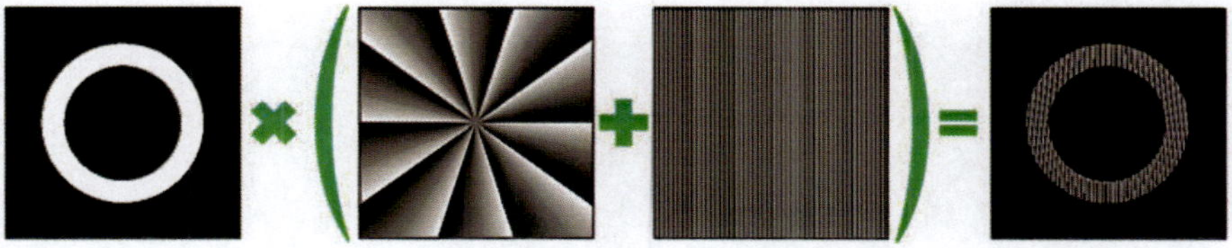

Fig. 3.59 Production of phase-only annular slit grating

place a SPP firstly to transform Gaussian beams into LG beams, and then transform LG beams into BG beams. Actually, the above procedure can also be accomplished by a phase-only grating. According to the schemes of imitating amplitude grating by phase-only grating discussed in Sect. 3.3.2, we can superpose both a blaze grating and a vortex grating on the transparent part of annular slit, to separate diffraction fields of the transparent and opaque part. Due to the superposed vortex grating, the diffractive beams will have helical wavefront and carry OAM. Hence when Gaussian beams are incident in, BG beams can be obtained in the far-field. Such phase-only grating is called phase-only annular slit grating. Figure 3.59 shows the procedure of how to produce phase-only annular slit grating, where combine blazed grating and vortex grating together firstly, and then multiply the transmittance function of annular slit:

$$T_{pr}(r, \varphi) = T_r(r, \varphi) \cdot \exp\left[i\left(\frac{2\pi x}{d} + l\varphi\right)\right] \tag{3.7.14}$$

Figure 3.60 gives several examples of carious phase-only annular slit grating and their corresponding simulated far-field diffraction patterns when Gaussian beams are incident. Note that here we just capture the field diffracted by transparent part. Similar with high order holographic axicon, phase-only annular slit grating can also be encoded on a SLM. In this scenario, BG beams with any orders can be generated very flexibly. Nevertheless, the generation efficiency of such method is lower than that of axicon method, because only rays incident in the transparent part are utilized, and most of the incident rays are loss. From the point of efficiency, generating BG beams through axicon method is more welcome.

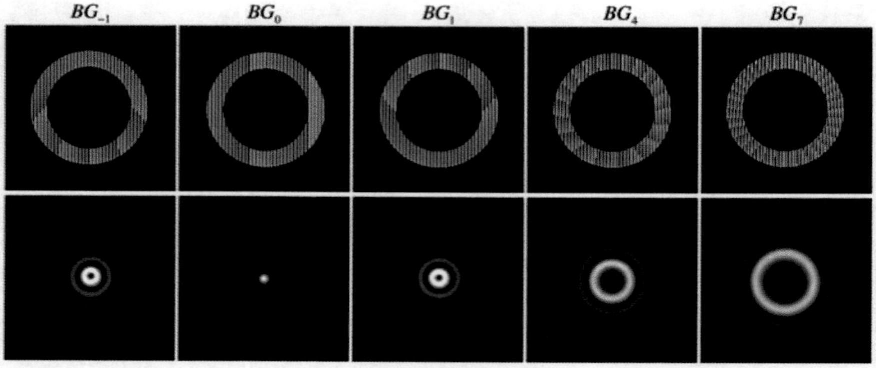

Fig. 3.60 Phase-only annular slit gratings with various orders and their corresponding simulated far-field diffractions when Gaussian beams are incident

References

1. Tamm C. Frequency locking of two transverse optical modes of a laser[J]. Phys Rev A. 1988;38(11):5960–3.
2. Tamm C, Weiss CO. Bistability and optical switching of spatial patterns in a laser[J]. J Opt Soc Am B: Opt Phys. 1990;7(6):1034–8.
3. Kim D J, Kim J W. Direct generation of an optical vortex beam in a single-frequency Nd:YVO4 laser.[J]. Optics Letters, 2015, 40(3):399–402.
4. Ngcobo S, Litvin I, Burger L, et al. A digital laser for on-demand laser modes.[J]. Nature Communications, 2013, 4(4):2289.
5. Beijersbergen M W, Allen L, H.E.L.O. van der Veen, et al. Astigmatic laser mode converters and transfer of orbital angular momentum[J]. Optics Communications, 1993, 96(1–3):123–132.
6. Gao C, Wei G, Weber H. Generation of the Stigmatic Beam with Orbital Angular Momentum[J]. Chin Phys Lett. 2001;18(6):771–3.
7. I. S. Gradshteyn, I. M. RYZHIK. Table of Integrals, Series, and Products[M]. Academic Press. Inc., San Diego, 1980: p. 1147.
8. Li F, Gao C, Liu Y, et al. Experimental study of the generation of Laguerre-Gaussian beam using a computer-generated amplitude grating [J]. Acta Physica Sinica. 2008;57(2):860–6.
9. Zhao D, Zhang H. Spatial light modulator [M]. Beijing: Beijing Institute of Technology Press, 1992. 110–135 (in Chinese).
10. Zhan H. Research of the performance measurement and wavefront correction application of phase-only liquid crystal spatial light modulator [D]. Harbin Institute of Technology, 2009 (in Chinese).
11. Soutar C. Determination of the physical properties of an arbitrary twisted-nematic liquid crystal cell[J]. Opt Eng. 1994;33(8):2704–12.
12. Wong DWK, Chen G. Redistribution of the zero order by the use of a phase checkerboard pattern in computer generated holograms[J]. Appl Opt. 2008;47(47):602–10.
13. Guo CS, Liu X, He JL, et al. Optimal annulus structures of optical vortices[J]. Opt Express. 2004;12(19):4625–34.
14. Guo CS, Liu X, Ren XY, et al. Optimal annular computer-generated holograms for the generation of optical vortices[J]. J Opt Soc Am A. 2005;22(2):385.
15. Xin J, Dai K, Zhong L, et al. Generation of optical vortices by using spiral phase plates made of polarization dependent devices[J]. Opt Lett. 2014;39(7):1984–7.
16. Marrucci L, Manzo C, Paparo D. Optical spin-to-orbital angular momentum conversion in inhomogeneous anisotropic media[J]. Phys Rev Lett. 2007;96(16): 163905.

17. Heebner J, Grover R, Ibrahim T. Optical microresonators : theory, fabrication, and applications[M]. Springer Science & Business Media: 2007.
18. Cai X, Wang J, Strain MJ, et al. Integrated compact optical vortex beam emitters[J]. Science. 2012;338(6105):363–6.
19. Lipson A, Lipson S G, Lipson H, Optical physics[M]. Cambridge University Press: 2010.
20. Liu YD, Gao C, Gao M, et al. Superposition and detection of two helical beams for optical orbital angular momentum communication[J]. Optics Communications. 2008;281(14):3636–9.
21. Gao C, Qi X, Liu Y, et al. Superposition of helical beams by using a Michelson interferometer[J]. Opt Express. 2010;18(1):72–8.
22. Goodman JW. Introduction to Fourier optics [M]. Beijing: Publishing house of electronics industry; 2006.
23. Lin J, Yuan XC, Tao SH, et al. Collinear superposition of multiple helical beams generated by a single azimuthally modulated phase-only element[J]. Opt Lett. 2005;30(24):3266–8.
24. Li S, Wang J. Adaptive power-controllable orbital angular momentum (OAM) multicasting[J]. Sci Rep. 2015;5:9677.
25. Zhu L, Wang J. Simultaneous generation of multiple orbital angular momentum (OAM) modes using a single phase-only element[J]. Opt Express. 2015;23(20):26221–33.
26. Durnin J. Exact solutions for nondiffracting beams. I. The scalar theory[J]. Journal of the Optical Society of America A, 1987, 4:651–654.
27. Arlt J, Dholakia K. Generation of high-order Bessel beams by use of an axicon[J]. Optics Communications. 2000;177(1–6):297–301.
28. Durnin J, Jr M J, Eberly J H. Diffraction-free beams.[J]. Physical Review Letters, 1987, 58(15):1499–1501.
29. Friberg AT. Stationary-phase analysis of generalized axicons[J]. J Opt Soc Am A. 1996;13(4):743–50.

Chapter 4
Vortices Lattices

Abstract Multiple vortex beams with various topological charges are usually essential for vortex beams system. An easy way to obtain multiple vortex beams, is to generate multiple single vortex beams, and then introducing them into one system. However, this scheme is complex and inconvenient. Such issue can be well addressed if we can generate multiple vortex beams simultaneously in one system. The basic idea is to design a special diffraction grating, which can diffract the incident Gaussian beams into various diffraction orders and meanwhile introduce various topological charges separately. Then in the far-field plane one can obtain multiple vortex beams in different positions, known as vortices lattices. Vortex beams lattices can be divided into three main classes, the dipole vortices lattice, the unipolar vortices lattice and the neither diploe nor unipolar vortices lattice. For unipolar vortices lattice, the topological charges of all the vortices in the lattice are identical; For dipole vortices lattice, the topological charges of vortex beams at each diffraction order is generally different, but that at the opposite diffraction order is opposite, and the sum of the topological charges of all the vortices in the lattice is 0. The neither diploe nor unipolar vortices lattice is the lattice that neither satisfies the unipolar condition nor the dipolar condition. In this chapter, various vortices lattices and there generating approaches are introduced from the point of designing diffraction gratings.

4.1 Basic Vortices Lattices

4.1.1 The Composite Fork-Shaped Grating and 3 × 3 Dipole Vortices Lattices

Firstly let's review the fork-shaped grating present in Sect. 3.2.3. When Gaussian beams are incident, the far-field diffraction has three main orders, where in the zeroth order is a Gaussian beam, in ± 1st orders are vortex beams with topological charges $\pm l$ respectively. This phenomenon indicates that a fork-shaped grating can be employed to generate three various vortex beams simultaneously (a Gaussian beam is the 0th order vortex beam indeed). And the far-field diffraction is a basic 1×3 vortices lattice.

S. Fu and C. Gao, *Optical Vortex Beams*, Advances in Optics and Optoelectronics,
https://doi.org/10.1007/978-981-99-1810-2_4

For another fork-shaped grating in y direction, obviously it can produce a 3×1 vortices lattice along y direction in the far-field.

The phase distribution of a fork-shaped grating with the fork number l_x in x direction reads from Eq. (3.2.16) as:

$$\phi_x(r, \varphi) = l_x\varphi + kr\cos\varphi\sin\theta_x \tag{4.1.1}$$

and its transmittance function is:

$$a(\phi_x) = \begin{cases} 1, & \mathrm{mod}(\phi_x, 2\pi) \le \pi \\ 0, & \text{others} \end{cases} \tag{4.1.2}$$

The phase distribution of a fork-shaped grating with the fork number l_y in y direction is:

$$\phi_y(r, \varphi) = l_y\varphi + kr\sin\varphi\sin\theta_y \tag{4.1.3}$$

and its transmittance function is:

$$a(\phi_y) = \begin{cases} 1, & \mathrm{mod}(\phi_y, 2\pi) \le \pi \\ 0, & \text{others} \end{cases} \tag{4.1.4}$$

If an amplitude grating is superposed by two fork-shaped gratings whose forks arranged along x and y directions separately, and has fork-shaped structure in both x and y direction, then its transmittance function is:

$$a(\phi) = a(\phi_x) \cdot a(\phi_y) \tag{4.1.5}$$

Such grating is called composite fork-shaped grating. Different from the common fork-shaped grating, the composite fork-shaped grating doesn't result from the beams' interference, but from the superposition of two fork-shaped grating of various fork directions. Figure 4.1 present the superposed process of a composite fork-shaped grating with $l_x = 1$ and $l_y = -3$ when $\theta_x = \theta_y$. Similar with the common fork-shaped grating, the composite fork-shaped gratings can also be made by laser etching, as shown in Fig. 4.2.

From Eq. (4.1.5), the transmittance function of a composite fork-shaped grating can be regarded as the superposition of that of two fork-shaped gratings with orthogonal fork directions. Therefore, the far-field diffractions of a composite fork-shaped grating can be acquired if we do the Fourier expansion for Eq. (4.1.5) by using the approach introduced in Sect. 3.2.3:

$$u(r', \varphi') = \sum_{b_x=-\infty}^{+\infty}\sum_{b_y=-\infty}^{+\infty} A_{b_x,b_y}\sum_{p=0}^{+\infty} c_p i^{2p+|b_x l_x + b_y l_y|} u_{p,b_x l_x + b_y l_y}(r', \varphi')$$

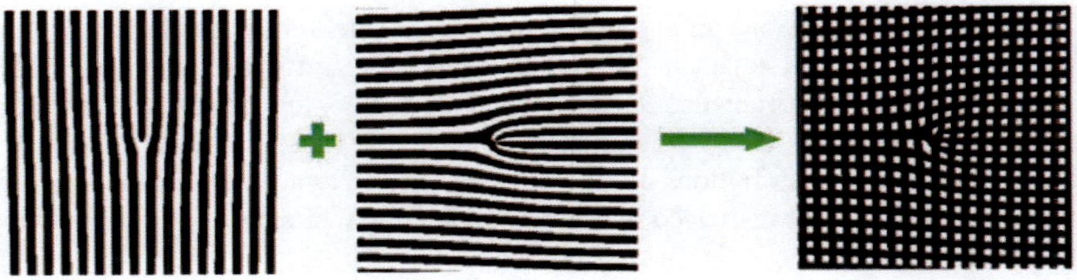

Fig. 4.1 The generation of composite fork-shaped grating with $l_x = 1$ and $l_y = -3$ providing that $\theta_x = \theta_y$

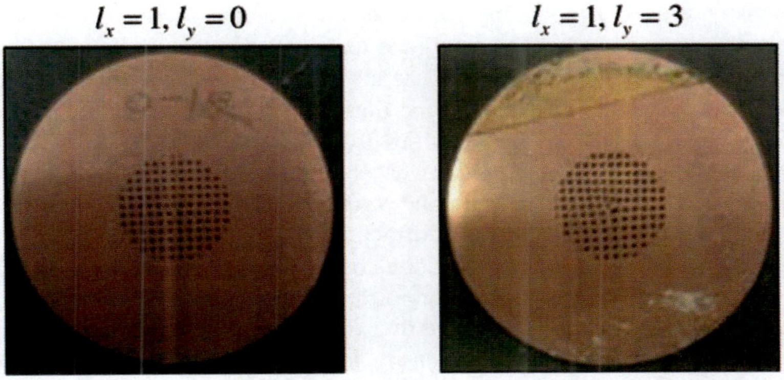

Fig. 4.2 The composite fork-shaped grating made by laser etching

$$\otimes\, \delta\left(f_x + \frac{b_x k \sin\theta_x}{2\pi} \right) \otimes \delta\left(f_y + \frac{b_y k \sin\theta_y}{2\pi} \right) \tag{4.1.6}$$

the parameters in which are identical with those in Eq. (3.2.29).

Equation (4.1.6) indicates that the far-field diffractions is a 3 × 3 dipole vortices lattice, which is the superposition of the diffraction orders b_x and b_y in x and y directions separately. And along both of the two directions there are position translations of $(b_x k\sin\theta_x)/(2\pi)$ and $(b_y k\sin\theta_y)/(2\pi)$ with respective to the center 0th diffraction order. In particular, when $\theta_x = \theta_y$, from Eq. (3.2.19) the grating constants in both x and y directions are identical, and by now the distance between the adjacent diffraction order in both of the two directions are identical too. In this book, each diffraction order in a vortices lattice is intuitively expressed in the form of two-dimensional coordinates as (b_x, b_y). For a composite fork-shaped grating, the topological charge of the vortex beam located in the diffraction order (b_x, b_y) is:

$$l_{b_x,b_y} = b_x l_x + b_y l_y \tag{4.1.7}$$

The above discussions imply that, for a composite fork-shaped grating, the distances between the adjacent diffraction orders in x and y directions are determined separately by parameters θ_x and θ_y, and the topological charge in diffraction order (b_x,b_y) is determined by parameters l_x and l_y. When $l_x = 1$ and $l_y = 3$, the topological charge distributions of nine different diffraction orders can be obtained through Eq. (4.1.7), as displayed in Fig. 4.3, indicating that nine vortex beams with topological charge from $-9\sim+9$ is produced.

Figure 4.4 shows the composite fork-shaped grating with $l_x = 1$, $l_y = 3$ and its far-field diffraction patterns when Gaussian beams are incident in. Obviously, \the incident power is mainly focus on the (0, 0) order, followed by the four diffraction orders (1, 0), (1, 0), (0, 1) and (0, 1). Meanwhile, with the increase of ($|b_x|+|b_y|$) value, the intensity of the corresponding vortex beam is decrease. Actually, the composite fork-shaped grating is a kind of amplitude rectangular grating with the ratio of slit width to grating constant is 0.5, and has all the features of amplitude rectangular grating. Therefore, the intensities of the four diffraction orders (1,0), (−1,0),(0,1) and (0,−1) are identical, and the intensities of the diffraction orders (−1,−1), (−1,1), (1,−1) and (1,1) are also identical but weaker than that of orders (1,0), (−1,0), (0,1) and (0,−1). However, the intensity of the vortex beams at the diffraction orders (0,1) and (0,−1) is weaker than that at the diffraction orderss (1,0) and (−1,0) intuitively. Such phenomenon is result from that, the absolute value of the topological charge of the vortices at the diffraction orders (0,1) and (0,−1) is larger than that at the diffraction orders (1,0) and (0,−1) and thus have larger beam size (one can refer to Sect. 1.4), which have lower power density. For the diffraction orders (−1,−1) and (1,1), their intensities are relatively weak, while their topological charges are the largest, so the power densities are the lowest, making these two diffraction orders almost invisible. In summary, in the vortices lattice produced from composite fork-shaped grating, the intensities of vortex beams located at the diffraction order that satisfy ($|b_x| + |b_y| = 2$) is the weakest, while the intensity of Gaussian beams at

Fig. 4.3 The topological charge distributions of 9 different diffraction orders of the composite fork-shaped grating with $l_x = 1$, $l_y = 3$

(−1,1) $l=2$	(0,1) $l=3$	(1,1) $l=4$
(−1,0) $l=-1$	(0,0) $l=0$	(1,0) $l=1$
(−1,−1) $l=-4$	(0,−1) $l=-3$	(1,−1) $l=-2$

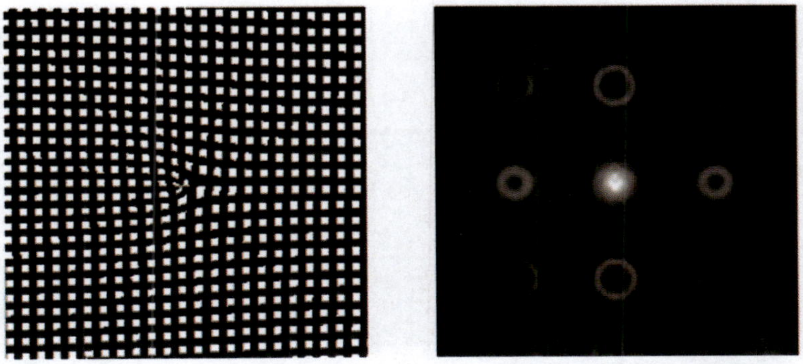

Fig. 4.4 The composite fork-shaped grating with $l_x = 1, l_y = 3$ and its far-field diffraction patterns when Gaussian beams are incident in

the diffraction order (0, 0) is highest. In other words, vortices in different diffraction orders may have different intensities, which has become the main defect of composite fork-shaped grating.

4.1.2 3 × 3 Unipolar Vortices Lattice

As for the 3 × 3 vortices lattices produced through illuminating a composite fork-shaped grating by Gaussian beams, if we want each diffraction orders have the same topological charge, namely, the unipolar vortices lattice, the condition that for arbitrary b_x and b_y the value $b_x l_x + b_y l_y$ is fixed must be satisfied. Thus there must be $l_x = l_y = 0$. By now the composite fork-shaped grating degenerates into a composite amplitude rectangular grating, as shown in Fig. 4.5. The composite fork-shaped grating with $l_x = l_y = 0$ has no fork-shaped structure indeed, thus it can't modulate the OAM dimensions of the incident beams. Therefore, when Gaussian beams are incident in, a 3 × 3 Gaussian beam lattice will present in the far-field. Similarly, if a vortex beam is incident, the far-field diffraction is a 3 × 3 vortices and each vortex in which have identical topological charges, as present in Fig. 4.5.

From the above discussion, two conditions need to be satisfied if we want to produce a 3 × 3 unipolar vortices lattice: ① a composite fork-shaped grating with $l_x = l_y = 0$ and ② incident vortex beams. In other words, when producing a 3 × 3 unipolar vortices lattice, a single mode vortex beam should be generated firstly according to the scheme discussed in Chap. 3, and then incident such vortex into the composite grating. In practical applications, the output of a common lasers is a Gaussian beam. Hence additional spiral phase plate, spatial light modulator or some other optical elements should be introduced firstly to generate vortex beams, which will increase the complexity and instability of the system to some extent. Although one cannot transform Gaussian beams into unipolar vortices lattices directly through a composite grating, some upgrades can be done for a composite fork-shaped grating, where

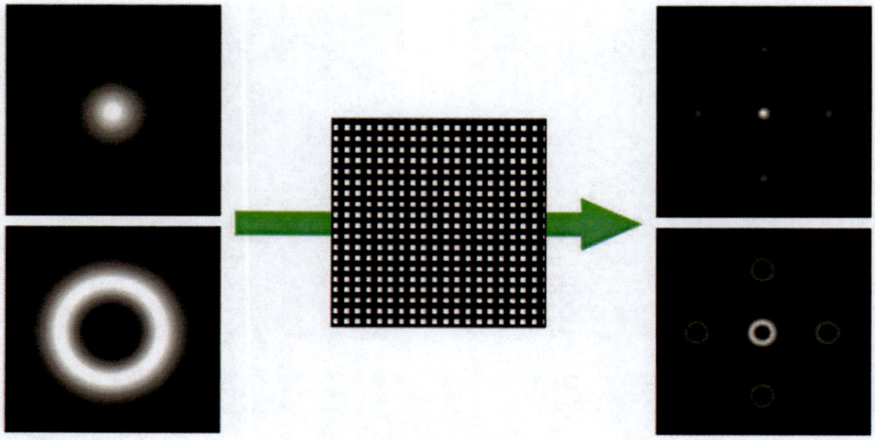

Fig. 4.5 The composite fork-shaped grating with $l_x = l_y = 0$ and its far-field diffraction patterns when Gaussian beams and vortex beams are incident in

both the helical phase and fork-shaped amplitude modulation can be introduced simultaneously.

The Sect. 3.3.3 of Chap. 3 have already discussed that the phase-only vortex grating can transform Gaussian beams into vortex beams with high efficiency. Therefore, the transmittance part of composite fork-shaped grating superposed by a phase-only vortex grating can realize the modulation of both helical phase and composite forked-shaped amplitude modulation simultaneously on one grating. Since the composite fork-shaped grating is a kind of amplitude grating, which means only by transforming it into phase grating can be combined perfectly with vortex grating. Here the scheme of blazed grating discussed in Sect. 3.3.2 is employed to imitate the function of composite fork-shaped grating, where a blazed grating is introduced in the transparent part to separate the transparent part from the opaque part. For the transparent part with blazed grating, an addition helical phase is superposed. When Gaussian beams are incident in, one can obtain a 3×3 unipolar vortices directly. Note that the superposed grating is phase-only, and should be encoded on a spatial light modulator in practical applications.

Figure 4.6 presents the producing process of the superposed phase-only grating when $l_x = l_y = 0$ and $l = 3$. Figure 4.7 displays the simulated far-field intensity and phase distributions when Gaussian beams passing through the grating in Fig. 4.6. One can find easily that a 3×3 unipolar vortices array where the topological charges of all the nine vortices are identical and equal to 3 is generated.

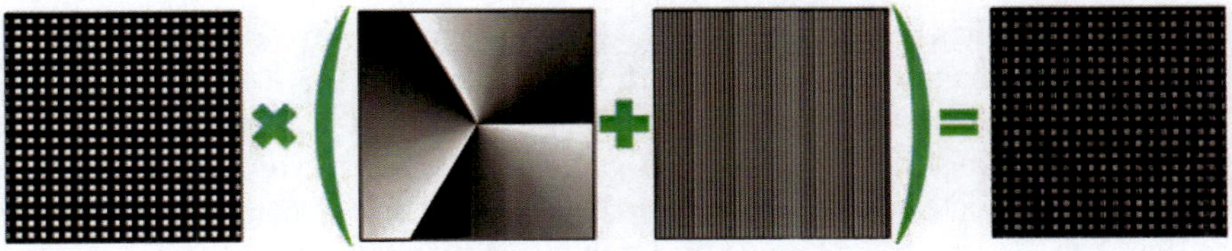

Fig. 4.6 The producing process of the phase-only grating that can produce unipolar vortices lattice based on composite fork-shaped grating ($l_x = l_y = 0, l = 3$)

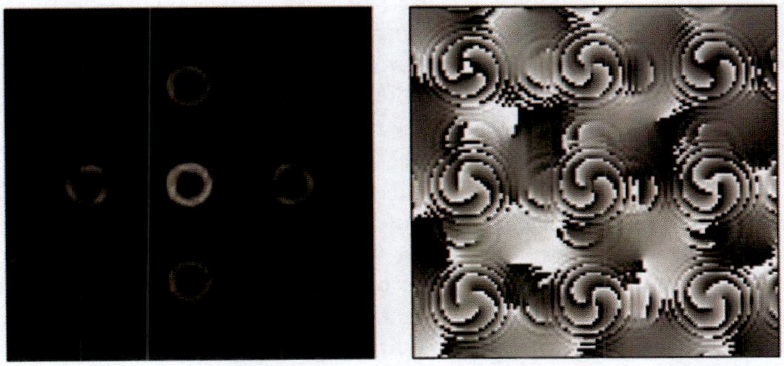

Fig. 4.7 The far-field intensity and phase distributions when Gaussian beams passing through the grating in Fig. 4.6

4.1.3 3 × 3 Vortices Lattice with Asymmetric Topological Charge Distribution

In the 3 × 3 vortices lattice generated by composite fork-shaped grating discussed above, the topological charges are a symmetric distribution with respect to the diffraction order (0, 0), that is, the topological charge in the diffraction order (0, 0) is 0, and the topological charge of the vortex beam of the diffraction order of the position symmetric with respect to (0, 0) is opposite to each other. Such phenomenon can be easily understood as that, the incident Gaussian beams is a 0th order vortex beam, thus the topological charge of the beam in diffraction order (0, 0) must be zero.

If the beam incident in the composite fork-shaped grating is not the fundamental Gaussian beam, but a l_0th order vortex beam $|l_0\rangle$, then according to the derivation done in Eqs. (3.2.22)~(3.2.29) one can obtain a similar form as Eq. (4.1.6) where the only difference is the additional initial helical phase term. By now one can acquire that the topological charge $l_{bx,\,by}$ of the vortex beam located on the diffraction order (b_x, b_y) is:

$$l_{b_x,b_y} = l_0 + b_x l_x + b_y l_y \tag{4.1.8}$$

indicating that when a vortex beam is incident into a composite fork-shaped grating, the topological charges of all the nine diffraction orders in the far-field 3 × 3 vortices lattice will add l_0 on the original basis. This means that for a specific composite fork-shaped grating, the topological charge distribution of the vortices lattice in the far-field diffraction can be determined by the order of the incident vortex beam.

Based on the scheme that imitating amplitude grating through phase-only grating introduced in sub-Sect. 4.1.2, the l_0th order vortex grating can be superposed with a composite fork-shaped grating to obtain a new integrated grating. When Gaussian beams incident in such integrated grating, one can obtain a 3 × 3 vortices lattice whose topological charge distribution matches Eq. (4.1.8) directly. Figure 4.8a presents one of such grating that integrated by a composite fork-shaped grating ($l_x = 1$, $l_y = 3$) and a vortex grating ($l_0 = 5$) accordance to this method. The transparent part is superimposed by the blazed grating and the vortex grating. According to Eq. (4.1.8), when Gaussian beams are incident in, the topological charge of the nine diffraction orders from bottom left to top right is + 1 ~ + 9, as shown in Fig. 4.8b. Figure 4.9 presents the far-field intensity and phase distributions when Gaussian beams passing through the grating in Fig. 4.8a, which fit well with the prediction on Fig. 4.8b.

In summary, the 3 × 3 vortices lattice with identical or asymmetric topological charge distribution can be produced through integrating a composite fork-shaped grating and a vortex grating. However, limitations still present. Firstly, the generated vortex beams lattice is diffracted by the first diffraction order of the blazed grating superposed in the transparent region, while half of the incident power is loss, leading to the low efficiency; Secondly, the integrated grating also has the defect of composite fork-shaped grating, that is, the uneven intensity distributions. The higher the diffraction order ($|b_x| + |b_y|$), the lower the intensity, which makes it can only produce 3 × 3 lattice. Therefore, one must design new diffraction gratings to produce vortices lattice with selective diffraction orders, topological charge and intensity distributions.

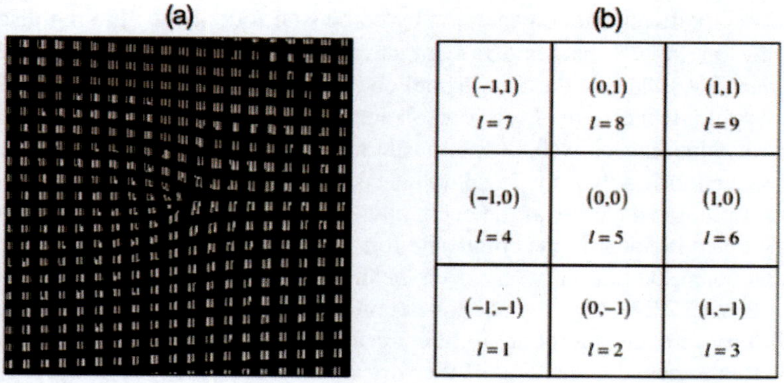

Fig. 4.8 **a** The phase-only grating result from the superposition of a composite fork-shaped grating ($l_x = 1$, $l_y = 3$) and a vortex grating ($l_0 = 5$); **b** The topological charge distributions of far-field diffractions when Gaussian beams are incident in

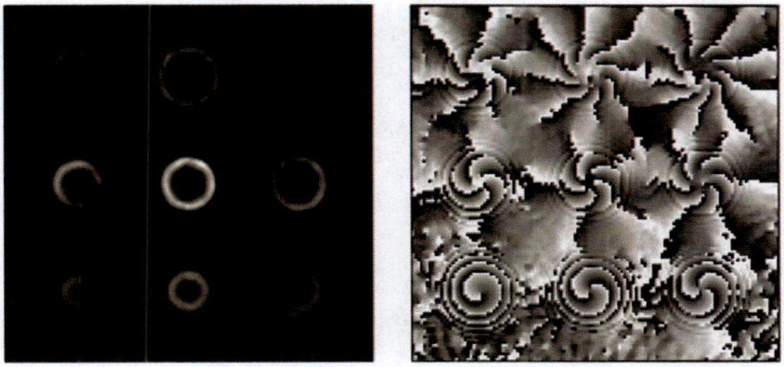

Fig. 4.9 The far-field intensity and phase distributions when Gaussian beams passing through the grating in Fig. 4.8a

4.2 Design and Optimization of Complex Optical Lattice

4.2.1 The Fourier Expansion of Diffraction Gratings

Generating vortices lattice through composite fork-shaped gratings is accomplished by amplitude modulation indeed. Although in the latter of Sect. 4.1 the topological charge control is done by the scheme of superposing additional phase-only vortex grating, it should be emphasized that the diffraction fields are also result from the amplitude modulation because the phase grating is used to imitate amplitude grating. Thus the high diffraction order can't be observed because the weak intensity compared with the diffraction order (0, 0) and only a 3×3 vortices lattice is present. Meanwhile, the topological charge distribution of the vortex beam in the lattice can only increase, decrease or be the same with the increasing of diffraction orders, and the vortices lattice with arbitrary topological charge distribution cannot be realized. Therefore, a new scheme of designing new gratings to generate vortices lattice with selective diffraction orders and topological charges must be developed.

Here we will discuss an approach of controlling the diffraction orders, intensities and topological charges in a lattice through the Fourier expansion of diffraction grating [1, 2]. Considering a one-dimensional lattice (the x direction in a Cartesian coordinates xOy), and defining the target lattice is:

$$s(x) = \sum_{a} \mu_a \exp(i\alpha_a) \exp(i\alpha\gamma x) \tag{4.2.1}$$

where a corresponds to the target diffraction orders, and doesn't contain non-target order as the missing orders. Parameters μ_a and α_a denote the amplitude and initial phase in order a respectively. γ is the spatial angular frequency representing the grating period and is defined as:

$$\gamma = \frac{2\pi}{T} \tag{4.2.2}$$

with T the grating constant.

Usually, $s(x)$ is a complex grating, whose phase-only form is:

$$T_P(x) = \exp[i\phi(x)] = \frac{s(x)}{|s(x)|} \tag{4.2.3}$$

Since additional harmonic components are introduced in the pure-phasing process of Eq. (4.2.3), the Fourier expansion of Eq. (4.2.3) can be written as:

$$\exp(i\phi(x)) = \sum_{b=-\infty}^{+\infty} c_b \exp(ib\gamma x) \tag{4.2.4}$$

with c_b the Fourier coefficient:

$$c_b = \frac{1}{2\pi/\gamma} \int_{-\frac{\pi}{r}}^{\frac{\pi}{r}} \exp[i\phi(x)] \exp(-ib\gamma x) dx \tag{4.2.5}$$

The Fourier coefficient c_b is a complex number, which can be regarded as the complex amplitude in diffraction order b:

$$c_b = |c_b| \exp(i\tau_b) \exp(il_b\varphi) \tag{4.2.6}$$

where $|c_b|$, τ_b and l_b denote the amplitude, initial phase and topological charge in diffraction b separately. The diffraction efficiency of the phase-only grating defined by Eq. (4.2.3) reads:

$$\eta = \frac{\sum_a |c_a|^2}{\sum_{b=-\infty}^{+\infty} |c_b|^2} \tag{4.2.7}$$

where the numerator and denominator denote the intensity sum of all the target diffraction orders and all the diffraction orders respectively, thus the efficiency $\eta \leq 1$.

In summary, one can design a phase-only grating by following the procedure below:

① Defining the target diffraction orders b;

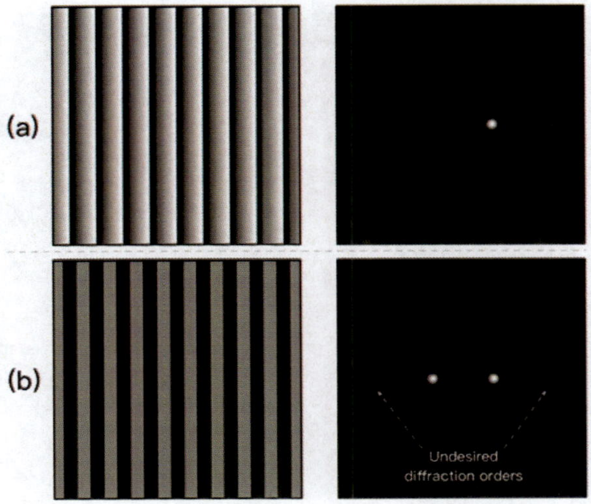

Fig. 4.10 The phase-only grating designed through Fourier expansion and its far-field diffraction patterns. **a** Only the 1st diffraction order is present; **b** Both the ±1st diffraction orders are present with equal intensities

② Defining the amplitude $|c_b|$, initial phase τ_b and topological charge l_b of the beam in the target diffraction order b. For the missing order, the amplitude $|c_b|$ is forced to 0.

Figure 4.10a and b displays the phase-only gratings designed according to the above procedures which can produce the Gaussian beams lattices that only have 1st diffraction order and have both the ±1st diffraction orders with equal intensities separately, and there corresponding far-field diffraction patterns. One can find obviously that, for the situation with only the 1st diffraction order, the designed grating is actually the blazed grating. While for the situation of ±1st diffraction orders with equal intensities, the designed grating is binary 0-π grating. Its far-field diffractions contain not only the desired ±1st diffraction orders, but also weak patterns in ±3rd orders, indicating that the phase-only grating designed according to the above procedure is imperfect.

Next let's consider a more complex situation, where three Gaussian beams with equal intensities are present in the −1st, +1st and +2nd diffraction orders, and the other orders are missed. Following the above procedure, the parameters are set as $|c_{-1}| = |c_{+1}| = |c_{+2}| = 1$, $\tau_{-1} = \tau_{+1} = \tau_{+2} = l_{-1} = l_{+1} = l_{+2} = 0$. And for the missing orders, $|c_b| = 0$. Then the obtained phase-only grating is present in Fig. 4.11a. Figure 4.11b shows its far-field diffraction patterns when Gaussian beams are incident in. And the transverse (along x axis) intensity curve of far-field diffractions is shown in Fig. 4.11c. Obviously the intensities of the desired three diffraction orders is different, which is different with our desire. Such difference will usually be more parent if the target diffraction orders are asymmetric with respect to the 0th diffraction order.

The reason of this phenomenon can be understood from properties of phase-only gratings. Considering the conditions of phase-only grating given by Eqs. (3.6.9), (4.2.3) should satisfy that $T(x)T^*(x)$ is a constant, namely:

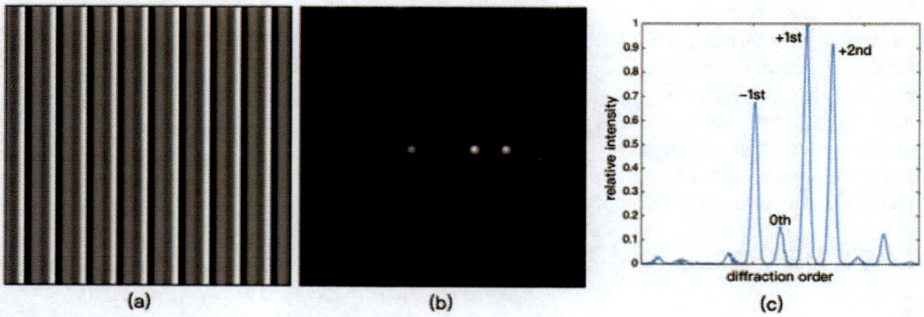

Fig. 4.11 **a** The phase-only grating designed based on Fourier expension to produce three Gaussian beams with equal intensities located on the −1st, + 1st and + 2nd diffraction orders; **b**, **c** are the diffraction patterns and transverse intensity distributions in the far-field when Gaussian beams passing through the grating in **a**

$$\sum_{b=-\infty}^{+\infty}\sum_{b'=-\infty}^{+\infty} c_b c_{b'}^* \exp\left[i\left(b'-b\right)\gamma x\right] = |C|^2 \tag{4.2.8}$$

which only valid for a unique b. Therefore, for the grating designed by this method, it is impossible to generate the ideal lattice by phase-only modulation. There must be undesired diffraction orders which contribute to the undesired intensity distributions. This indicates that optimization algorithms must be used to optimize the designed diffraction grating on the basis of the above two steps, so as to minimize the intensity of the undesired diffraction order and satisfy our desire.

4.2.2 *The Gerchberg–Saxton Algorithm and Grating Optimization*

Gerchberg-Saxton (GS) algorithm is firstly proposed by Gerchberg and Saxton in 1970s, whose nature is the process of iteration calculation between the input and output plane and thus to design special diffractive optical elements [3]. GS algorithm makes it available to obtain the phase distribution of the input plane from the already known and desired amplitude distribution in the input and output planes separately, so that the incident beams can be modulated according to our desire. Here the GS algorithm can be employed to optimize the grating introduced in Sect. 4.2.1, where the incident power is focused on the target diffraction orders, and the intensity proportion also matches our desire. As for the non-target diffraction orders, $|c_b| \to 0$.

The flow chart of GS algorithm is present in Fig. 4.12. The amplitude of the incident beam E_i (in the transmit plane) and the beam after diffraction E_o (in the receiver plane) are already known. Firstly set an initial phase φ_0, and then such initial phase together with the amplitude of the incident beam E_i constitutes the complex amplitude of the input beam $E_i\exp(i\varphi_0)$; Next calculate the complex amplitude in the receiver plane $E_m\exp(i\varphi_m)$ by employing diffraction theories introduced in Chap. 1; Thirdly replace the amplitude of $E_m\exp(i\varphi_m)$ by the already known amplitude E_o, thus obtain $E_o\exp(i\varphi_m)$; Fourthly, doing the transverse diffraction calculation for $E_o\exp(i\varphi_m)$, and obtain $E_m'\exp(i\varphi_m')$; Fifthly replace the amplitude E_m' by the amplitude of incident beam E_i, and obtain the new incident complex amplitude $E_i\exp(i\varphi_m')$ that for the next iterated calculation. Finally, after undergoing many iterations until a certain iteration condition is met, out of the loop, and the phase distribution of the phase in the transmit plane is delivered.

When producing vortices lattice, what we concern is the optical fields in the far-field, thus actually the Fraunhofer diffraction is employed in GS algorithm here. From

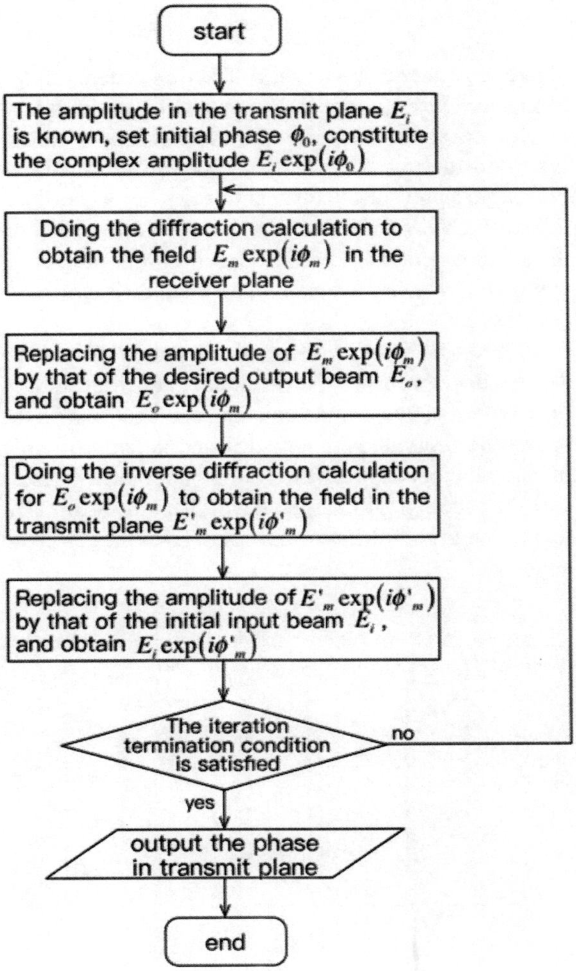

Fig. 4.12 The flow chart of GS algorithm

Sect. 1.4.5, the Fraunhofer diffraction can be accomplished through Fourier transformation directly, and similarly the inverse Fraunhofer diffraction can be accomplished through inverse Fourier transformation.

GS algorithm is an iterative algorithm, which can quickly converge to an optimal solution, and its operation speed is very fast. However, since GS algorithm is also a local optimization algorithm, it is easy to converge to the local optimal solution, and the desired results cannot be obtained. When calculating optical phase, the convergence of GS algorithm can be improved by selecting initial phase reasonably. When optimizing a grating, the initial phase is usually set as [4]:

$$\phi_0 = \arg\left[\sum_{b\in B} c_b \exp(ib\gamma x)\right] \tag{4.2.9}$$

where B is the all the target diffraction orders, $\arg(\varsigma)$ denotes taking the argument of the complex number ς. By now, GS algorithm has good convergence.

Next two cases are present to show the optimization performance of GS algorithm. First considering the case that each desired diffraction orders has equal intensity. Figure 4.13 presents when generating a 1×9 Gaussian beams lattice with nine equal-intensity Gaussian beams in $-4 \sim +4$ diffraction orders, the phase distribution of the grating optimized by GS algorithm in one grating constant (T). Meanwhile the simulated intensity distribution of each diffraction order is also present, which meet our desire.

Figure 4.14 shows the optimized grating that can generate nine vortex beams in $-4 \sim +4$ diffraction orders with equal intensities (a 1×9 vortices lattice), whose phase distribution in one grating constant is identical with that in Fig. 4.13. Figure 4.14 also shows the simulated far-field diffraction patterns when Gaussian beams are incident in, and all the 9 desired diffraction patterns are present. Note that vortex beams with various topological charges have different beam size, thus in Fig. 4.14 the beams located in higher diffraction orders have bigger beam size and result in low power densities.

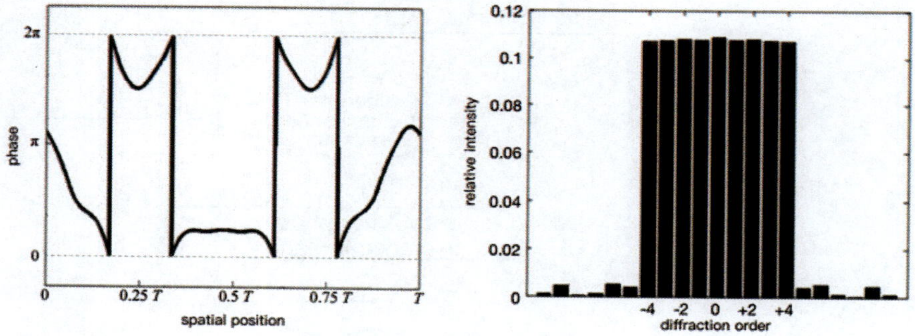

Fig. 4.13 The phase distribution of the optimized grating versus spatial positions in one grating constant, and the simulated intensity distribution of each diffraction order [4]

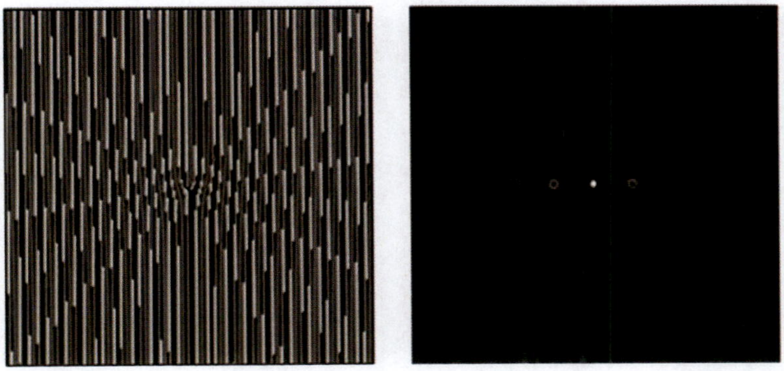

Fig. 4.14 The optimized grating that can generate nine vortex beams in −4 ~ + 4 diffraction orders with equal intensities, and the simulated far-field diffraction patterns when Gaussian beams are incident in

GS algorithm also shows good performance when generating the lattice with non-equal intensity distributions. Figures 4.15 and 4.16 show the one-dimensional phase-only grating to generate four beams in 0th, +1st, +2nd and +3rd diffraction orders with intensity proportion 1:2:1:2. Figure 4.16 also gives the simulated far-field diffraction patterns when Gaussian beams are incident in. In Figs. 4.13, 4.14, 4.15 and 4.16, using GS algorithm to optimize diffraction gratings can be understood as, the diffraction order that cannot be displayed by ordinary fork-shaped grating is displayed in a specific proportion, so that the topological charge distribution of vortex beams satisfies the rule given in Sect. 3.2.3. Actually, this method can also be applied in the generation of vortices lattice with arbitrary topological charge distributions, which will be discussed in detail in the following sub-Sect. 4.3.2.

In summary, on the basis of Sect. 4.2.1, the introduction of GS algorithm can optimize diffraction gratings to make the diffraction fields meet our desire. However,

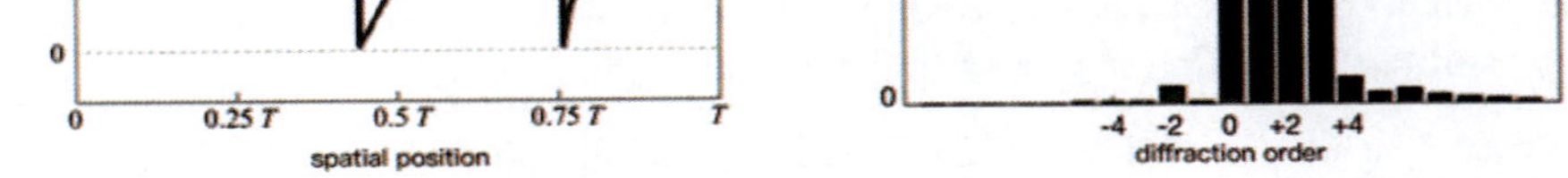

Fig. 4.15 The phase distribution of the optimized grating that can generate four beams in 0th ~ + 3rd diffraction orders with the intensity proportion 1:2:1:2 versus spatial positions in one grating constant, and the simulated intensity distribution of each diffraction order [4]

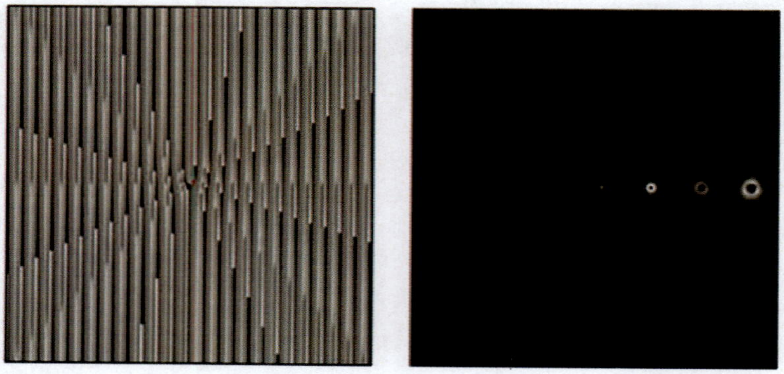

Fig. 4.16 The optimized grating that can generate four vortex beams with topological charge 0 ~ + 3 in 0th ~ + 3rd diffraction orders separately with the intensity proportion 1:2:1:2, and the far-field diffraction patterns when Gaussian beams are incident in

the optimization process needs constant iteration, which increases the calculation time and complexity of diffraction grating design to a certain extent. When the desired intensity of each diffraction order is equal, it can be achieved by a simpler grating design method, namely the Dammann grating, which will be discussed in the next section.

4.2.3 Dammann Grating

Dammann grating was firstly proposed by Dammann et al. in 1971, which is a 0-π phase-only grating to produce parallel beam array [5]. Dammann grating can produce beam lattice with any arrangement, and the uniformity and other parameters of the grating are independent of the incident beams. Recently, Dammann gratings have find applications in producing equal-intensity beam lattices [6] and optical coupling [7] and so on. The structure of a Dammann grating can be obtained by optimal design, where the essence of the design process is to find an optimal set of parameters to make the generated beam evenly distributed. Such optimization algorithm can be directly solved by analytic algorithm, and optimized by iterative algorithm, for instance, the local search algorithm, the global search algorithm and so on.

Supposing that the transmittance function of a one-dimensional Dammann grating is $\phi_D(x)$, since the phase distribution of a Dammann grating is 0-π binary, the transmittance function in one grating constant T can be sketched in Fig. 4.17 as:

where x_κ is the coordinate of inflection point. In the inflection point $x_\kappa T$, the value of $[\phi_D(x)]$ will jump, that is, from 0 to π, or from π to 0. From Eqs. (4.2.4) ~ (4.2.6), the transmittance function of a diffractive grating can be Fourier expanded, where the Fourier coefficient of each term c_b denotes the optical fields in diffraction order b. Doing the Fourier expansion for the Dammann grating in Fig. 4.16, one can obtain

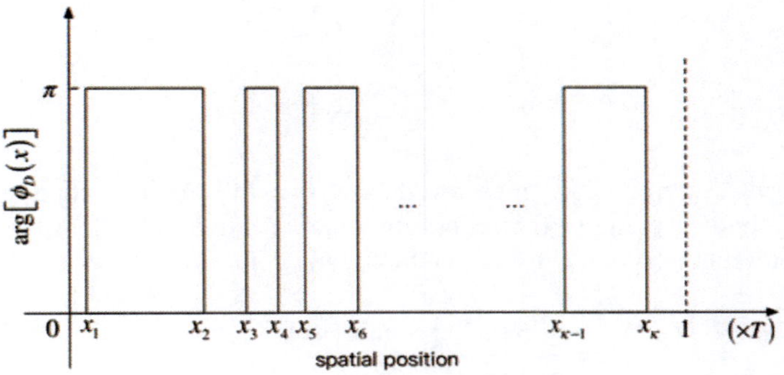

Fig. 4.17 The transmittance function of one-dimensional Dammann gratings in one grating constant T

the field distribution of 0th diffraction order as [6]:

$$c_0 = 2\sum_{\kappa=1}^{N}(-1)^{\kappa}x_{\kappa} \tag{4.2.10}$$

where N is the total number of inflection points within a grating constant, and must be even. For the non-zero diffraction orders, the optical field distribution is:

$$c_b = \frac{1}{\pi b}\sum_{\kappa=1}^{N}(-1)^{\kappa}\exp(-i2\pi b x_{\kappa}) \tag{4.2.11}$$

The intensity of the beam in diffraction order b is $P_b = |P_b|^2$. From Eqs. (4.2.10) and (4.2.11), one can acquire that the intensity of different diffraction orders in the diffraction fields of Dammann gratings are independent of the grating constant T, but associate with parameters N and x_{κ} of the inflection points. And the grating constant T only determine the distance between the adjacent diffraction orders. Let all the desired diffraction orders constitute a set B, if we want all the desired diffraction orders have the same intensities, the condition:

$$\begin{cases} |c_0|^2 = |c_b|^2, & b \in B \\ |c_b|^2 = 0, & b \notin B \end{cases} \tag{4.2.12}$$

must be satisfied.

From Eq. (4.2.12), the GS algorithm introduced in Sect. 4.2.2 can be employed to calculate the number N and the coordinate x_{κ} of inflection points, thus to produce Dammann gratings. The diffraction efficiency of a Dammann grating is defined as:

$$\eta = \frac{\sum\limits_{b\in B} |c_b|^2}{\sum\limits_{b=-\infty}^{+\infty} |c_b|^2} \tag{4.2.13}$$

indicating that such efficiency must less than or equal to 1. Another important parameter of a Dammann grating is the uniformity, which represents the intensity uniformity of each present beam in the diffraction fields, and defined as:

$$U = \frac{\max\{|c_b|^2\} - \min\{|c_b|^2\}}{\max\{|c_b|^2\} + \min\{|c_b|^2\}}, \quad b \in B \tag{4.2.14}$$

where $\max\{\varsigma_b\}$ and $\min\{\varsigma_b\}$ denote the maximum and minimum of sequence $\{\varsigma_b\}$. The closer the uniformity value is to 0, the more uniform the beam intensity distribution in different diffraction orders.

The ultimate goal of optimizing a Dammann grating is to reach the highest diffraction efficiency η and smallest uniformity U. Usually such process is complex. To simplify such process, the numerical solutions of inflection points x_κ of various one-dimensional Dammann gratings are present here [8], which can be employed to design Dammann gratings directly.

In Table 4.1, each line omits the first inflection point with the value 0. Additionally, for the Dammann grating that can produce lattice with even beams number, only the inflexion point coordinates in the first half of the grating constant (0 ~ 0.5 T) are listed. The distribution of inflection points in the second half of the grating constant (0.5 T ~ T) is the same as that in the first half of the grating period, which can be acquired as that in the first half of the grating constant plus 0.5 [9]. It should be noticed that, for the Dammann gratings that can produce the lattice with even beam number ($1 \times M, M = 2\,m, m \in$ » $_+$), the even diffraction order in the far-field is missing, in other words, the present diffraction orders is $-2\,m + 1, -2\,m + 3, \ldots, 2\,m - 3, 2\,m - 1$, leading to the equal distance between the adjacent present diffraction orders.

Taking the generation of 1 × 5 and 1 × 6 Gaussian beams lattices as example, the designing processes of Dammann grating that can produce the lattices with odd and even beam numbers separately are introduced. For the generation of 1 × 5 lattice, from Table 4.1 there are four inflection points, whose coordinates are: $x_1 = 0$, $x_2 = 0.03863$, $x_3 = 0.39084$ and $x_4 = 0.65552$ respectively. The binary 0-2π sub-grating of each inflection points reads:

$$\arg[\phi_\kappa(x)] = \begin{cases} 2\pi, & \mathrm{mod}(x, T) \le x_\kappa T \\ 0, & \mathrm{mod}(x, T) > x_\kappa T \end{cases} \tag{4.2.15}$$

where mod(ς, ε) denotes the taking the remainder of ε from ς. Therefore, according to Fig. 4.17 the Dammann grating can be regarded as their linear superposition:

$$\arg[\phi_D(x)] = \frac{1}{2}\sum_{\kappa=1}^{N}(-1)^{\kappa}\arg[\phi_{\kappa}(x)] \tag{4.2.16}$$

As for the Dammann grating that can produce 1 × 5 Gaussian beams lattice, the number of inflection points is $N = 4$, and the designing process is sketched

Table 4.1 The numerical solutions of inflection points of one-dimensional Dammann gratings [8]

Lattice number	Coordinate of inflection point x_κ	Efficiency η	Uniformity U
1 × 2	0.5	0.8106	0.00001
1 × 3	0.73526	0.6642	0.00002
1 × 4	0.22057, 0.44563	0.7063	0.00001
1 × 5	0.03863, 0.39084, 0.65552	0.7738	0.00001
1 × 6	0.11444, 0.20897	0.8452	0.00001
1 × 7	0.23191, 0.42520, 0.52571	0.7863	0.00001
1 × 8	0.06185, 0.17654, 0.20858, 0.31797	0.7615	0.00004
1 × 9	0.06668, 0.12871, 0.28589, 0.45666, 0.59090	0.7249	0.00004
1 × 10	0.10838, 0.11857, 0.19679, 0.23559, 0.31524, 0.37300	0.7590	0.00044
1 × 11	0.15015, 0.36389, 0.54103, 0.55344, 0.71318, 0.76612, 0.91107	0.7664	0.00067
1 × 12	0.01969, 0.08713, 0.12696, 0.18922, 0.24877, 0.35609	0.7997	0.00030
1 × 13	0.17765, 0.31352, 0.41244, 0.49846, 0.57633, 0.70857, 0.73041	0.7962	0.00045
1 × 14	0.05199, 0.12988, 0.19138, 0.24210, 0.27299, 0.31191, 0.35922, 0.49499	0.8008	0.00127
1 × 15	0.18240, 0.27424, 0.58581, 0.67967, 0.71917, 0.82217, 0.90642	0.8318	0.00033
1 × 16	0.14083, 0.17569, 0.22147, 0.26700, 0.35651, 0.39536, 0.43934, 0.45235	0.8152	0.00007
1 × 17	0.12271, 0.36812, 0.38508, 0.46844, 0.56010, 0.72061, 0.80310, 0.86000, 0.93712	0.8085	0.00037
1 × 18	0.03713, 0.06531, 0.11324, 0.11804, 0.13544, 0.18221, 0.25235, 0.39999, 0.42891, 0.46460	0.8124	0.00039
1 × 19	0.08570, 0.14057, 0.35809, 0.44535, 0.50140, 0.50781, 0.61168, 0.65143, 0.74179, 0.89057, 0.93947	0.7966	0.00080
1 × 20	0.04610, 0.15487, 0.17904, 0.21314, 0.24548, 0.27122, 0.30407, 0.38182, 0.39422, 0.44093	0.8193	0.00204
1 × 21	0.07505, 0.22656, 0.47886, 0.53661, 0.61683, 0.63009, 0.68194, 0.73870, 0.80083, 0.84469, 0.92196	0.8201	0.00078
1 × 22	0.05594, 0.15405, 0.19038, 0.21583, 0.24074, 0.27138, 0.30809, 0.31684, 0.35007, 0.36146, 0.43254, 0.46241	0.7987	0.00097

(continued)

Table 4.1 (continued)

Lattice number	Coordinate of inflection point x_κ	Efficiency η	Uniformity U
1 × 23	0.10435, 0.15704, 0.21912, 0.25653, 0.26153, 0.27828, 0.31707, 0.36536, 0.42446, 0.55744, 0.65260, 0.89953, 0.94099	0.8134	0.00076
1 × 24	0.02429, 0.05414, 0.06357, 0.09289, 0.11691, 0.14255, 0.16572, 0.20359, 0.23521, 0.34309, 0.41865, 0.44707	0.8226	0.00427
1 × 25	0.15387, 0.16452, 0.27298, 0.33211, 0.38120, 0.44909, 0.52349, 0.64396, 0.76823, 0.82082, 0.85980, 0.91639, 0.95832	0.8283	0.00030
1 × 26	0.02297, 0.04900, 0.06575, 0.09419, 0.10844, 0.21426, 0.27527, 0.30298, 0.30778, 0.33468, 0.36283, 0.41160, 0.43772, 0.47229	0.8105	0.00277
1 × 27	0.02859, 0.08692, 0.13812, 0.24035, 0.36942, 0.43961, 0.48655, 0.53940, 0.59754, 0.76160, 0.76925, 0.84337, 0.88257, 0.92116, 0.97417	0.8035	0.00179
1 × 28	0.01644, 0.04317, 0.06095, 0.11648, 0.14444, 0.21911, 0.23895, 0.26930, 0.28823, 0.32389, 0.38798, 0.39853, 0.42512, 0.47821	0.8238	0.00183
1 × 29	0.01582, 0.11494, 0.23203, 0.26929, 0.36868, 0.40786, 0.43497, 0.44045, 0.47031, 0.52654, 0.68721, 0.74326, 0.79553, 0.82470, 0.86563, 0.90709, 0.94619	0.7846	0.00046
1 × 30	0.01771, 0.04688, 0.07217, 0.07782, 0.08859, 0.13849, 0.18676, 0.21026, 0.22839, 0.24944, 0.27177, 0.28582, 0.31038, 0.35288, 0.37314, 0.39557	0.8101	0.00249
1 × 31	0.09241, 0.14299, 0.19031, 0.23982, 0.34695, 0.45454, 0.48263, 0.63078, 0.67371, 0.71256, 0.75291, 0.79316,0.82855, 0.85338, 0.85886, 0.92831, 0.94824	0.8184	0.00805
1 × 32	0.05540, 0.08900, 0.11010, 0.13340, 0.17320, 0.19580, 0.21090, 0.23060, 0.24870, 0.33010, 0.34820, 0.40100, 0.43320, 0.44110, 0.46530, 0.48400	0.8303	0.00707
1 × 64	0.01041, 0.01743, 0.02951, 0.04058, 0.04768, 0.06088, 0.07056, 0.08398, 0.09737, 0.10275, 0.12850, 0.14324, 0.21278, 0.24661, 0.26268, 0.28020, 0.28895, 0.29665, 0.31767, 0.32416, 0.33771, 0.36176, 0.38281, 0.39622, 0.41280, 0.42374, 0.43336, 0.43816, 0.44859, 0.45944, 0.4669, 0.47427, 0.48392, 0.48935	0.8072	0.00534

in Fig. 4.18. Firstly generating four 0-2π binary sub-gratings through the coordinates of inflection points and Eq. (4.2.15). Next linear superimposing them based on Eq. (4.2.16) and doing 0-π binarization. Finally the Dammann grating is generated. Figure 4.19 gives the far-field diffraction patterns when Gaussian beams are incident in, where there are five Gaussian beams locate in $-2 \sim +2$ diffraction orders separately with equal intensities.

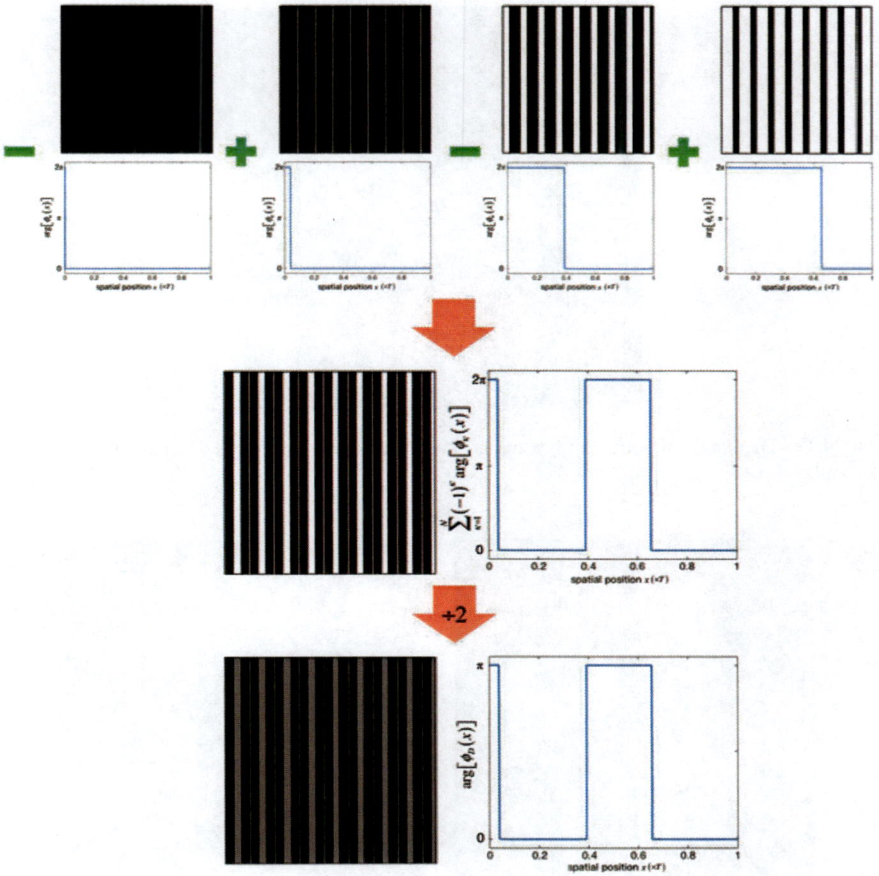

Fig. 4.18 The designing process of 1 × 5 Dammann grating

For generating 1 × 6 lattice, there are six inflection points from Table 4.1, whose value are: $x_1 = 0$, $x_2 = 0.11444$, $x_3 = 0.20897$, $x_4 = 0.5$, $x_5 = 0.61444$ and $x_5 = 0.70897$. Then following the procedures similar with Fig. 4.18, one can obtain the 1 × 6 Dammann grating as shown in Fig. 4.20a. And Fig. 4.20b presents the far-field diffraction patterns when Gaussian beams are incident in.

By now, the designing procedures of a one-dimensional Dammann grating are listed as follows:

① Confirming the number of beams in the desired lattices;
② Obtaining the coordinate of inflection points x_κ refer to Table 4.1;
③ Generating the 0-2π binary sub-gratings through Eq. (4.2.15);

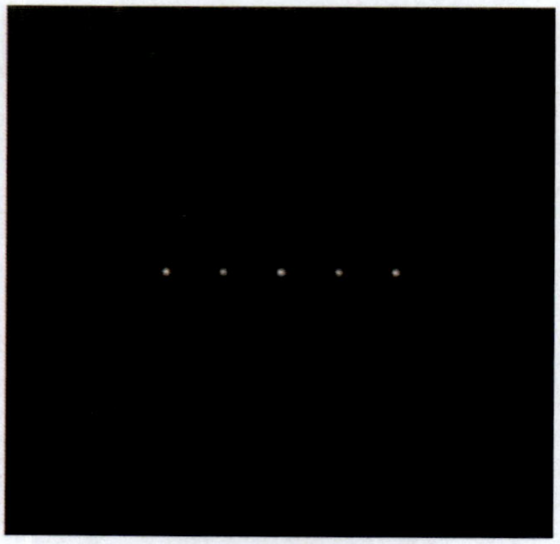

Fig. 4.19 The far-field diffraction patterns when Gaussian beams are incident in the 1 × 5 Dammann grating

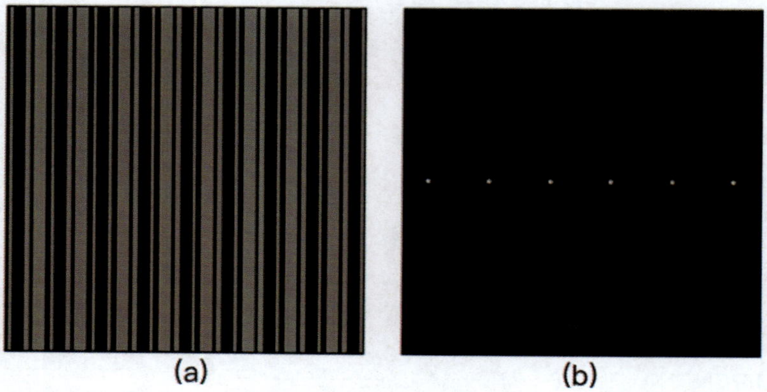

Fig. 4.20 **a** 1 × 6 Dammann grating. **b** The far-field diffraction patterns when Gaussian beams are incident in

④ Superimposing the sub-gratings based on Eq. (4.2.16) to obtain the final one-dimensional Dammann grating.

Obviously, for a 0-π binary Dammann grating, the grating constant won't affect the intensity of each diffraction order and only determine the distance between adjacent orders, thus a Dammann grating can be generated from the inflection points directly. When designing a grating to produce beams lattice with equal intensity, the procedures of generating Dammann gratings are much simpler that other gratings, which makes them of great significance in generating new structured fields as beams lattice and so on.

On the basis of one-dimensional Dammann grating, one can also design two-dimensional Dammann grating. The so-called two-dimensional Dammann grating can produce a two-dimensional rectangular beams lattice in the far-field when Gaussian beams are incident in. The two-dimensional Dammann grating can be regarded as the superposition of two one-dimensional Dammann grating with orthogonal directions. Supposing that a one-dimensional Dammann in x direction can produce a $1 \times n$ beams lattice, and another one-dimensional Dammann in y direction can produce a $m \times 1$ beams lattice, there their superposition results in a two-dimensional Dammann grating that can produce a $m \times n$ two-dimensional beams lattice. The process of superposition is a logical calculus and defined as:

$$\alpha \odot \beta = \begin{cases} \pi, & \alpha = \beta \\ 0, & \alpha \neq \beta \end{cases} \tag{4.2.17}$$

where both α and β are binary numbers and can be 0 or π. When $\alpha = \beta$, the calculate results is π. And when $\alpha \neq \beta$, the calculate results is 0. Thus the transmittance function of a two-dimensional Dammann grating $\phi_{D2}(x, y)$ reads:

$$\arg[\phi_{D2}(x, y)] = \arg\left[\phi_{D_x}(x)\right] \odot \arg\left[\phi_{D_y}(y)\right] \tag{4.2.18}$$

where $\phi_{D_x}(x)$ and $\phi_{D_y}(y)$ are the transmittance function of one-dimensional Dammann gratings in x and y directions respectively.

Next we will introduce the generating process of a two-dimensional Dammann grating taking the generation of 6×7 beams lattice as instance. As shown in Fig. 4.21a, firstly designing a 6×1 Dammann grating in y direction, whose designing procedures are identical with that in x direction except that the x coordinate is replaced by y coordinate. Form the point of the relationship between coordinate axes, Fig. 4.21a can be acquired by counterclockwise rotating 90° around its center of Fig. 4.20a. Secondly designing a 1×7 Dammann grating in x direction as shown in Fig. 4.21b. And finally after combining the two one-dimensional gratings through Eq. (4.2.18) then the final 6×7 two-dimensional Dammann grating is obtained, as shown in Fig. 4.21c. Figure 4.21 also shows the simulated far-field diffraction patterns when Gaussian beams are incident in the designed 6×7 two-dimensional Dammann grating. One can find easily a 6×7 Gaussian beams lattice is present, which fit well with our desire.

In Fig. 4.21, the distance between the adjacent diffraction orders in vertical (y) and horizontal (x) directions of the generated 6×7 Gaussian beams lattice are different. This is because in vertical directions the −5th, −3rd, −1st, +1st, +3rd and +5th diffraction orders are present and the −4th, −2nd, 0th, +2th and +4th diffraction orders are missed. Therefore the distance between the adjacent diffraction orders in vertical direction is twice as much as that in horizontal direction, leading to the non-uniform spatial distribution of the beams, which is unfavorable to some of practical applications as array lighting and something must be done to compensate. As mentioned previously, the grating constant T determines the distance between adjacent orders. Thus one can set various grating constant in horizontal and vertical directions to accomplish uniform spatial distribution.

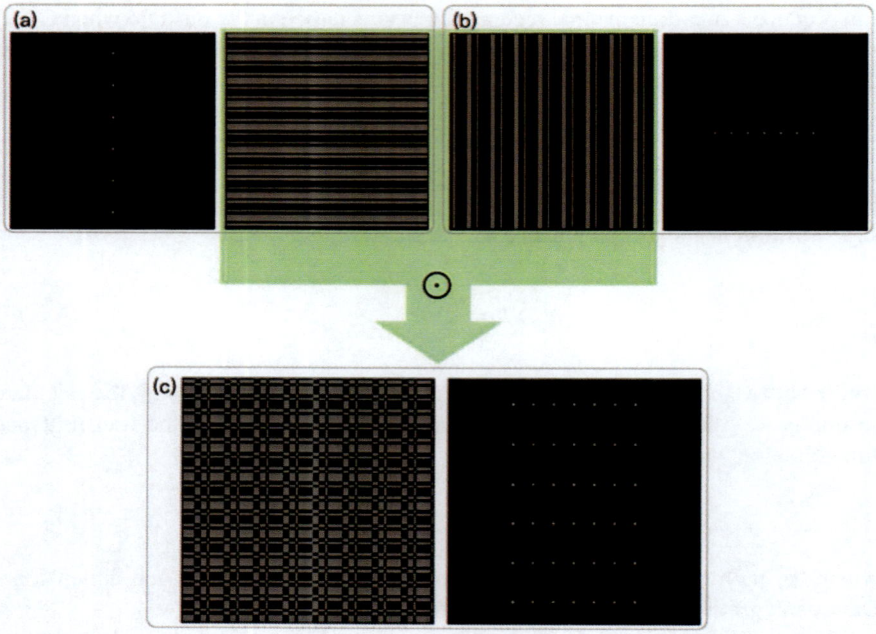

Fig. 4.21 **a** 6 × 1 one-dimensional Dammann grating and the far-field diffractions when Gaussian beams are incident in; **b** 1 × 7 one-dimensional Dammann grating and the far-field diffractions when Gaussian beams are incident in; **c** The 6 × 7 two-dimensional Damman grating superposed by the two one-dimensional grating shown in **a** and **b**, and the far-field diffractions when Gaussian beams are incident in

Considering a one-dimensional Dammann grating, α_b is the diffraction angle of diffraction order b, under paraxial approximation $\sin\alpha_b = \alpha_b$, then from the grating equation under normal incidence:

$$b\lambda = T \sin \alpha_b \tag{4.2.19}$$

one can obtain the diffraction angle:

$$\sin \alpha_b = \frac{b\lambda}{T} \tag{4.2.20}$$

If d_f denotes the distance between the Dammann grating and the receiving plane, from geometry when α_b is small enough the distance between the bth and 0th diffraction orders is:

$$d_b = d_f \alpha_b = \frac{d_f b\lambda}{T} \tag{4.2.21}$$

Thus the distance between adjacent diffraction orders is:

Fig. 4.22 The 6 × 7 two-dimensional Dammann grating with $T_y = 2T_x$ and its far-field diffraction patterns when Gaussian beams are incident in

$$\Delta d = d_{b+1} - d_b = \frac{d_f \lambda}{T} \tag{4.2.22}$$

As for the situation shown in Fig. 4.21, in order to accomplish the uniform spatial beam distribution, the distance between adjacent diffraction orders in x and y directions Δd_x and Δd_y should meet with $2\Delta d_y = \Delta d_x$, thus one can acquire from Eq. (4.2.22) that $T_y = 2T_x$. Therefore, here by setting the grating constant in the vertical (y) direction to be twice as much as that in the horizontal (x) direction, the uniform position distribution of beams can be obtained.

Figure 4.22 gives the 6 × 7 two-dimensional Damman grating with $T_y = 2T_x$, and the far-field diffraction patterns when Gaussian beams are incident in. Compared with Fig. 4.21c, the produced Gaussian beam lattice has a uniform spatial beam distribution, where the distance between adjacent diffraction orders in both x and y directions are identical.

In summary, a Dammann grating can produce the desired rectilinear and rectangular Gaussian beams lattice. If some improvements are done for Dammann gratings, they can also generate vortex beams lattice with equal intensity distributions, which will be discussed detailly in next section.

4.3 Two-Dimensional Vortices Lattice

4.3.1 Dammann Vortex Grating and Basic Rectangular Lattice

Now that a two-dimensional Gaussian beam lattice can be produced from a Dammann grating, can we generate two-dimensional vortex beam lattice providing that the structure of Dammann grating is adjusted properly? The answer is surly YES. Similar

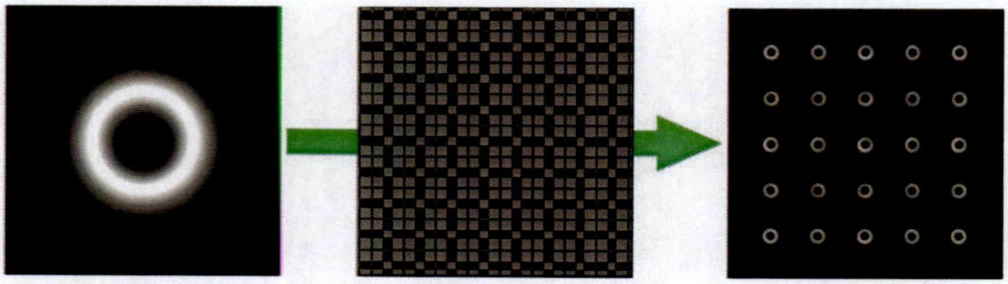

Fig. 4.23 The far-field diffraction patterns when am vortex beam (|+3⟩) is incident into a 5 × 5 Dammann grating

with the composite fork-shaped grating in Sect. 4.1, one can improve the Dammann grating in two main ways. One is to superposed with other phase-only grating. The other is to introduce fork-shaped structure and thus to compute a new kind of binary 0-π phase-only grating, namely the Dammann vortex grating [10, 11].

The discussion in Sect. 4.2.3 only focus on the situation when Gaussian beams are incident in. When a vortex beam is incident, there is no modulation on the topological charge since there is no fork-shaped structure in the Damman grating. And the far-field diffraction is a unipolar vortices lattice, as displayed in Fig. 4.23.

A common lasers usually output Gaussian beams. If we introduce a phase element in front of the Dammann grating to transform Gaussian beams into vortex beams firstly, the complexity of the system will increase to some extent. Therefore, we aim to produce vortex beam lattices when a Gaussian beam is incident, which require to introduce both the helical phases and beam splitting abilities. Here one can employ the approach that similar with that introduced in Sect. 4.1.2, as superposing a vortex grating on a Dammann grating. Different from composite fork-shaped grating, Dammann grating is phase-only, thus it can be combined together with vortex gratings directly, as shown in Fig. 4.24. Figure 4.25 presents the far-field diffraction patterns and the corresponding phase distributions when a Gaussian beam is incident in the superposed gratings shown in Fig. 4.24. Obviously, the beam in each diffraction order has helical phase, the topological charges of all of which are identical with the order of the superposed vortex grating. In other words, the two-dimensional unipolar vortices lattice is obtained.

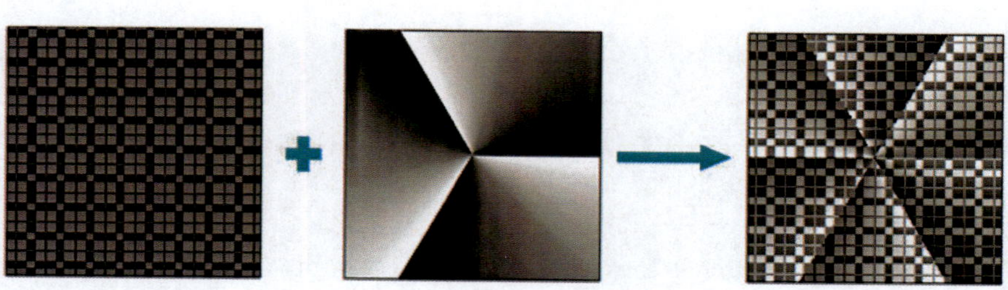

Fig. 4.24 The superposition of a Dammann grating and a 3rd order vortex grating

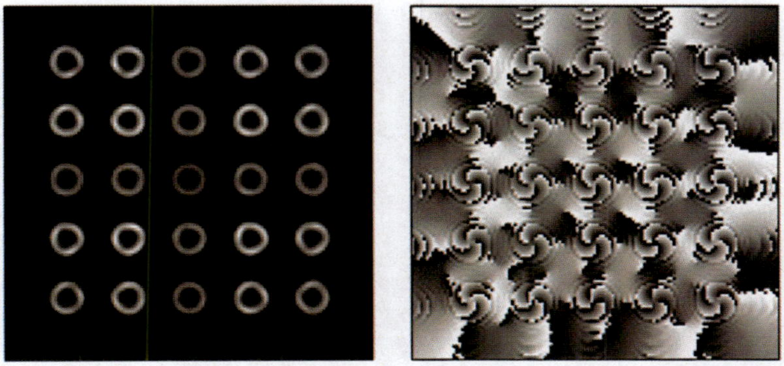

Fig. 4.25 The far-field diffraction patterns and the corresponding phase distributions when a Gaussian beam is incident in the superposed grating in Fig. 4.24

The grating shown in Fig. 4.24 is from the superposition of a vortex grating and a Dammann grating, and the topological charge of vortex beam in each diffraction orders is only determined by the order of the vortex grating, thus the diffraction order distributions are only determined by the Dammann grating. According to numerical solution of the inflection points x_κ given in Table 4.1, one can design arbitrary $m \times n$ two-dimensional unipolar vortices lattice, as displayed in Fig. 4.26.

An interesting phenomenon that, in the first column of Fig. 4.26 we generate 2 × 2 two-dimensional unipolar vortices lattice with topological charges of all the four beams are +2. While from the corresponding phase distributions the undesired diffraction orders still present helical phase structure. The reason is that Dammann gratins are a kind of binary 0-π gratings with optimization, and there must be weak beams present in undesired diffraction orders. We superpose a +2nd order vortex grating on a 2 × 2 Dammann grating, which actually is a modulation for the incident Gaussian beams. Therefore the weak beams present in undesired diffraction orders also have helical phase distributions.

The scheme that combined vortex grating and Dammann grating together can ideally generate vortex beams lattice, however, the topological charge in various diffraction orders are identical, which still doesn't meet the demands in some applications. At following we will introduce a new grating, which is based on Dammann grating, called Dammann vortex grating. The topological charge distributions in its far-field diffraction are associated with the diffraction orders.

Similar with the ordinary Dammann gratings, the Dammann vortex grating can also regarded as the linear combination of multiple 0-2π sub-gratings, where the sub-grating with inflection points x_κ reads:

$$\arg[\phi_\kappa(x)] = \begin{cases} 2\pi, & \mathrm{mod}[S(x), 2\pi] \le 2\pi x_\kappa \\ 0, & \mathrm{mod}[S(x), 2\pi] > 2\pi x_\kappa \end{cases} \tag{4.3.1}$$

with:

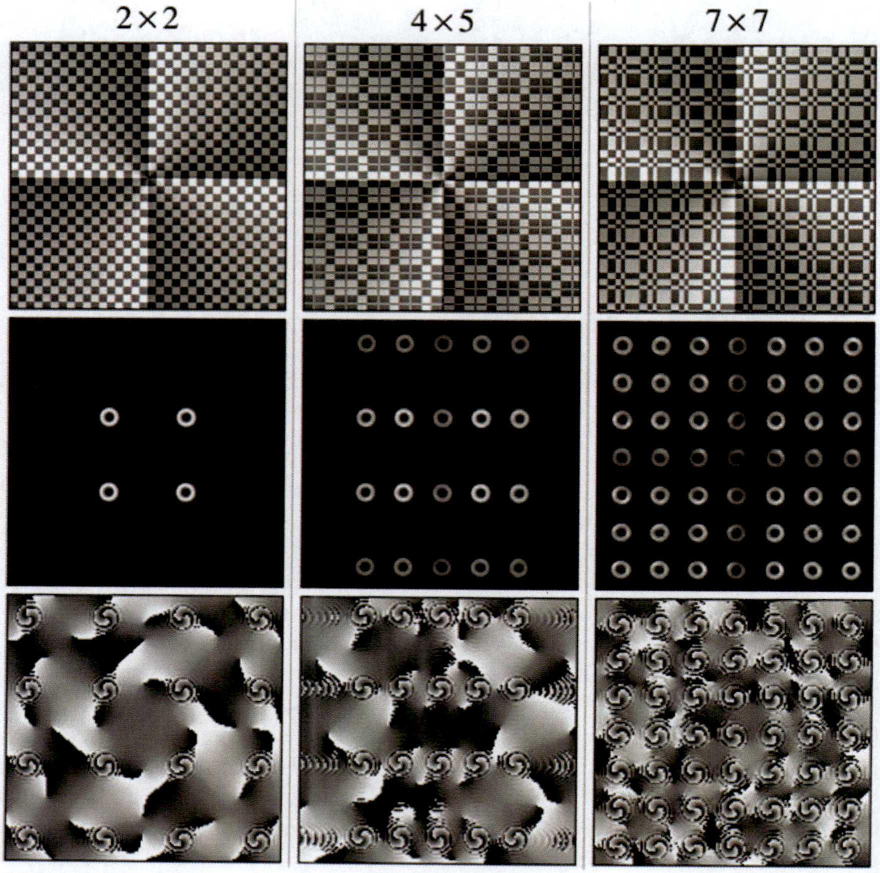

Fig. 4.26 Some of the phase-only gratings that superposed by + 2nd order vortex grating and various Dammann grating, and their corresponding $m \times n$ two-dimensional unipolar vortices lattice when Gaussian beams are incident in

$$S(x) = l_x\varphi + \frac{2\pi x}{T} \tag{4.3.2}$$

where l_x denotes the order of a Dammann vortex grating, φ denotes the azimuthal angle and T denotes the grating constant. When $l_x = 0$, substitute Eq. (4.3.2) into Eq. (4.3.1), the obtained equation is identical with Eq. (4.2.15). In other words, Dammann vortex grating degenerates into Dammann grating, and Dammann grating is a special kind of Dammann vortex grating with the order $l_x = 0$.

The design procedure of Dammann vortex grating is also identical with that of Dammann grating. Firstly generating all the N various binary 0-2π sub-gratings according to the inflection points $x_\kappa (\kappa \in \{1, 2, 3, ..., N\})$ and Eq. (4.3.1). And then combining them linearly through Eq. (4.2.16). Figure 4.27 presents the design procedure, where takes 1×5 first order ($l_x = 1$) Dammann vortex grating as instance.

From Fig. 4.27, one can find easily that Dammann vortex gratings have similar fork-shaped structure compared with fork-shaped gratings. In addition, Eq. (4.3.2) is very close to Eq. (3.2.16), implying that the diffraction properties of Dammann vortex gratings are similar to those of ordinary amplitude forked gratings. Figure 4.28 presents the far-field diffraction patterns when Gaussian beams are incident in a Dammann vortex grating shown in Fig. 4.27, which is a 1×5 vortices lattice with the topological charges from left to right are $-2, -1, 0, +1$, and $+2$ respectively. Such phenomena imply that, the modulation introduce for the incident Gaussian beam of Dammann vortex grating is similar with that of amplitude fork-shaped grating, where the topological charge of bth diffraction order reads bl_x. Meanwhile, different from fork-shaped grating, Dammann vortex grating is a binary 0-π phase-only diffractive grating, which make the undisplayed diffraction order of the fork-shaped grating appear through reasonably setting the inflection point, and finally generate vortices lattices with the intensity of all the diffraction orders same. Therefore, Dammann

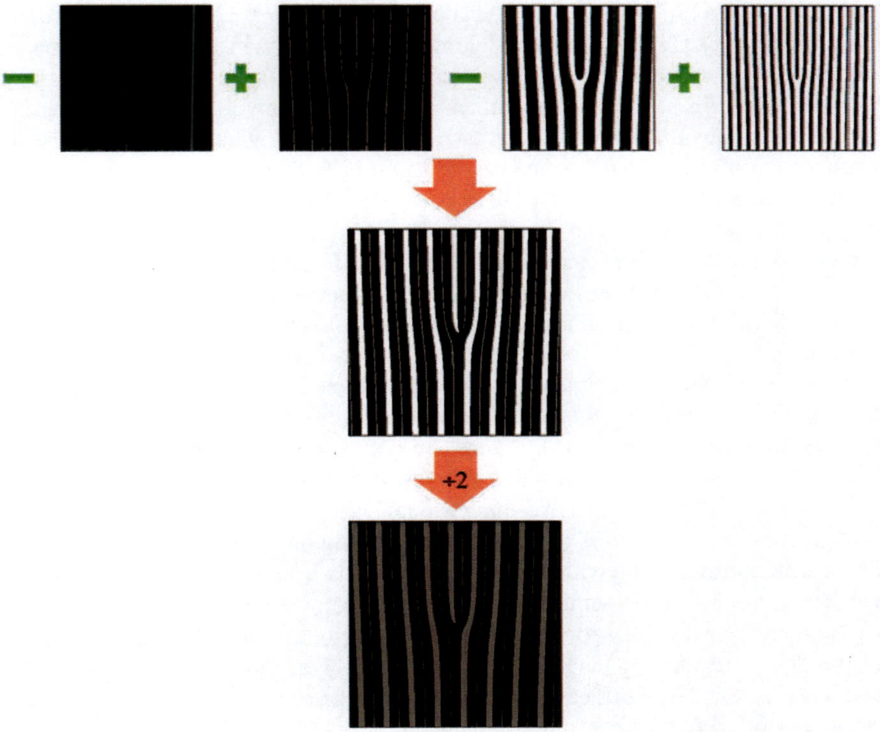

Fig. 4.27 The design procedure of the first order ($l_x = 1$) 1×5 one-dimensional Dammann vortex grating

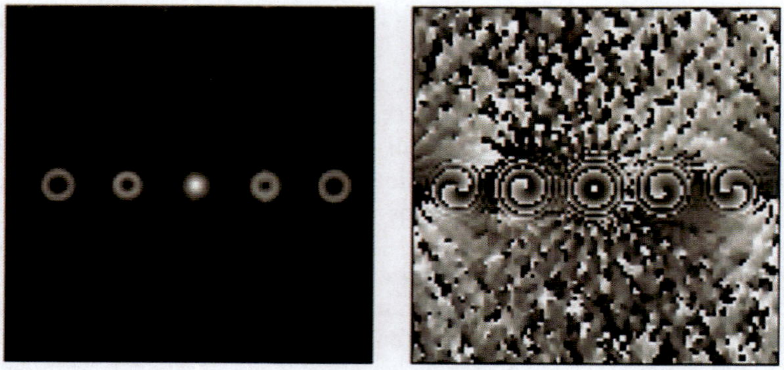

Fig. 4.28 The far-field diffraction patterns and corresponding phase distributions when Gaussian beams are incident in the Dammann vortex grating in Fig. 4.27

vortex gratings are kinds of optimized fork-shaped gratings, and also the extending of fork-shaped gratings.

According to Eq. (4.2.17), one can superpose a transverse one-dimensional $1\times n$ Dammann vortex grating with another longitudinal one-dimensional $m\times 1$ Dammann vortex grating, and then obtain a two-dimensional $m\times n$ Dammann vortex grating, as sketched in Fig. 4.29. For a one-dimensional Dammann vortex grating, the vortex beams in various diffraction orders have different topological charges. Hence the obtained two-dimensional Dammann vortex grating will present more complex topological charge distributions. The theoretical derivation of such process is rather complicated, and readers can derive it by themselves. Here, the conclusion is given directly as: the topological charge of the beams in diffraction order (b_x, b_y) reads $(b_x l_x + b_y l_y)$, which is consistent with the composite fork-shaped grating [Eq. (4.1.7)]. Such conclusion indicates that the two-dimensional Dammann vortex grating is an extending of the composite fork-shaped grating, where the desired diffraction orders can be present with equal intensities through reasonable design. In addition, since the topological charge of the vortex beam with the opposite diffraction order that symmetric about the center (0, 0) must be opposite, the total topological charge of the vortices lattice is 0, indicating that the vortex beams lattices generated from Gaussian beam illuminating a two-dimensional Dammann vortex grating are dipolar.

Specially, according to Figs. 4.27 and 4.29, designing a 5×5 two-dimensional Dammann vortex grating with parameters setting as $l_x = 1$, $l_y = 5$, and the grating constant $T_x = T_y$, as shown in Fig. 4.30a. The topological charge distributions can be acquired based on the topological charge distribution law of two-dimensional Dammann vortex grating mentioned above, as displayed in Fig. 4.30b. Figure 4.30c and d are the far-field diffraction patterns and the corresponding phase distributions separately when Gaussian beams are incident in. Figure 4.30 indicates that, when a Gaussian beam passes through a 5×5 two-dimensional Dammann vortex grating with $l_x = 1$ and $l_y = 5$, the far-field diffractions contain 25 different diffraction orders, the topological charge of which are $-12 \sim +12$ orderly. In other words,

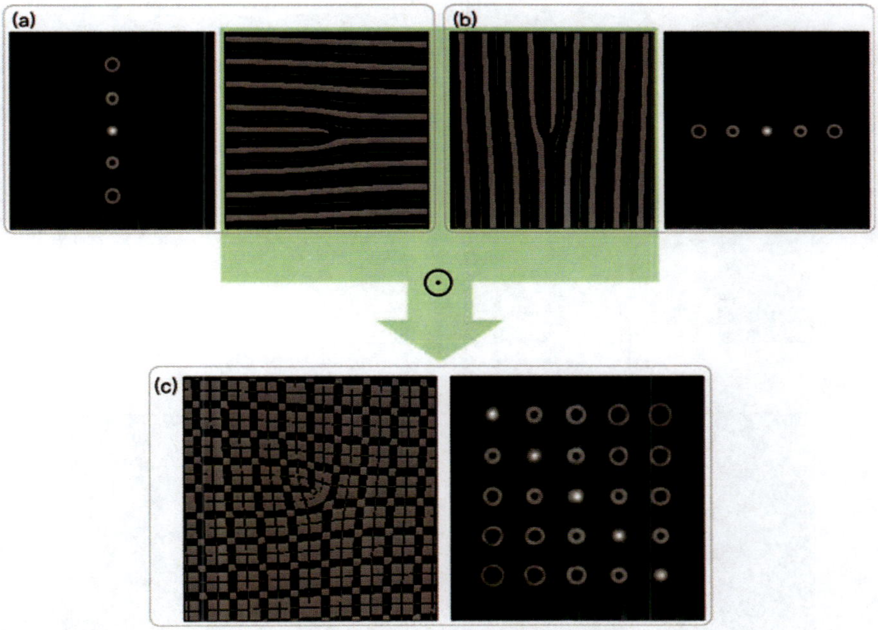

Fig. 4.29 The generation of two dimensional Dammann vortex grating. **a** One dimensional 5 × 1 Dammann vortex grating with the order $l_x = 1$, and its far-field diffraction patterns when Gaussian beams are incident in; **b** One dimensional 1 × 5 Dammann vortex grating with the order $l_y = 1$, and its far-field diffraction patterns when Gaussian beams are incident in; **c** The finally generated two dimensional 5 × 5 Dammann vortex grating and its far-field diffraction patterns when Gaussian beams are incident in

25 various vortex beam beams with topological charge −12 ~ + 12 are generated simultaneously. Such two-dimensional Dammann vortex grating called standard 5 × 5 Dammann vortex grating, and has already found lots of applications in the detection of multiplexed vortex beams [11], the OAM spectra analysis [12] and so on, which will be discussed in detail in Chap. 5.

The standard 5 × 5 Dammann vortex grating can be extended to a more general cases, namely the standard $m \times m$ Dammann vortex grating, which is defined as: m is odd, $l_x = 1$, $l_y = \mathrm{m}$, and the grating constants in the two orthogonal directions are the same. When a Gaussian beam passing through a standard $m \times m$ Dammann vortex grating, the far-field diffractions contain m^2 vortices beams, and the topological charges from bottom left to top right are—$(m^2 - 1)/2$ ~ $(m^2 - 1)/2$ in sequence. The design of standard Dammann vortex grating is also simple, where only four conditions should be satisfied as: ① m is odd; ② $l_x = 1$; ③ $l_y = \mathrm{m}$; ④ $T_x = T_y$. The significance of standard Dammann vortex grating is that, the diffraction field contains all the vortex beams with topological charges from $-(m^2 - 1)/2$ to $(m^2 - 1)/2$ simultaneously, and their intensities are equal.

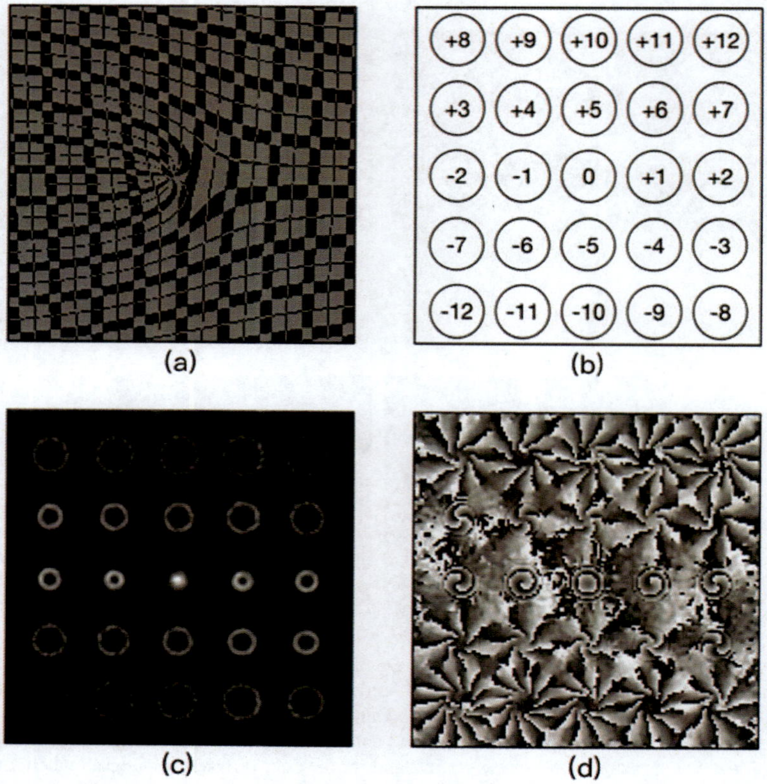

Fig. 4.30 Standard 5 × 5 Dammann vortex grating (**a**) and the topological charge distribution (**b**), intensity profile (**c**) and phase distribution (**d**) of its far-field diffractions

On the basis of standard $m \times m$ Dammann vortex grating, one can obtain a new grating as integrated Dammann vortex grating [13], which is the superposition of Dammann vortex grating and $+ (m^2 - 1)/2$th order or $-(m^2 - 1)/2$th order vortex gratings, as sketched in Fig. 4.31. When a Gaussian beam is incident, because of the component of a vortex grating, all the topological charges of vortices in diffraction fields are shifted, and finaly obtain m^2 vortex beams with the topological charges ranging from $-(m^2 - 1)$ to 0, or from 0 to $(m^2 - 1)$, as shown in Fig. 4.32. The integrated Dammann vortex grating can accomplish the diagnostic of N-fold multiplexed OAM states, which will be discussed in detail in Chap. 5.

This section introduces the approaches of generating vortex beams lattice through Damman vortex grating and its extending, which has the features of high diffraction efficiency and equal intensity distribution compared with the composite fork-shaped grating discussed in Sect. 4.1. The design and computing of Dammann vortex grating is also easy, which can be accomplished through the numerical solution of the inflection point given in Table 4.1. However, this method also has limitations as, the

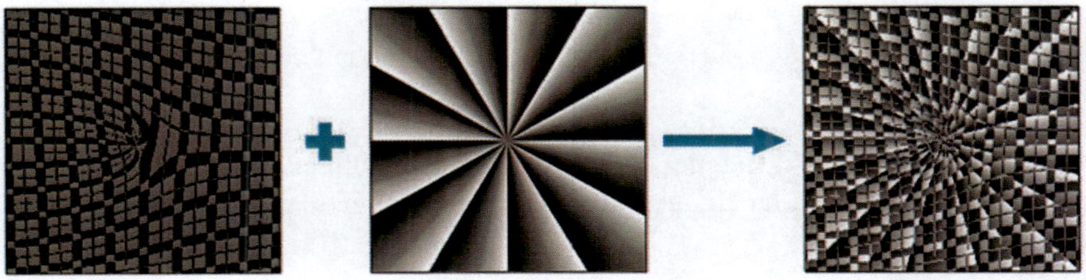

Fig. 4.31 Integrating Dammann vortex grating and vortex grating together, where takes $m = 5$ and $+12$th order vortex grating as example

Fig. 4.32 The far-field diffraction patterns and phase distributions when a Gaussian beam pass through the integrated Dammann vortex grating given in Fig. 4.30

topological charges in the same direction of the vortices lattice must increase or decrease with the diffraction order. Additionally, it is impossible to generate vortex beam lattices with inhomogeneous intensity distributions.

4.3.2 Special Rectangular Vortices Lattice

Dammann vortex gratings demonstrate good performances in producing vortex beams with identical intensities. However, they still powerless if vortices lattices with unequal intensity distributions, or arbitrarily distributed topological charges (none increasing or decreasing) are desired. Thus new gratings must be developed to achieve this goal.

The transmittance function $\exp[i\phi(x, y)]$ of a two-dimensional phase-only diffractive gratings under Cartesian coordinates can be Fourier expanded as:

$$\exp[i\phi(x, y)] = \sum_{a,b=-\infty}^{+\infty} c_{a,b} \exp\left[i\left(a\gamma_x x + b\gamma_y y\right)\right] \tag{4.3.3}$$

where a and b are the diffraction orders along x and y directions separately. γ_x and γ_y are the spatial angular frequency along x and y directions, which represent the grating constants:

$$\begin{cases} \gamma_x = \frac{2\pi}{T_x} \\ \gamma_y = \frac{2\pi}{T_y} \end{cases} \tag{4.3.4}$$

with T_x and T_y the grating constants along x and y directions respectively. The Fourier coefficient $c_{a,b}$ denotes the complex amplitude of diffraction order (a, b):

$$c_{a,b} = \left|c_{a,b}\right| \exp\left(i\tau_{a,b}\right) \exp\left(il_{a,b}\varphi\right) \tag{4.3.5}$$

with $|c_{a,b}|$, $\tau_{a,b}$ and $l_{a,b}$ the amplitude, initial phase and topological charge of diffraction order (a, b) respectively, φ the azimuthal angle. Equations (4.3.3)~(4.3.5) indicate a way to design gratings that one can preset the amplitudes, initial phases, topological charges and other parameters firstly and then accumulate them together. Such process is identical with that of the one-dimensional gratings discussed in Sect. 4.2.1.

Nevertheless, such scheme is non-perfect as discussed in Sect. 4.2.1, since when the number of desired diffraction orders N is greater than one, there must be extraneous diffraction orders, leading to the whole diffraction efficiency:

$$\eta = \frac{\sum_{(a,b)\in B} \left|c_{a,b}\right|^2}{\sum_{a,b=-\infty}^{+\infty} \left|c_{a,b}\right|^2} \tag{4.3.6}$$

must lower than 1. Note that in Eq. (4.3.6) B corresponds to a set of desired diffraction orders.

According to the designing approaches of diffractive grating given by Eqs. (4.3.3)~(4.3.5), here an example of producing a special 3 × 3 rectangular vortices lattice is present. The target is that, among the nine diffraction orders, the + 1st order, + 2nd order, + 3rd order and another + 3rd order vortex beams are located on the (−1,1)th, (1,0)th, (0,−1)th and (1,−1)th diffraction orders respectively with the intensity proportion 0.5:1:1:1. And the other five diffraction orders are missed. Hence, from Eqs. (4.3.3) and (4.3.5), the parameters are setting as:

$\left|c_{-1,1}\right| = \sqrt{0.5} = 0.707$;
$\left|c_{1,0}\right| = \left|c_{0,-1}\right| = \left|c_{1,-1}\right| = 1$;
All the other $\left|c_{a,b}\right| = 0$;
For $\forall a, b \in \{-1, 0, 1\}, \tau_{a,b} = 0$;

$l_{-1,1} = 1$;
$l_{1,0} = 2$;
$l_{0,-1} = l_{1,-1} = 3$.

Then the designed diffractive grating is given in Fig. 4.33. Figure 4.34 presents the far-field diffraction patterns and the intensity proportions of various diffraction orders. One can find easily that, the diffraction field comprise four evident "doughnuts", and the topological charges of all the present orders meet well with our desire. However, there is a weak distribution of intensity in the position of the missing order, and the energy ratio of the four diffraction orders is not strictly 0.5:1:1:1. The reason of this phenomenon is the same as that of one-dimensional diffractive gratings in Sect. 4.2.1. It is an approximate process to approximate the ideal diffraction grating function by a pure phase transmittance function in the form of Fourier superposition. Therefore, for the two-dimensional diffraction grating designed though Fourier expansion, optimization algorithms as GS algorithm introduced in Sect. 4.2.2 are still needed.

The optimized diffraction grating is shown in Fig. 4.35. Its far-field diffraction patterns and the intensity proportion of various diffraction orders when Gaussian beams are incident in are displayed in Fig. 4.36. Compared Fig. 4.33 with Fig. 4.35, they show little difference with naked eyes. While from Fig. 4.36, the optimized intensity proportion of various diffraction orders satisfies well with the desire. Nevertheless, after the optimization, diffraction orders that should be missed still present weak intensities, indicating that the results optimized through GS algorithm are very close, but cannot be completely equal, to the ideal one.

Fig. 4.33 The two-dimensional diffraction grating from Fourier expansion

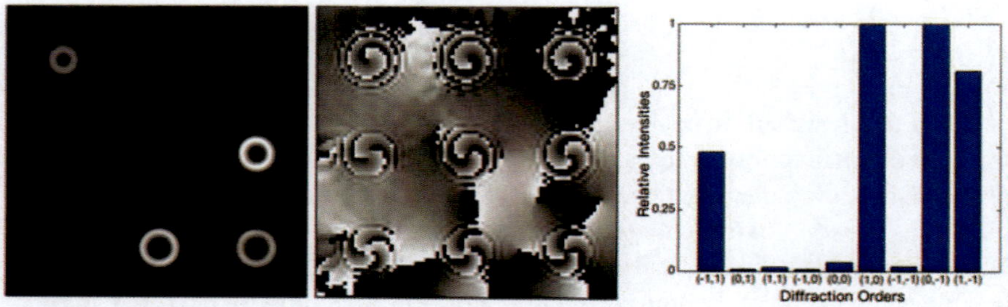

Fig. 4.34 The far-field diffraction patterns and the corresponding phase distributions and intensity proportion of various diffraction orders when Gaussian beams pass through the grating in Fig. 4.33

Fig. 4.35 The optimized grating of Fig. 4.33

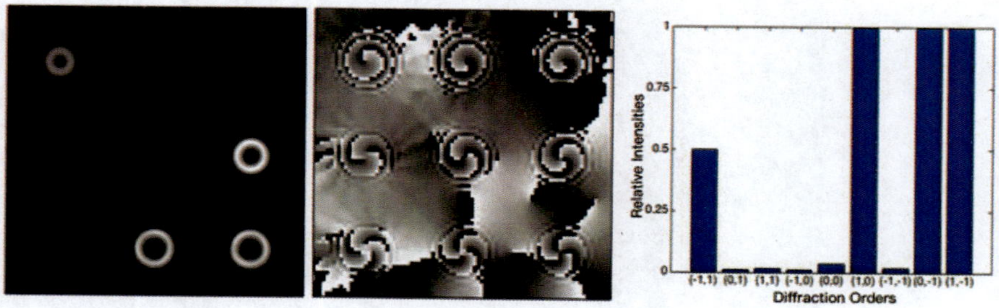

Fig. 4.36 The far-field diffraction patterns and the corresponding phase distributions and intensity proportion of various diffraction orders when Gaussian beams pass through the grating in Fig. 4.35

In summary, when designing two-dimensional diffraction gratings, the specific values of the parameters in the Fourier expansion coefficient should be determined according to the expected diffraction fields, and then the grating should be optimized through optimization algorithms as GS algorithm and so on. Since the phase-only grating cannot be strictly equal to the ideal diffraction grating, the optimized diffraction fields are very close, but cannot be exactly the same, to the ideal expectation. In other words, there must be independent and undesired diffraction orders. Compared with Dammann vortex grating, diffraction gratings designed by such method can generate rectangular vortices lattices with arbitrary spatial positions, intensity and topological charge distributions, but also accompanied by complicated design process.

4.3.3 Annular Vortices Lattices

Multiple vortex beams with various topological charges located along an annular path make up annular vortices lattices, which may find applications in OAM encoding [14] and so on. As the annular vortex beam lattice looks like a "necklace" surrounded by "pearls" (vortex beams), it is also known as "pearl necklace beams" vividly. Each vortex beam in the lattices is located on the same circle, thus they are all equidistant from the center. Reviewing the 2 × 2 rectangular lattice in Fig. 4.26, the contained four vortex beams are located on the (−1,−1)th, (−1,1)th, (1,−1)th and (1,1)th diffraction orders respectively, and are all equidistant from the center, indicating that the 2 × 2 rectangular lattice can also be regarded as an annular vortices lattice. Similarly, the 1 × 2 or 2 × 1 linear vortices lattice is also a special annular vortices lattice.

The above mentioned three special annular vortices lattices can be obtained through the methods introduced previously. While for generating more complex annular vortices lattices, for instance, with $2N$ ($N \in$ »$_{+}$) vortex beams, a new grating design approach must be developed. Considering an optical field contain multiple vortex beams with equal intensities:

$$s(r, \varphi) = \sum_{b=-N}^{N-1} \exp(il_b\varphi) \exp[ikr\rho_b \cos(\varphi - \theta_b)] \tag{4.3.7}$$

with (ρ_b, θ_b) the center coordinate, l_b the topological charge of diffraction order b respectively, if for any diffraction order b, ρ_b is a constant, and θ_b is the equal interval value in the interval [0, 2π), then Eq. (4.3.7) describes $2N$ vortex beams with topological charge l_b that evenly distributed on a circle with the radius of ρ_b. For Eq. (4.3.7), the phase-only is carried out, and then the transmittance function reads:

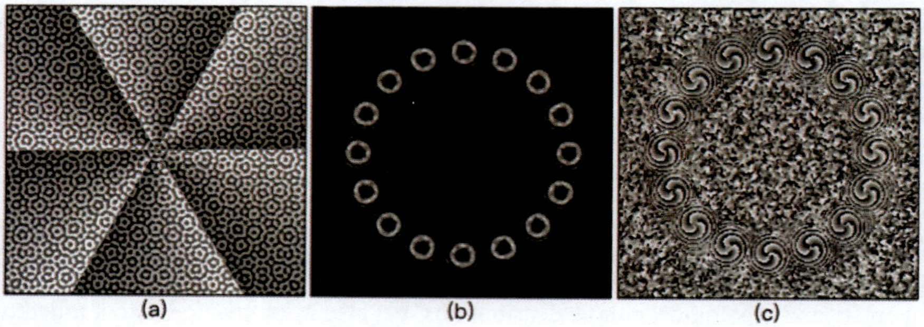

Fig. 4.37 The diffraction modulation plate (**a**) that can generate annular vortices lattice with all the topological charge + 3, and its far-field diffraction patterns (**b**) and corresponding phase distributions (**c**)

$$T(r,\varphi)=\exp[i\phi(r,\varphi)]=\frac{s(r,\varphi)}{|s(r,\varphi)|} \tag{4.3.8}$$

As discussed in Sect. 4.2.1, a phase-only grating contains nothing about amplitude information, leading to the unexpected diffraction fields. Hence GS algorithm or other optimal algorithms should be employed for $\phi(r,\varphi)$ to make the diffraction fields infinity closed to $s(r,\varphi)$. Usually, the optimized grating that can produce annular vortices lattices called diffraction modulation plate.

When design a diffraction modulation plate, if l_b in Eq. (4.3.7) is fixed, we can produce unipolar annular vortices lattice, as shown in Fig. 4.37. If $l_b = b$, then the topological charges of the vortex beams in the lattices associate with locations (diffraction orders). Figure 4.38 gives three various diffraction modulation plates and their corresponding annular vortices lattices, which contain 8, 12 and 18 vortex beam beams with topological charges $\pm 1 \sim \pm 4$, $\pm 1 \sim \pm 6$ and $\pm 1 \sim \pm 9$ separately. And by now, dipole annular vortices lattices are obtained. When doing the design, l_b can be any value, thus to generate annular vortices with arbitrary topological charge distributions. The amplitude factor can also be added to Eq. (4.3.7) to control the intensity of each diffraction order. The method is exactly the same as that of the rectangular lattice introduced in Sect. 4.3.2, which will not be repeated here.

4.4 Three-Dimensional Vortices Lattices

The $m \times n \times p$ three-dimensional vortices lattice is a kind of spatial distribution structure, which is made up of p $m \times n$ two-dimensional lattices on various locations, as shown in Fig. 4.39. In order to generate a three-dimensional vortices lattice, an optical element should be designed first, which can focus the incoming beam and generate multiple focused spots at different positions of the optical axis of the diffraction field. Such goal can be achieved through Fresnel zone plate (FZP), it can

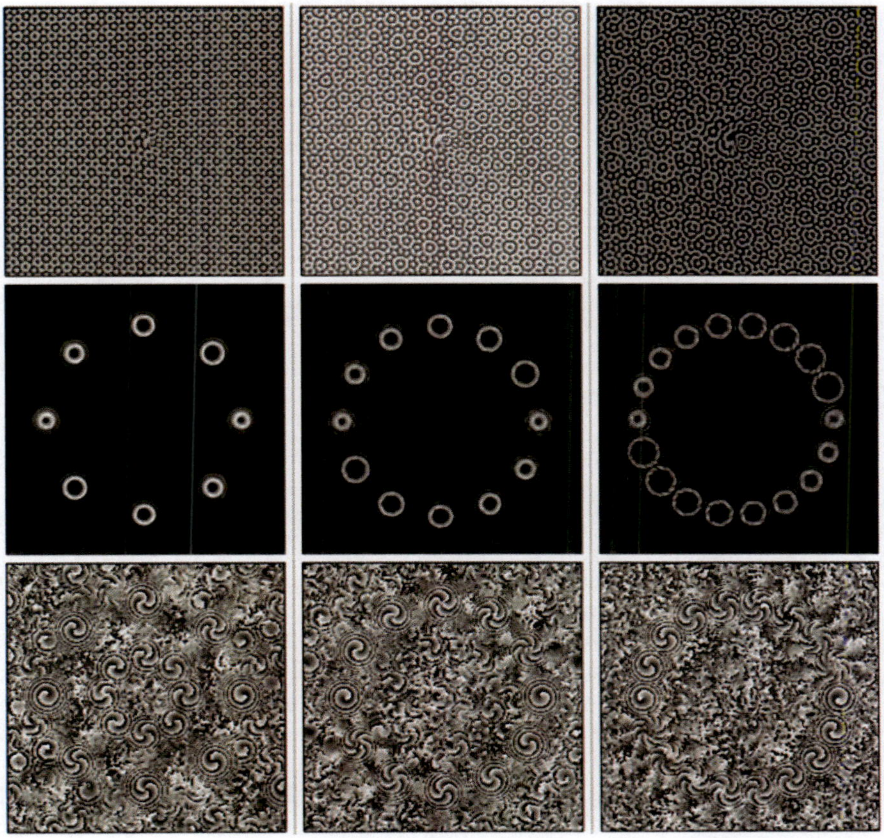

Fig. 4.38 Three various diffraction modulation plates, and their far-field diffraction patterns as well as the corresponding phase distributions

focus the incident beams into an infinite number of points on the optical axis of the diffraction field, in which the intensity of the main focus is the highest, and the intensity of the other focuses declines rapidly with the increase of the diffraction order. In other words, FZP can produce multiple focal points, but only the main focal point can be employed, thus works as a lens. G. Saavedra et al. demonstrated a fractal zone plates, where in diffraction fields series of fractal focal points similar to side lobes appear near each focal point, forming an axial multi-focal light field, but the intensity distribution of each focal point is nonuniform [15]. Afterwards, the demonstration of Dammann zone plate (DZP) addresses effectively the issue of nonuniform-intensity focal points [16]. DZPs are based on the binary 0-π Dammann grating, The traditional binarization of phase-only FZP is optimized, so as to obtain multiple equal intensity focal spots in different axial positions, which paves the way for three-dimensional vortice lattices with equal intensity distributions.

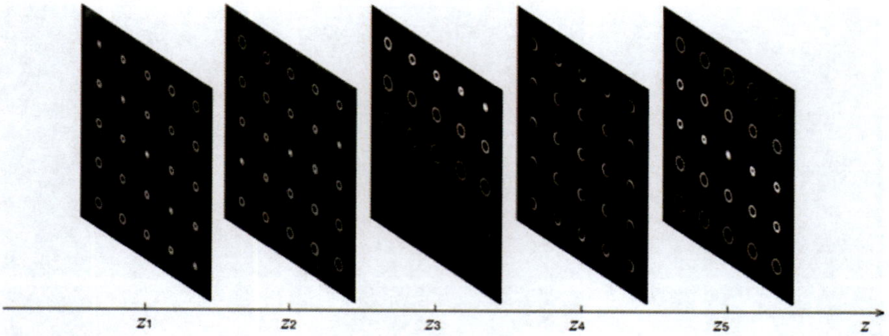

Fig. 4.39 Three-dimensional vortices lattices

4.4.1 Dammann Zone Plate

Dammann zone plate (DZP) is a new kind of binary 0-π FZP, where phase modulation details (inflection points) with respect to the square of radial coordinates are added in each grating period, so as to generate any number of focal spots with equal axial intensity in a certain range of diffraction fields [16], as sketched in Fig. 4.40. Figure 4.41 gives the distribution of the transmittance function ($\phi_{DZP}(r)$) of a DZP with normalized radial coordinates. One can find easily that, DZPs have similar binary structure compared with Dammann gratings introduced in Sect. 4.2.3, where multiple inflection points are present, and the phase value shifts from 0 to π or from π to 0 at each inflection point.

The model of generating any number of focal spot distribution with equal axial intensity using a DZP is shown in Fig. 4.42. The incident beams are phase-only modulated by the DZP firstly, and then are focalized by a lens. Finally there will be

Fig. 4.40 The structure of DZP

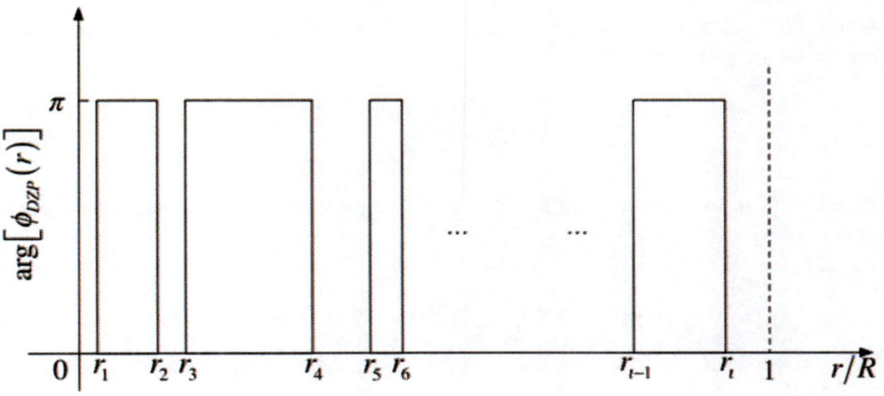

Fig. 4.41 The transmittance function of DZPs

multiple focal points with equal intensities on both sides of the image focus of the lens. A cylindrical coordinate is established with the focus as the center of the circle and the z axis as the direction of the optical axis. From the diffraction theory, the diffraction fields near the focus is [16]:

$$E(u, v) = 2\int_0^1 \phi_{DZP}(r) J_0(vr_\iota) \exp\left(\frac{-iur_\iota^2}{2}\right) r_\iota dr_\iota \quad (4.4.1)$$

where r_ι denotes the normalized inflection points as $r_\iota = r/R$ with r the radial coordinates and R the pupil radius of the system respectively. $J_0(\varsigma)$ is the 0-th order Bessel function of the first kind. u and v are defined as:

$$u = \frac{2\pi (\mathrm{NA})^2 z}{\lambda} \quad (4.4.2)$$

$$v = \frac{2\pi (\mathrm{NA})\rho}{\lambda} \quad (4.4.3)$$

Fig. 4.42 The model of generating any number of focal spot distribution with equal axial intensity using a DZPs [16]

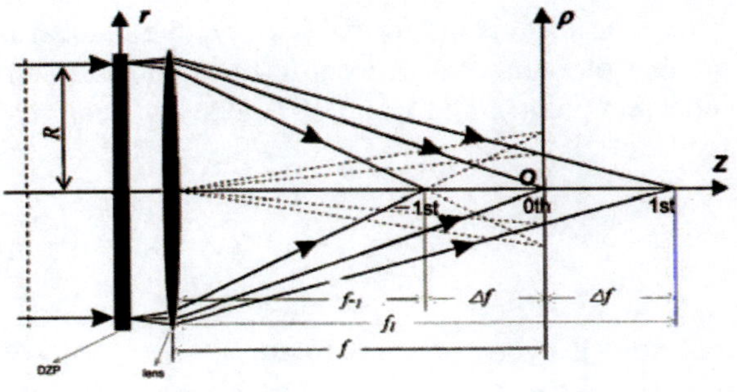

where NA is the numerical aperture, ρ is the radial coordinates of the plane where the focus locates. Making:

$$\xi = \frac{r_\kappa^2 - 0.5}{4\pi} \tag{4.4.4}$$

then from Eq. (4.4.1) one can obtain the axial light field distributions on the optical axis ($v = 0$) is:

$$E(u, 0) = 4\pi \int_{-\infty}^{+\infty} T(\xi) \exp(-i2\pi u\xi) d\xi \tag{4.4.5}$$

where,

$$T(\xi) = \begin{cases} \phi_{DZP}(r), -\frac{1}{8\pi} < \xi < \frac{1}{8\pi} \\ 0, \text{others} \end{cases} \tag{4.4.6}$$

Equation (4.4.5) indicates that, the axial light field distribution behind the DZP and a lens can be regarded as the Fourier transformation of the transmittance function of the DZP. According to the characteristics of Fraunhofer diffraction integral, the far-field diffractions of Dammann gratings introduced in Sect. 4.2.3 are the Fourier transformation of the transmittance function actually when plane waves are incident, indicating that the DZPs are consistent with the Dammann gratings. Similar with Dammann gratings, for DZPs a series of inflection points are added in each period of the transmittance function with respect to the square of the radial coordinates, namely a fixed number of phase modulation rings of different radii are added to each concentric ring of equal area in the traditional FZP, and finally produce multiple axial focal points with identical intensities.

Here the focal position of diffraction order b is present directly [16]:

$$z_b = f + \frac{2N_p b\lambda}{(\text{NA})^2} \tag{4.4.7}$$

with f the focal length of the lens, N_p the number of DZP's period known as the number of equal-area concentric rings contained in the binary 0-π FZP with no inflection point. From Eq. (4.4.7), one can acquire the distance between adjacent focuses:

$$\Delta z = z_{b+1} - z_b = \frac{2N_p\lambda}{(\text{NA})^2} \tag{4.4.8}$$

indicating that the distance between adjacent focuses are associated with the number of period N, the wavelength λ, and the numerical aperture NA. Similar with Dammann grating, the diffraction efficiency and uniformity are defined separately as:

$$\eta = \frac{\sum\limits_{b\in B} I_b}{\sum\limits_{b=-\infty}^{+\infty} I_b} \tag{4.4.9}$$

$$U = \frac{\max\{I_b\} - \min\{I_b\}}{\max\{I_b\} + \min\{I_b\}}, b \in B \tag{4.4.10}$$

In Eqs. (4.4.9) and (4.4.10), I_b corresponds to the focus intensity of diffraction order b, and B is the set of desired diffraction orders. Under the condition of setting the the number of diffraction focal spot p, period number N_p and other parameters, the zone plate was optimized by optimization algorithms, so as to achieve the highest diffraction efficiency and the lowest uniformity value, and finally the DZP was obtained.

Since the consistance of DZPs and Dammann gratings, the normalized inflection points of DZPs r_ι can be calculated directly from that of Dammann gratings (x_κ) given in Table 4.1. It must be noticed that, the DZP introduces a series of inflection points in each period Δr^2 with respect to the square of the radial coordinates r^2, which is different from Dammann gratings. However, the actual design is based on the transmittance function with respect to the radial coordinate, so it is impossible to calculate the inflection points within a grating period like that of Dammann grating, but all inflection points within the interval of [0, R] need to be calculated. This also indicates that when the number of generated diffraction focal spots is fixed, if the period number N_p contained in the interval [0,R] changes, the normalized value of inflection points r_ι will also change.

When calculating r_ι, one should confirm the number of axial focuses p firstly, and then find the inflection points x_κ and the total number of inflection points N of the desired $1 \times p$ array by refering to Table 4.1. The computed DZP has NN_p inflection points in total, which reads:

$$r'_\iota = q + x_\kappa, \begin{cases} q = 0, 1, 2, ..., N_p - 1 \\ \kappa = 1, 2, 3, ..., N \end{cases} \tag{4.4.11}$$

and correspond to the inflection points in each period of the radial coordinate squared r^2. Therefore, the normalized inflection points along radial directions are:

$$r_\iota = \sqrt{\frac{r'_\iota}{N_p}} \tag{4.4.12}$$

By now, one can realize that, when calculating the normalized inflection points r_ι, it is not necessary to optimize the zone plate through complex algorithms, but can calculate directly from the inflection points x_κ of one-dimensional Dammann grating given in Table 4.1 along with Eqs. (4.4.11) and (4.4.12). Table 4.2 presents the normalized inflection points r_ι of DZPs in some of the cases, which can be employed directly when designing DZPs.

Figure 4.43 presents the simulated results of the distribution of the focused fields along the z-axis of the DZP designed when the number of focuses p is 5 and the

Table 4.2 The normalized inflection points r_ι of DZPs [16]

The number of axial focuses p	The number of periods N_p	The inflection points r_ι	Diffraction efficiency η	Uniformity U
2	10	0, 0.2236, 0.3162, 0.3873, 0.4472, 0.5000, 0.5477, 0.5916, 0.6325, 0.6708, 0.7071, 0.7416, 0.7746, 0.8062, 0.8367, 0.8660, 0.8944, 0.9220, 0.9487, 0.9747	0.8130	9.3×10^{-6}
2	20	0, 0.1581, 0.2236, 0.2739, 0.3162, 0.3536, 0.3873, 0.4183, 0.4472, 0.4743, 0.5000, 0.5244, 0.5477, 0.5701, 0.5916, 0.6124, 0.6325, 0.6519, 0.6708, 0.6892, 0.7071, 0.7246, 0.7416, 0.7583, 0.7746, 0.7906, 0.8062, 0.8216, 0.8367, 0.8515, 0.8660, 0.8803, 0.8944, 0.9083, 0.9220, 0.9354, 0.9487, 0.9618, 0.9747, 0.9874	0.8112	3.8×10^{-6}
3	10	0, 0.2712, 0.3162, 0.4166, 0.4472, 0.5230, 0.5477, 0.6112, 0.6325, 0.6881, 0.7071, 0.7573, 0.7746, 0.8207, 0.8367, 0.8795, 0.8944, 0.9346, 0.9487, 0.9867	0.6679	4.2×10^{-3}
3	15	0, 0.2214, 0.2582, 0.3401, 0.3651, 0.4270, 0.4472, 0.4990, 0.5164, 0.5619, 0.5774, 0.6183, 0.6325, 0.6701, 0.6831, 0.7181, 0.7303, 0.7631, 0.7746, 0.8056, 0.8165, 0.8460, 0.8563, 0.8845, 0.8944, 0.9214, 0.9309, 0.9569, 0.9661, 0.9911	0.6658	1.9×10^{-3}
4	5	0, 0.2100, 0.2985, 0.3162, 0.3796, 0.4349, 0.4472, 0.4941, 0.5377, 0.5477, 0.5866, 0.6238, 0.6325, 0.6664, 0.6994, 0.7071, 0.7376, 0.7675, 0.7746, 0.8026, 0.8301, 0.8367, 0.8626, 0.8883, 0.8944, 0.9188, 0.9429, 0.9487, 0.9717, 0.9945	0.7126	4.4×10^{-2}

(continued)

Table 4.2 (continued)

The number of axial focuses p	The number of periods N_p	The inflection points r_ι	Diffraction efficiency η	Uniformity U
4	10	0, 0.1485, 0.2111, 0.2236, 0.2684, 0.3075, 0.3162, 0.3494, 0.3802, 0.3873, 0.4148, 0.4411, 0.4472, 0.4712, 0.4945, 0.5000, 0.5216, 0.5427, 0.5477, 0.5675, 0.5870, 0.5916, 0.6100, 0.6281, 0.6325, 0.6497, 0.6668, 0.6708, 0.6871, 0.7033, 0.7071, 0.7225, 0.7379, 0.7416, 0.7563, 0.7711, 0.7746, 0.7887, 0.8028, 0.8062, 0.8198, 0.8334, 0.8367, 0.8497, 0.8629, 0.8660, 0.8787, 0.8914, 0.8944, 0.9067, 0.9190, 0.9220, 0.9338, 0.9458, 0.9487, 0.9602, 0.9719, 0.9747, 0.9859, 0.9973	0.7079	1.4×10^{-4}
5	5	0, 0.0879, 0.2796, 0.3621, 0.4472, 0.4558, 0.5274, 0.5754, 0.6325, 0.6385, 0.6915, 0.7288, 0.7746, 0.7796, 0.8235, 0.8550, 0.8944, 0.8987, 0.9371, 0.9649	0.8032	2.9×10^{-2}
5	10	0,, 0.0622, 0.1977, 0.2560, 0.3162, 0.3223, 0.3729, 0.4069, 0.4472, 0.4515, 0.4890, 0.5153, 0.5477, 0.5512, 0.5823, 0.6046, 0.6325, 0.6355, 0.6626, 0.6823, 0.7071, 0.7098, 0.7342, 0.7520, 0.7746, 0.7771, 0.7994, 0.8158, 0.8367, 0.8390, 0.8597, 0.8750, 0.8944, 0.8966, 0.9160, 0.9304, 0.9487, 0.9507, 0.9691, 0.9826	0.7812	7.6×10^{-3}

number of periods N_p is 10. Obviously, the intensities of the five focuses are almost same.

On the basis of DZPs, one can also introduce helical phase $\exp(il_{\text{DZP}}\varphi)$ to acquire Dammann vortex zone plates (DVZPs), where l_{DZP} is the order of a DVZP, also known as basic topological charge. DVZPs have similar characteristics to one-dimensional Dammann gratings except that they produce focal spots on the optical axis rather than he diffraction plane. Due to the helical phases, the focused fields of DVZPs are vortex beams carrying OAM, and the topological charge of diffraction order b is bl_{DZP}. Other parameters such as the distance between adjacent focuses, the diffraction efficiency, the uniformity and the design method of the DVZP are the same as those of the ordinary DZP, which won't be repeated here. Figure 4.44 presents various DVZPs with $p = 7$ and $N_p = 8$, their orders l_{DZP} are from -1 to $+\,4$. Obviously

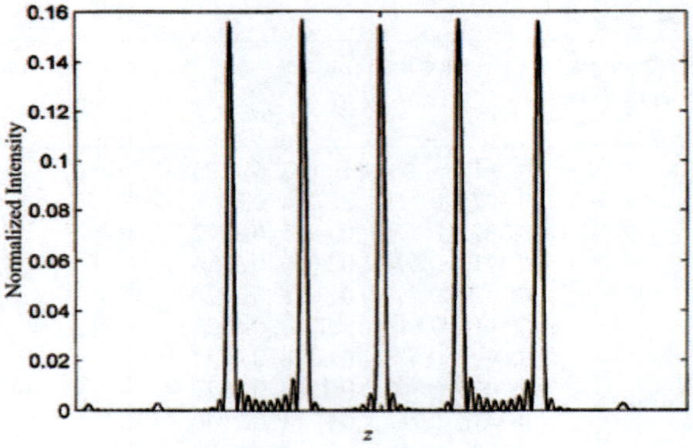

Fig. 4.43 The simulated results of the distribution of the focused fields along the z-axis of a DZP [16]

the 0-th DVZP is the common DZP indeed, and DVZPs with opposite order have opposite helices.

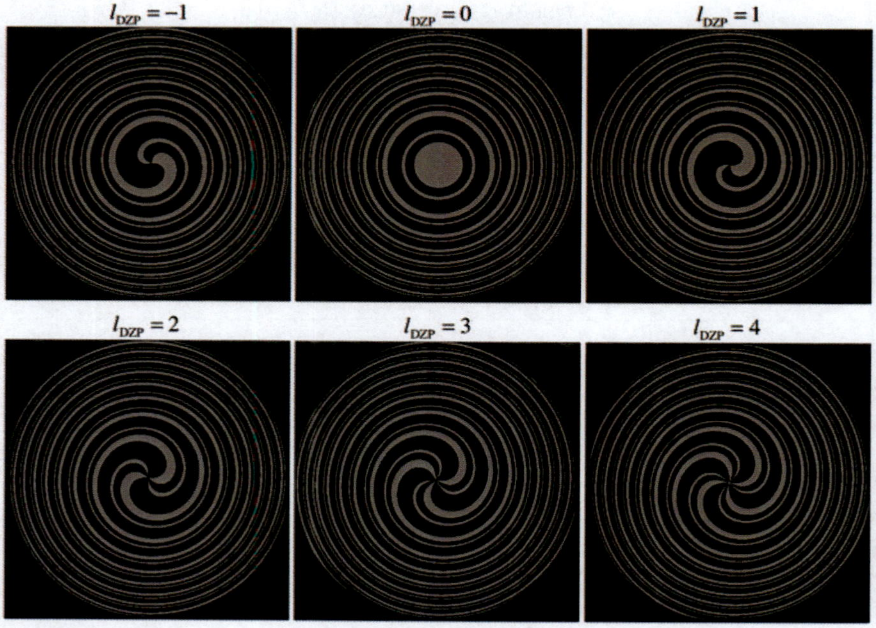

Fig. 4.44 Various DVZPs

4.4.2 The Generation of Three-Dimensional Vortices Lattices

DZPs make it possible to produce multiple axial focuses along z direction in diffraction fields. And multiple vortex beams with various topological charges can also present in each focus if the helical phase is added in DZPs. The Dammann vortex grating introduced in Sect. 4.3.1 can produce $m \times n$ two-dimensional vortices lattices in xOy plane when Gaussian beams are incident in. Therefore, one can imagine that a feasible way to achieve three-dimensional vortices lattices is to combine a DVZP and Dammann vortex grating together to gain a new grating, where the diffraction fields contain multiple focusing planes, each focusing plane has an $m \times n$ lattice, so that the overall fields is represented as an $m \times n \times p$ three-dimensional vortices lattices. Such binary 0-π phase-only gratings called three-dimensional Damman vortex gratings [17–19].

A two-dimensional Dammann vortex grating can be Fourier expanded as:

$$T_D(x, y) = \sum_{a,b=-\infty}^{+\infty} c_{a,b} \exp\left[i\left(a\gamma_x x + b\gamma_y y\right)\right] \exp\left[i\left(al_x + bl_y\right)\varphi\right] \tag{4.4.13}$$

with a, b the diffraction orders along x and y axes separately, $c_{a,b}$ the Fourier coefficient, l_x, l_y the order of the grating, φ the azimuthal angle. γ_x and γ_y are the grating's spatial angular frequency, which represent the grating constant and are defined by Eq. (4.3.4).

A DVZP can be Fourier expanded as:

$$T_{DZP}(r, \varphi) = \sum_{q=-\infty}^{+\infty} c_q \exp\left[iq\gamma_\xi \xi\right] \exp(iql_z\varphi) \tag{4.4.14}$$

where (r,φ) denotes the normalized polar coordinates of the pupil plane, l_z denotes the order of DZP, c_q is the Fourier coefficient, q is the diffraction orders, and ξ reads:

$$\xi = \sqrt{1 - (r \sin\alpha)^2} \tag{4.4.15}$$

with α half aperture angle. In Eq. (4.4.14), γ_ξ is the spatial angular frequency with respect to ξ:

$$\gamma_\xi = \frac{2\pi}{T_\xi} \tag{4.4.16}$$

where T_ξ is the spatial period with respect to ξ. Combining the two-dimensional Dammann vortex grating in Eq. (4.4.13) and the DVZP in Eq. (4.4.14) together, the transmittance function of the obtained grating is:

$$T_{TD}(r,\varphi)=\sum_{a,b=-\infty}^{+\infty}c_{a,b}\exp\left[i\left(a\gamma_x x+b\gamma_y y\right)\right]\exp\left[i\left(al_x+bl_y\right)\varphi\right]$$
$$\cdot\sum_{q=-\infty}^{+\infty}c_q\exp\left[iq\gamma_\xi\xi\right]\exp(iql_z\varphi)$$
$$=\sum_{a,b,q=-\infty}^{+\infty}c_{a,b}c_q\exp\left[i\left(a\gamma_x r\cos\varphi+b\gamma_y r\sin\varphi+q\gamma_\xi\xi\right)\right]$$
$$\exp\left[i\left(al_x+bl_y+ql_z\right)\varphi\right] \tag{4.4.17}$$

In Eq. (4.4.17), the first exponential term determines the position distribution of each vortex beam in the three-dimensional lattice, and the second exponential term determines the order of each vortex beams in each diffraction order, namely:

$$l_{a,b,q}=al_x+bl_y+ql_z \tag{4.4.18}$$

Next discussing the distance between adjacent diffraction orders along the three various orthogonal directions (x, y and z axes). In the plane xOy, the distance between adjacent diffraction orders are determined by the grating constants of two-dimensional Dammann vortex grating T_x and T_y respectively. For instance, in x axis, from Eq. (4.2.22), such distance reads:

$$\Delta x=\frac{d_f\lambda}{T_x} \tag{4.4.19}$$

If N_x corresponds to the number of the grating period along x directions, a_x is the linearity of the grating in the x-axis, then $T_x=a_x/N_x$. As the half aperture angle is small, the linearity a_x and d_f satisfy $a_x=2d_f\sin$. Thus Eq. (4.4.19) can be written as:

$$\Delta x=\frac{N_x\lambda}{2\sin\alpha} \tag{4.4.20}$$

Similarly, the distance between adjacent diffraction orders in y directions reads:

$$\Delta y=\frac{N_y\lambda}{2\sin\alpha} \tag{4.4.21}$$

with N_y the number of grating period along y axis. If N_ξ corresponds to the number of spatial periods of the superposed DVZPs with respect to ξ, then from Eq. (4.4.8):

$$\Delta z=\frac{2N_\xi\lambda}{(\mathrm{NA})^2} \tag{4.4.22}$$

From the definition of numerical aperture:

$$\mathrm{NA} = n_0 \sin\alpha \tag{4.4.23}$$

considering that the refractive index behind the grating $n_0 \approx 1$, by substituting Eq. (4.4.23) into Eq. (4.4.22) and then simplifying, the distance between adjacent diffraction orders in the along z-axis is obtained as:

$$\Delta z = \frac{N_\xi \lambda}{1 - \cos\alpha} \tag{4.4.24}$$

By now, we have already the distance between adjacent diffraction orders of all the three orthogonal directions. They have uniform form, only associate with the wavelength, half aperture angle and grating periods.

Since both the Dammann vortex grating and DVZP are binary 0-π gratings, to make the superposed three-dimensional Dammann vortex grating 0-π binarization, one still need to combine them through the operation of Eq. (4.2.17), where such process is sketched in Fig. 4.45.

Figure 4.46 presents the optical system to generate three-dimensional $m \times n \times p$ vortices lattices through a three-dimensional Dammann vortex grating. As mentioned in the previous section, the DZP (DVZP) needs to be matched with a lens to achieve the function of producing multi-focus. In practice, a thin convex lens should be placed behind the Dammann vortex grating to focus, so that the two-dimensional lattice in the far-field can be observed at a limited distance. Far-field diffractions can be observed on the back focal plane of a thin convex lens when a colorimetric Gaussian beam (approximately parallel or plane wave) is incident. Therefore, it is necessary to place a lens behind the three-dimensional Dammann vortex grating. Some works [18] also integrate the transmittance function of convex lens [Eq. (2.5.7) in chapter two] and three-dimensional Dammann vortex grating together, as shown in Fig. 4.47. It should be noticed that the focal length of the lens should be reasonably designed to conform to the NA value of the DVZP. In this way, the three-dimensional vortices lattice can be generated directly through the combined grating without a lens being placed behind.

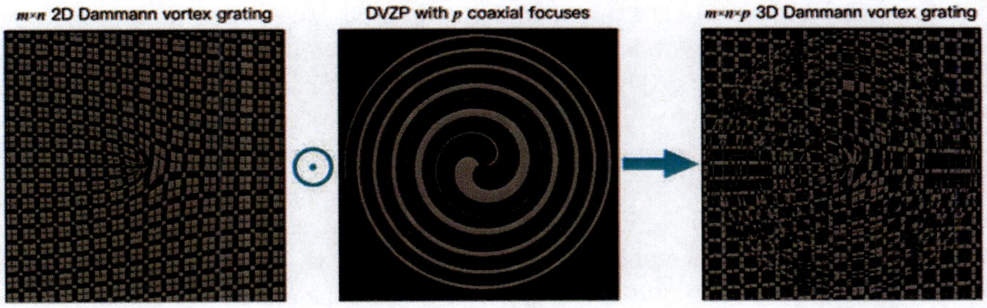

Fig. 4.45 The three-dimensional 5 × 5 × 5 Dammann vortex grating from a two-dimensional 5 × 5 Dammann vortex grating and a DVZP with five focuses

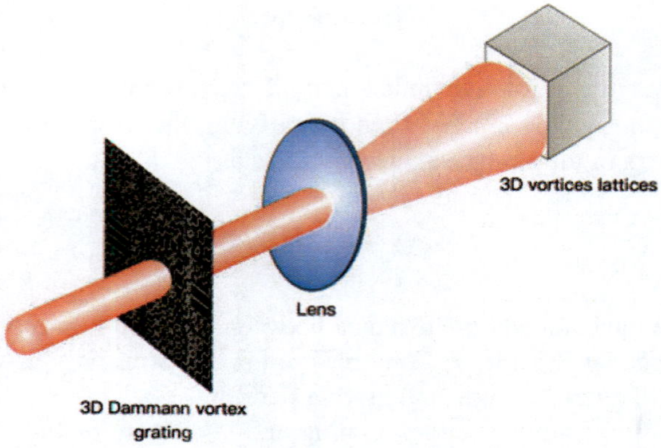

Fig. 4.46 The optical system to generate three-dimensional vortices lattices through a three-dimensional Dammann vortex grating

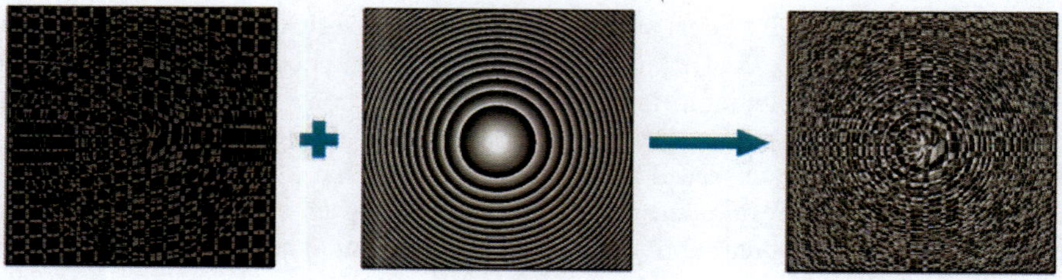

Fig. 4.47 The superposition of a three-dimensional Dammann vortex grating and a lens

Actually, the three-dimensional Dammann vortex grating is the superposition of a two-dimensional Dammann vortex grating and a DVZP, where the phase information contained increases greatly, and the grating details become very messy. Usually such kind of complex phase grating is difficult to make, thus liquid–crystal spatial light modulator (SLM) is employed to simulate. However, limited by the resolution of SLM, the three-dimensional Dammann vortex grating can only be used in the case of focusing field of lens with low numerical aperture. While in some practical applications (such as photonic crystal fabrication), the numerical aperture of the focusing lens is usually large enough to provide high energy density and micron or even submicron focused spots. Although in theory one can employ both the three-dimensional Dammann vortex grating and high numerical aperture objective to achieve this goal, the aperture of the objective is usually a few millimeters, which limits the period number and the total pixel number of the three-dimensional Dammann vortex grating on the SLM within the aperture, thus greatly reducing the uniformity and efficiency of the generated three-dimensional lattice. Therefore, for the case of high numerical aperture, it is imperfect to adopt the three-dimension Darmmann vortex grating. The optical system to generate the three-dimensional vortices lattice usually consists of a two-dimensional Dammann vortex grating, a confocal lens group, a DZP, and the aberration focused objective lens [19]. The confocal lens group can compress or enlarge the divergence Angle of the 2-dimensional daman grating according to the needs. In addition, a spatial filter can be added to the spectrum surface of the confocal lens group to change the distribution of the daman array according to the requirements. During the optical path construction, the center of DZP should be aligned with the center of the focusing objective lens, and on this premise, an $m \times n$ two-dimensional Dammann grating designed for the focusing objective lens should be placed in front of the confocal lens group, to obtain a three-dimensional focused spot lattice in the focusing field of the objective lens with high numerical aperture.

References

1. Romero LA, Dickey FM. Theory of optimal beam splitting by phase gratings. I. One-dimensional gratings. J Opt Soc Am Opt Image Sci Vis. 2007;24(8):2280–95.
2. Romero LA, Dickey FM. The mathematical theory of laser beam-splitting gratings. Prog Opt. 2010;54(10):319–386.
3. Gerchberg RW, Saxton WO. A practical algorithm for the determination of phase from image and diffraction plane pictures. Optik. 1972;35:237–50.
4. Qi XQ, Gao CQ, Liu YD. Generation of helical beams with pre-determined energy distribution based on phase modulation gratings. Acta Phys Sin. 2010;59(1):264–70.
5. Dammann H, Görtler K. High-efficiency in-line multiple imaging by means of multiple phase holograms. Opt Commun. 1971;3(5):312–5.
6. Dammann H, Klotz E. Coherent optical generation and inspection of two-dimensional periodic structures. J Mod Opt. 1977;24(4):505–15.
7. Di C, Zhou C. Dammann garing-based dynamic optical coupler. Acta Opt Sin. 2007;27(7):1275–8.
8. Zhou C, Liu L. Numerical study of Dammann array illuminators. Appl Opt. 1995;34(26):5961–9.
9. Morrison RL, Walker SL, Cloonan TJ. Beam array generation and holographic interconnections in a free-space optical switching network.. Appl Opt. 1993;32(14):2512–2518.
10. Moreno I, Davis JA, Cottrell DM, et al. Encoding generalized phase functions on Dammann gratings. Opt Lett. 2010;35(10):1536–8.
11. Zhang N, Yuan XC, Burge RE. Extending the detection range of optical vortices by Dammann vortex gratings. Opt Lett. 2010;35(20):3495–7.
12. Fu S, Zhang S, Wang T, et al. Measurement of orbital angular momentum spectra of multiplexing optical vortices. Opt Express. 2016;24(6):6240–8.
13. Fu S, Wang T, Zhang S, et al. Integrating 5 × 5 Dammann gratings to detect orbital angular momentum states of beams with the range of -24 to +24. Appl Opt. 2016;55(7):1514–7.
14. Gao C, Qi X, Liu Y, et al. Superposition of helical beams by using a Michelson interferometer. Opt Express. 2010;18(1):72–78.
15. Saavedra G, Furlan WD, Monsoriu JA. Fractal zone plates. Opt Lett. 2003;28(12):971–3.
16. Zhou C, Yu J. Dammann zone plate. Chinese patent: 201010585480.4, 2011-05-18.
17. Davis JA, Moreno I, Martínez JL et al. Creating three-dimensional lattice patterns using programmable Dammann gratings. Appl Opt 2011;50(20):3653–3657.
18. Huang L, Song X, Reineke B, et al. Volumetric generation of optical vortices with metasurfaces. ACS Photonics. 2017;4(2):338–46.
19. Yu J, Zhou C, Jia W, et al. Three-dimensional Dammann array. Appl Opt. 2012;51(10):1619–1630.

Chapter 5
Diagnosing Orbital Angular Momentum for Vortex Beams

The detection of vortex beams mainly refers to the measurement of the OAM component contained in the vortex beam and its specific gravity. As the characteristic value of a vortex beam, topological charge l determines the distribution of its helical wave front and the magnitude of the OAM carried by each photon. Hence, the mode diagnostic of OAM of a vortex beam is a measurement of its order or topological charge l in fact. For single-mode vortex beams, the aim is simply to determine the l value. For multiplexed vortex beams, in addition to determining the OAM components, the intensity ratio of each component, namely, the OAM spectrum, should also be measured. Accurate measurement of OAM components and OAM spectra is of great significance for the practical application of vortex beams. This chapter will focus on several scheme for vortex beam OAM detection, including interferometry, diffractometry and polarimetry.

5.1 Basic Schemes for OAM Measurement

5.1.1 Measuring OAM Through Torque

The torque measurement method [1, 2] is accomplished by the device mechanical rotation result from the OAM carried by the vortex beams. A typical measurement system of this scheme is displayed in Fig. 5.1a. The system consists of a cylindrical lens CL, two reflecting mirrors M1 and M2, a fixed rectangular bracket and a hanging thin wire. Among them, the cylindrical lens CL and the mirror M2 are fixed together by the rectangular bracket and suspended by thin wires. Meanwhile, M1 is fixed by an external structure. The incident Hermite Gaussian beam enters the cylindrical lens CL from below, and is reflected by the M1 after propagation for a certain distance, and passes through the cylindrical lens again to form a vortex beam carrying OAM. Since the carried OAM is conserved during the beam transmission, the cylindrical lens will be twisted slightly, which will further lead to the twist of M2. The twisting

S. Fu and C. Gao, *Optical Vortex Beams*, Advances in Optics and Optoelectronics,
https://doi.org/10.1007/978-981-99-1810-2_5

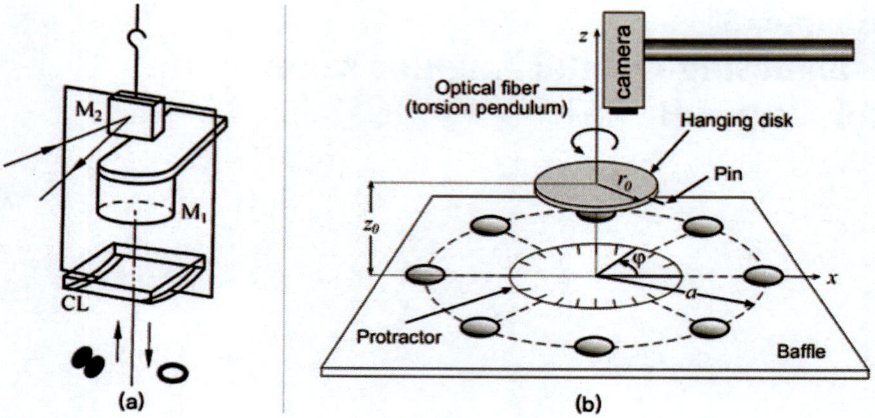

Fig. 5.1 Experimental setup of measuring OAM through torque [1, 2]

angle can be determined by the reflection angle change of another incident beam that reflected by M2. After the derivation of mechanical relationships, the OAM carried by the vortex beam can be calculated through the changing torque.

As shown in Fig. 5.1b, another torque OAM measurement scheme is to measure the rotation angle of the suspended rubber-wrapped acrylic wafer during the generating a vortex beam.

5.1.2 Secondary Intensity Moment

Any beam can be characterized by the secondary intensity moment, including vortex beams [3]. The secondary intensity moment contains terms associated with angular momentum, so it can be used to measurement the OAM of a beam. In general, any beam has at most ten independent parameters of secondary intensity moments, which can be expressed by a 4 × 4 matrix V as:

$$V = \begin{bmatrix} \langle x^2 \rangle & \langle xy \rangle & \langle x\theta_x \rangle & \langle x\theta_y \rangle \\ \langle xy \rangle & \langle y^2 \rangle & \langle y\theta_x \rangle & \langle y\theta_x \rangle \\ \langle x\theta_x \rangle & \langle y\theta_x \rangle & \langle \theta_x^2 \rangle & \langle \theta_x\theta_y \rangle \\ \langle x\theta_y \rangle & \langle y\theta_y \rangle & \langle \theta_x\theta_y \rangle & \langle \theta_y^2 \rangle \end{bmatrix} \tag{5.1.1}$$

In Eq. (5.1.1), the physical significance of each parameter is given by

$\langle x^2 \rangle$ and $\langle y^2 \rangle$ are proportional to the square of the beam width in both x and y directions.

$\langle \theta_x^2 \rangle$ and $\langle \theta_y^2 \rangle$ are proportional to the square of the far-field divergence angle in both x and y directions.

$\langle x\theta_x \rangle$ and $\langle y\theta_y \rangle$ are inversely proportional to the radius of curvature of beam equiphase surface in the x and y directions.
$\langle xy \rangle$ and $\langle \theta_x\theta_y \rangle$ indicate the orientation of the beam in near and far fields, respectively.
$\langle x\theta_y \rangle$ and $\langle y\theta_x \rangle$ are the twisting parameter of the beam.

The relationship between the OAM of a beam and the secondary intensity moment is discussed as follows. The secondary moment parameters $\langle x\theta_y \rangle$, $\langle y\theta_x \rangle$ of a field $E(x,y)$ are defined as [4]:

$$\langle x\theta_y \rangle = \frac{1}{2P(-ik)} \iint xE(x,y)\frac{\partial}{\partial y}E^*(x,y)dxdy + c.c \tag{5.1.2}$$

$$\langle y\theta_y \rangle = \frac{1}{2P(-ik)} \iint yE^*(x,y)\frac{\partial}{\partial x}E(x,y)dxdy + c.c \tag{5.1.3}$$

For a beam propagating along the z-axis, its electric field intensity reads:

$$\vec{E}(x,y,z,t) = E_0 \cdot u(x,y,z) \cdot \vec{e}(x,y,z) \cdot \exp[i(kz - \omega t)] \tag{5.1.4}$$

where k is the wave number and ω is the angular frequency. $E_0 \cdot u(x,y,z)$ represents the amplitude of the electric field intensity, where $u(x,y,z)$ is a normalized complex scalar function to describe the amplitude distribution of the field. $u(x,y,z)$ satisfies wave equation under paraxial approximation. $\vec{e}(x,y,z)$ represents the unit polarization vector. Substituting Eq. (5.1.4) into Eqs. (5.1.2) and (5.1.3), we get

$$\langle x\theta_y \rangle = \frac{i\lambda}{4\pi} \iint xu\frac{\partial u^*}{\partial y}dxdy + c.c \tag{5.1.5}$$

$$\langle y\theta_x \rangle = \frac{i\lambda}{4\pi} \iint yu\frac{\partial u^*}{\partial x}dxdy + c.c \tag{5.1.6}$$

Subtract Eq. (5.1.5) by Eq. (5.1.6):

$$\langle x\theta_y \rangle - \langle y\theta_x \rangle = \frac{i\lambda}{4\pi} \iint \left(xu\frac{\partial u^*}{\partial y} - yu\frac{\partial u^*}{\partial x} \right)dxdy + c.c \tag{5.1.7}$$

Comparing Eq. (1.2.35) with Eq. (5.1.7), the relationship between the OAM J_{zl} and the secondary intensity moment is expressed as:

$$J_{zl} = \frac{P}{c}\left(\langle x\theta_y \rangle - \langle y\theta_x \rangle\right) \tag{5.1.8}$$

where P is the total power of the beam and c is the light velocity in vacuum. Equation (5.1.8) shows that as long as the secondary moment of a beam is measured, the OAM it carried can be measured. Then information such as the topological charge

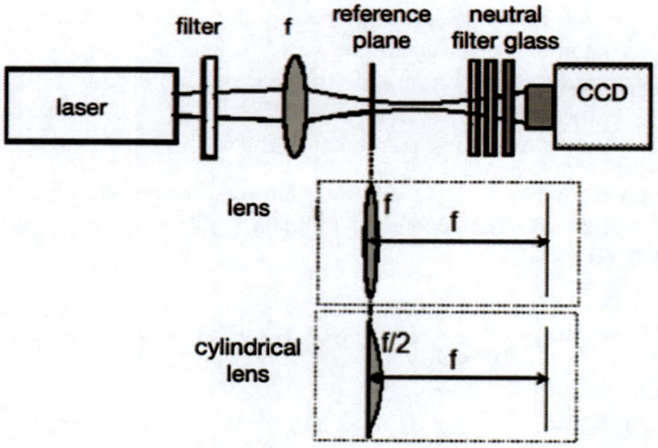

Fig. 5.2 Experimental setup for measuring OAM of a vortex beam using the secondary intensity moment

can be inferred. The experimental device for measuring beam's OAM using the secondary intensity moment is shown in Fig. 5.2.

① Firstly, choosing an arbitrary reference plane and measure the secondary moments at this reference plane $\langle xy_{ref}\rangle$:

$$\langle xy_{ref}\rangle = \frac{1}{P}\iint xyI(x,y,z)dxdy \tag{5.1.9}$$

where P is the beam power and $I(x,y,z)$ is the intensity distribution at the reference surface.

② Placing a lens with focal length f on the reference plane. Measuring the secondary moment $\langle xy_{sph}\rangle$, $\langle xy_{sph}\rangle$ at the focal plane of the lens and the crossing moment with the original beam $\langle \theta_x\theta_y\rangle_{ref}$ satisfy:

$$\langle xy_{sph}\rangle = f^2 \cdot \langle \theta_x\theta_y\rangle_{ref} \tag{5.1.10}$$

③ After step ②, replacing the lens by a cylindrical lens with focal length $f/2$ and measure the transverse moment $\langle xy_{eyl}\rangle$ of the beam in the plane $z = f$. The relationship between $\langle xy_{eyl}\rangle$ and the secondary moment of the initial beam meets:

$$\langle xy_{eyl}\rangle = -\langle xy_{ref}\rangle + f\cdot\langle y\theta_x\rangle_{ori} - f\cdot\langle x\theta_y\rangle_{ori} + f^2\langle \theta_x\theta_y\rangle_{ref} \tag{5.1.11}$$

Thus, from Eq. (5.1.11),

$$\langle y\theta_x\rangle_{ori} - \langle x\theta_{y1}\rangle_{ori} = \frac{1}{f}\left(\langle xy_{eyl}\rangle - \langle xy_{sph}\rangle + \langle xy_{ref}\rangle\right) \tag{5.1.12}$$

Here, substituting Eq. (5.1.12) into Eq. (5.1.8), the OAM value of the vortex beam is acquired.

5.1.3 Measuring OAM Using the Rotational Doppler Effect

Doppler effect is a well-known physical phenomenon which can be understood as the frequency shift of a wave if there is relative motion between the source and the observer. Doppler shift exists in both mechanical and electromagnetic waves. For light waves, the relative motion between the source and the observer results in a frequency shift Δf of the wave: $\Delta f = f_0 v/c$, with f_0 the initial optical frequency, v the motion velocity and c the light velocity. Doppler effect plays an important role in speed measurement, fluid detection and other applications.

The transverse rotation of a vortex beam with a helical wavefront is equivalent to a longitudinal translation. Therefore, the rotation of the vortex beam will also result in the frequency shift of the beam, an effect known as rotational Doppler effect. For vortex beams with various topological charge, the frequency shift is different because of the different rotation speed of the wavefront. The relation between the frequency shift Δf of the vortex beam and its angular frequency Ω, topological charge l is given directly [5]:

$$\Delta f = \frac{l\Omega}{2\pi} \tag{5.1.13}$$

Based on the rotational Doppler effect of the vortex beam, Vasnetsov et al. gave a scheme to measure the OAM of a vortex beam [6]. After the fundamental Gaussian beam has passed through a beam splitter, one beam is used as the reference beam and the other beam is diffracted through a grating to simulate the signal light. The reference beam is modulated by a reflector driven by a piezoelectric ceramic to produce a side frequency of $f = 10$ Hz, which is then superimposed on the signal light. Usually to observe the rotational Doppler effect, relative angular motion must be generated between the beam and the observer. The beam can generally be rotated using a Dove prism (Dove prism), or the detector can be rotated.

The main shortcomings of this scheme are as follows: firstly, for multiplexed vortex beams superimposed by multiple non-zero radial indices p, the point intensity cannot be used to distinguish the OAM of the beam; secondly, for multimode mixed single-loop vortex beams, when the modulation frequency f is not large enough, other frequency component interference terms appear and the OAM component corresponding to the low frequency cannot be distinguished. Finally, this scheme requires a fundamental mode Gaussian beam as a reference beam, which makes the setup more complex and therefore difficult to use in practical vortex beam detection.

5.1.4 Measuring OAM Using the Conjugate Relation Between OAM and the Mean Value of Rotation Operator

As mentioned in Sect. 2.3.2, OAM spectrum and the mean of rotation operator in the optical wave function are in the relation of Fourier transformation. Such relation can be employed to measure OAM of a vortex beam. If the wave function of the incident beam is ψ and the rotation angle θ of the beam can be expressed as $\exp\left(i\theta\hat{l}\right)$, then the average value of the rotation operator is [7]:

$$M(\theta) = \frac{\iint d^2x\psi^* \exp\left(i\theta\hat{l}\right)\psi}{\iint d^2x|\psi|^2} \tag{5.1.14}$$

Here $M(\theta)$ is a complex number. Through Fourier transformation, the OAM spectral can be obtained:

$$P_l = \frac{1}{2\pi}\int M(\theta)\exp(-i\theta l)d\theta \tag{5.1.15}$$

This suggests that the detection of vortex beams can be carried out by measuring the average value of the rotation operator firstly and then doing a Fourier transformation to obtain its OAM spectrum.

An experimental system for measuring the OAM distribution of a vortex beam using such approach is shown in Fig. 5.3 [8]. In Fig. 5.3, the output intensity difference of the two output beams is proportional to the real part of the average value $M(\theta)$ of the rotation operator. And the intensity difference of the two output beams after inserting the half wave plate (HWP) is proportional to the imaginary part of the average value $M(\theta)$ of the rotation operator. After rotating the Dove prism to different angle θ and obtaining all the real and imaginary parts of $M(\theta)$, the OAM components of the beam and the OAM spectrum can be calculated directly using Eq. (5.1.15).

This approach can be employed to measure the OAM spectrum of any space-shaped beam, and has no requirement for the radial index p and the purity of the incident beam. For symmetric OAM spectrum ($P_l = P_{-l}$), no additional half-wave plate is needed, while for asymmetric OAM spectrum ($P_l \neq P_{-l}$) is difficult to carry out. In addition, the scheme also faces the problem of how to ensure the coaxial interference of the beam and the difference between the two arms remain stable when the Dowell prism rotates.

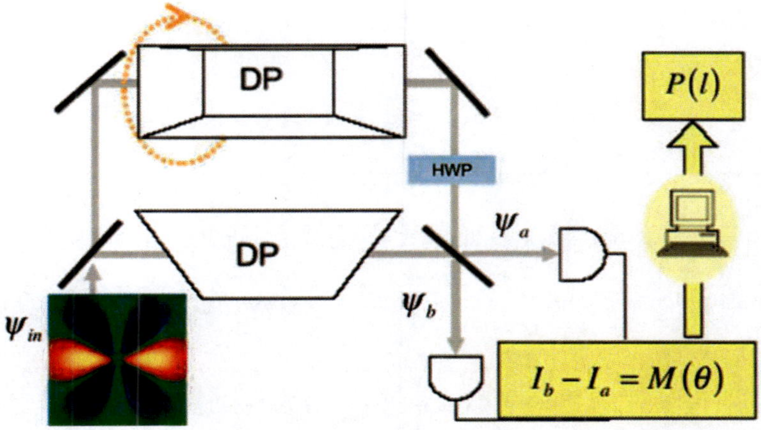

Fig. 5.3 Experimental system for measuring the OAM distribution of vortex beams using the mean of rotation operators. DP: Dove prism; HWP: half wave plate [8]

5.2 Interferometry

Measuring OAM through interferometry is to introduce a reference beam to interfere with the vortex beam, or splitting the vortex beams and then interfering with each other. The topological charge can be determined by interferometric field properties.

5.2.1 Interference Vortex Beams with Plane Waves

There are two interference forms between a plane wave and a vortex beam, parallel interference and non-parallel interference. Parallel interference means that the optical axis of the vortex beam is parallel to that of the plane wave, otherwise it is non-parallel interference. Since the vortex beam has a helical wave front, the two types of interference give completely different interference fields.

Let's consider the non-parallel interference firstly. To simplify the operation, we consider only the helical phase and the propagation distance $z = 0$. By now a vortex beams can be expressed in polar coordinates (r,φ) as

$$E_{OV}(r, \varphi) = \exp(il\varphi) \tag{5.2.1}$$

where l is the topological charge. When a plane wave whose wave vector in the xOz plane is at an angle θ to the optical axis (z-axis) in the Cartesian coordinate at $z = 0$:

$$E_0(x, y) = \exp(ikx \sin\theta) \tag{5.2.2}$$

where $k = 2\pi/\lambda$ is the wave number. Then the intensity of their interference field reads:

$$I = |E_{OV} + E_0|^2 = (E_{OV} + E_0)(E_{OV} + E_0)* \tag{5.2.3}$$

Substituting Eqs. (5.2.1) and (5.2.2) into Eq. (5.2.3):

$$I = [\cos(l\varphi) + \cos(kx \sin\theta)]^2 + [\sin(l\varphi) + \sin(kx \sin\theta)]^2 \tag{5.2.4}$$

From trigonometric function and letting $x = r\cos\varphi$, Eq. (5.2.4) can be simplified to

$$I(r, \varphi) = 4\cos^2\left\{\frac{1}{2}[l\varphi - kr\cos\varphi\sin\theta]\right\} \tag{5.2.5}$$

Equation (5.2.5) is essentially the same as Eq. (3.2.18), showing that the interference of a tilted plane wave with a vortex beam results in a fork-shaped intensity distributions, as shown in Fig. 5.4. The number of center forks is identical with the absolute value of the topological charge $|l|$, and the fork opening direction is determined by the sign of l. When $l = 0$, the interference field degenerates to the common interference fringe obtained by two plane wave interference.

Figure 5.4 indicates that, when diagnosing OAM state of a vortex beam, a plane wave can be used to interfere non-parallelly with the vortex beam. The topological charge can be determined by the number of center forks and the opening direction of forks.

However, since the vortex beam has a hollow ring structure, there is no intensity at the beam center. The interference field center only contains plane wave components but no vortex beam component. In other words, there is no vortex wavefront at the field center to interfere with the plane wave. Previous analysis illustrates that, the fork structure of the interferometric field exists at its center. Hence in practice, although the interferometric field has a fork structure, the sign of the topological charge can be determined by the fork opening direction, but fork number cannot be accurately

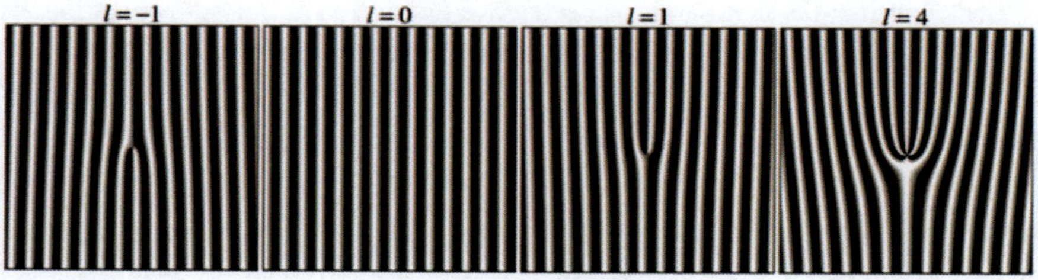

Fig. 5.4 Intensity distribution when a vortex beam interferent non-parallelly with a tilted plane wave

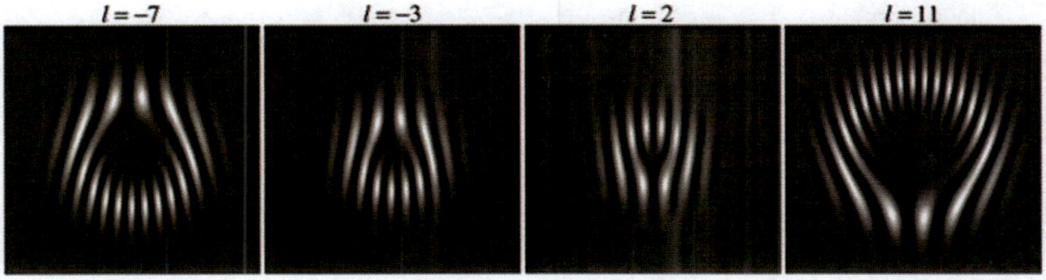

Fig. 5.5 Intensity distributions of interference field of vortex beams and tilted plane waves

acquired because vortex beam non-existent here. The actual field of a vortex beam interfering with a tilted plane wave is shown in Fig. 5.5.

From Eq. (5.2.5), the spatial period of the interference forked fringe is:

$$T = \frac{\lambda}{\sin\theta} = \lambda \csc\theta \tag{5.2.6}$$

It is revealed that for a fixed wavelength, T is only determined by the tilt angle θ. The smaller θ is, the larger the spatial period T is. When $\theta = 0$, the optical axis of the vortex beam is parallel to the wave vector of the plane wave. In other words, they interfere in parallel, and $T \to \infty$, the interference field shows no period intensity distributions.

When $\theta = 0$, Eq. (5.2.5) is

$$I(r, \varphi) = 4\cos^2\left[\frac{1}{2}l\varphi\right] = 2\{1 + \cos[l\varphi]\} \tag{5.2.7}$$

Equation (5.2.7) shows that the interference field has a periodic cosine intensity distribution along the azimuthal, showing a petal-like shape, and the number of petals is the same as the absolute value of the topological charge l, as shown in Fig. 5.6.

The center of the interferometric field is still free of vortex beams, but we can read the absolute value of the topological charge of vortex beams directly from the number

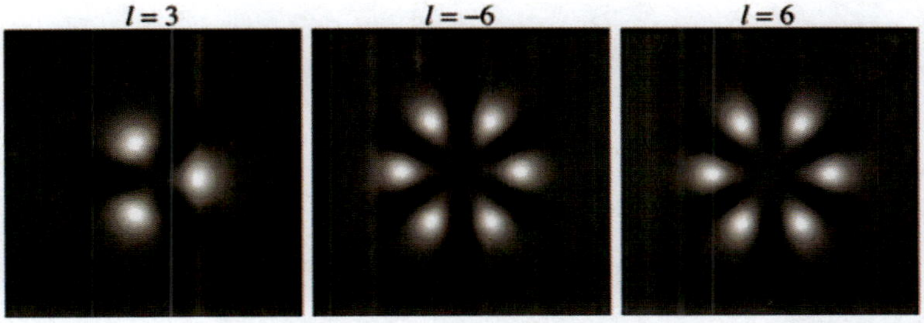

Fig. 5.6 Intensity distribution of parallel interference field between vortex beams and a plane wave

of petals it presents. Parallel interference with plane waves is a good solution to the problem as the exact value of the fork number in non-parallel interference cannot be clearly read. However, the sign of the topological charge cannot be determined at this point.

When using this method to detect vortex beams, the Gaussian beam is generally used as the reference beam. Although the Gaussian beam can be seen as a plane wave at the beam waist and at infinity, the wave front at other locations is a curved surface with a small curvature, distributed in a regular pattern and related to position z. As a higher-order mode form of the fundamental Gaussian mode, the wavefront at the non-waist is also a superposition of a helical wavefront and a surface of small curvature, as shown in Fig. 5.7. Then when a vortex beam is not in the beam waist ($z = 0$), its coaxial interference field with the fundamental Gaussian beam can present the sign information of the topological charge, due to the tiny curvature of the curved wavefront it contains. Note that by coaxial interference, we mean that the optical axis of the vortex beam coincides with that of the Gaussian beam, which is also a form of parallel interference. Figure 5.8 shows the intensity distribution of the interferometric field when vortex beams interfere with Gaussian beams coaxially at $z \neq 0$, illustrating that a vortex beam can be detected by interfering coaxially with the fundamental Gaussian beam.

Interfering with a Gaussian beam to detect a vortex beam is a relatively simple method, requiring only an additional fundamental mode Gaussian beam and a beam

Fig. 5.7 Helical phases of the vortex beam $|+2\rangle$ at $z = 0$ and $z \neq 0$

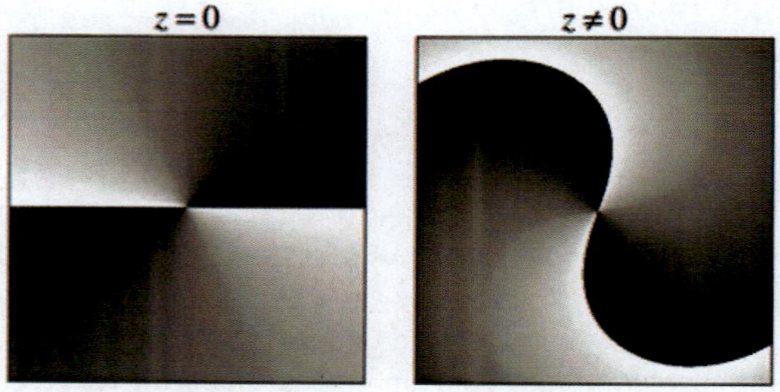

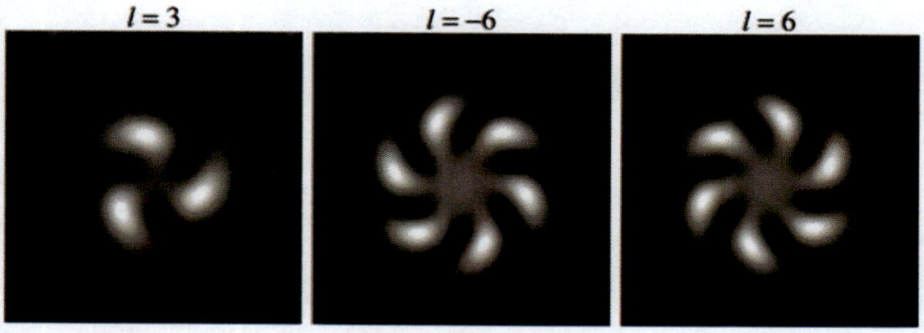

Fig. 5.8 Interferometric intensity distribution of coaxial interference between a vortex beam and a Gaussian beam

combining element such as a splitting prism. However, readers should be aware that the size of the fundamental mode Gaussian beam must be larger than that of the vortex beam, otherwise good interference cannot be obtained. This approach also has the disadvantage that the introduction of an additional reference beam makes the detection system more complex. Meanwhile, the need of ensure co-axial interference between the Gaussian beam and the vortex beam makes the optical path adjustment more difficult.

5.2.2 Interferometry for Helical Phase Measurement

The topological charge l, the eigen value of a vortex beam, determines its helical phase or wavefront distribution. Once we know the phase distribution of a vortex beam, we can directly obtain the topological charge or OAM it carries. This subsection describes an interferometric method of measuring the phase distribution of vortex beams that is applicable to both vortex beams and other structured optical fields.

By introducing a fundamental mode Gaussian beam interfere with a vortex beam, the helical phase of the vortex beam can be calculated by image processing [9].

Considering only two beams, which are denoted as

$$E_1 = |E_1| \cos \phi_1 \tag{5.2.8}$$

$$E_2 = |E_2| \cos \phi_2 \tag{5.2.9}$$

where $|E_1|$, $|E_2|$ are their amplitudes and ϕ_1, ϕ_2 are their phases, respectively. Note that the phase referred to here is the total phase, including helical phases. When E_1 and E_2 interfere coaxially, the intensity distribution of the interference field is:

$$|E_1 + E_2|^2 = |E_1|^2 + |E_2|^2 + 2|E_1||E_2| \cos(\phi_1 - \phi_2) \tag{5.2.10}$$

If an additional phase π/2 is introduced to the beam E_2, the intensity distribution of their interference fields becomes:

$$\left| E_1 + E_2 \exp\left(\frac{i\pi}{2}\right) \right|^2 = |E_1|^2 + |E_2|^2 + 2|E_1||E_2| \sin(\phi_1 - \phi_2) \tag{5.2.11}$$

From Eqs. (5.2.10) and (5.2.11), we have

$$\tan(\phi_1 - \phi_2) = \frac{\left|E_1 + E_2 \exp(\frac{i\pi}{2})\right|^2 - |E_1|^2 - |E_2|^2}{|E_1 + E_2|^2 - |E_1|^2 - |E_2|^2} \tag{5.2.12}$$

When the phase ϕ_2 of E_2 is known, the phase of E_1 can be found as follows.

$$\phi_1 = \phi_2 + \arctan\left[\frac{\left|E_1 + E_2\exp\left(\frac{i\pi}{2}\right)\right|^2 - |E_1|^2 - |E_2|^2}{|E_1 + E_2|^2 - |E_1|^2 - |E_2|^2}\right] \tag{5.2.13}$$

Note that the terms in the inverse tangent function in Eq. (5.2.13) are all intensity distributions, including intensities of E_1 and E_2, intensity of the E_1 and E_2 interferometric fields, and intensity of the E_1 and E_2 interferometric fields with an additional $\pi/2$ phase, all of which can be obtained with array detectors such as a CCD camera. By defining E_1 as the beam to be measured and E_2 as the reference beam, the phase distribution of the beam to be measured can be obtained directly by measuring the intensity of the two beams when they interfere coaxially.

When measuring the phase distribution of a vortex beam, the expanded fundamental Gaussian beam is usually chosen as the reference beam. Since the phase can be approximated as a plane, ϕ_2 in Eq. (5.2.13) is equate to 0. Thus, only the four intensity distributions in Eq. (5.2.13) need to be measured. The schematic diagram of the experimental system for measuring the phase distribution of a vortex beam is shown in Fig. 5.9. The vortex beam E_1 and the reference beam E_2 are combined coaxially to obtain the intensity distributions of the two interferometric fields $|E_1 + E_2|^2$ and $|E_1 + E_2\exp(i\pi/2)|^2$ respectively. The phase delay here can be achieved by a liquid crystal retarder.

Figure 5.10 gives an example of phase measurement through interferometry. The intensity distributions of the vortex beam ($|+5\rangle$) at the beam waist position ($z = 0$) $|E_1|^2$ and the reference Gaussian beam $|E_2|^2$, as well as the interferometric field intensity distributions $|E_1 + E_2|^2$, $|E_1 + E_2\exp(i\pi/2)|^2$ are given in Fig. 5.10a, respectively.

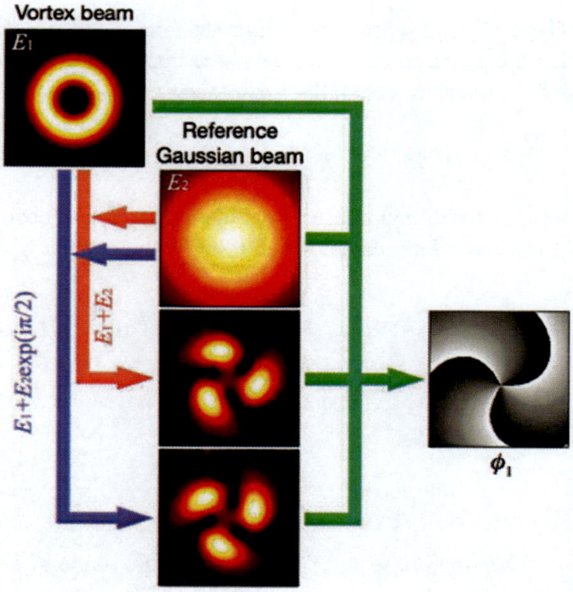

Fig. 5.9 Measuring the phase of a vortex beam by interferometry

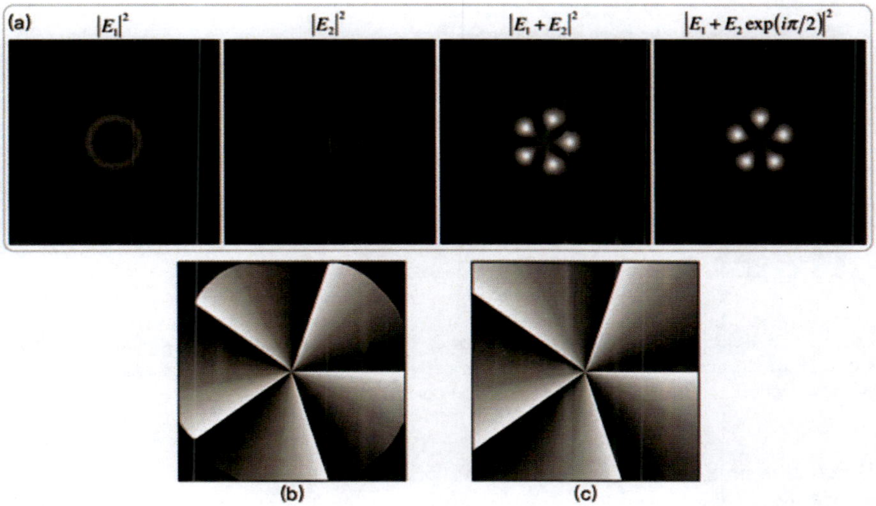

Fig. 5.10 Simulation results of the interferometric measurement of the vortex beam phase

The interferometric field pattern is rotated after an additional π/2 phase is introduced to the Gaussian beam. In addition, the intensity distributions of $|E_1|^2$ and $|E_2|^2$ here are normalized to the intensity maximum of the interferometric field and therefore appear slightly darker. Figure 5.10b gives the measured helical phase of the vortex beam through the intensity distribution given in (a) and Eq. (5.2.13); Fig. 5.10c shows the actual helical phase of the vortex beam. Figure 5.10 shows that the helical phase of the vortex beam calculated by Eq. (5.2.13) is in good agreement with the theoretical one.

5.2.3 *Young's Double-Slit Interference of Vortex Beams*

Young's double-slit interference is a very famous optical experiment. It shows that when a spherical wave from a point source passes through a double-slit, it produces bright and dark interference fringes on the receiver and the intensity distribution $I(x)$ of the fringes satisfies:

$$I(x) \infty \cos^2\left(\frac{\pi a x}{\lambda d}\right) \tag{5.2.14}$$

where λ is the wavelength of the light, a is the distance between the double slits and d is the distance from the double slits to the receiver. If the incident wave is a vortex beam with a helical wave front, there will be a phase difference along y-direction through the part of the double slit as shown in Fig. 5.11, contributing to interference fringe distortion. Note that the slit width is enlarged in Fig. 5.11 for clarity of display.

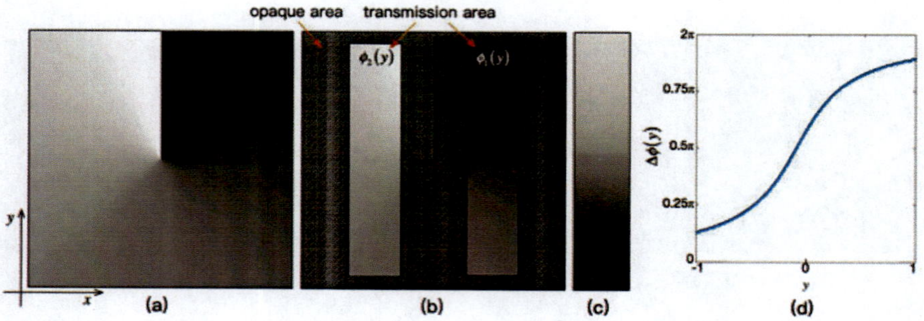

Fig. 5.11 **a** Helical phase of a +1st order vortex beam. **b** Transmitted helical phase of a double slit. **c** Phase difference in the y-axis direction between the double slits. **d** Phase difference in the y-axis direction as a function of the normalized position y

In actual double-slit interference the slit width should be taken to be very small. The magnitude of the phase difference in y direction is determined by the tilt degree of the helical wave front results from topological charge l of the vortex beam. Thus, the distortion degree of the interference fringe can be regarded as a function of the topological charge l. This suggests that topological charges can be diagnosed from the distortion degree of the interference fringe when a vortex beam is incident in a double slit [10, 11].

When a vortex beam illuminates the double slit, an additional phase difference in y-direction $\Delta\phi(y)$ result from the helical phase should be added to Eq. (5.2.14):

$$\Delta\phi(y) = \phi_2(y) - \phi_1(y) \tag{5.2.15}$$

From Fig. 5.12, it is easy to understand that the magnitude of $\Delta\phi(y)$ associated with coordinate y satisfies

$$\Delta\phi(y) = 2l\theta = 2l \arctan\frac{a}{2y} \tag{5.2.16}$$

Considering Eq. (5.2.14), the intensity distribution of Young's double slit interference of a vortex beam is obtained as

$$I(x, y) \propto \cos^2\left(\frac{\pi a x}{\lambda d} + \frac{\Delta\phi(y)}{2}\right) = \cos^2\left(\frac{\pi a x}{\lambda d} + l \arctan\frac{a}{2y}\right) \tag{5.2.17}$$

In Fig. 5.12, when $y \to +\infty$, $\theta \to 0$, the interference fringe distribution is the same as that when the Gaussian beam is incident. When $y \to -\infty$, $\theta \to \pi$, from Eq. (5.2.16) $\Delta\phi(y) = 2\pi l$, namely, an additional phase $2\pi l$ is introduced. According to the Yang's interference fringe conditions and Eq. (5.2.17), if introduce additional 2π phase to $\Delta\phi(y)$, the fringe has to be shifted by one position period, namely, a misalignment is created. Therefore, the fringe has to be shifted by l fringe when $y \to +\infty$, and l misalignments are created at the center of the interference fringe.

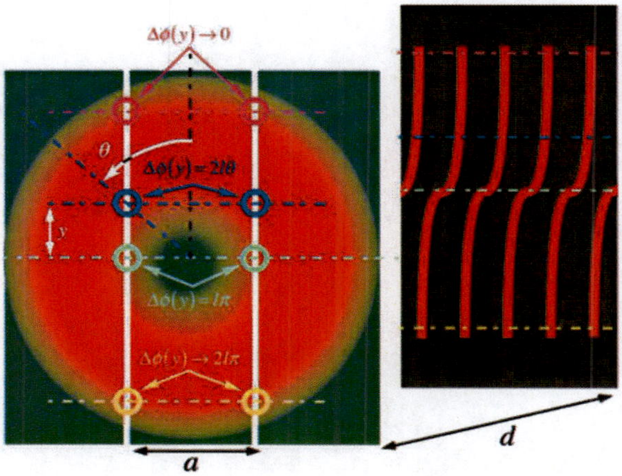

Fig. 5.12 Relationship between phases in double slit and position y when a l-th order vortex beam is incident in [11]

This means that the value of the topological charge l of a vortex beam can be obtained from the intensity distribution of the interferometric field by means of the relation given in Eq. (5.2.17).

Figure 5.13 shows the intensity distribution of the interferometric field of the vortex beam after incident in a double slit. A clear misalignment present and the number of misaligned stripes is equal to the absolute value of the topological charge. In addition, since vortex beams with opposite topological charges transmitted by the double slit have opposite phase differences along y-axis, the misalignment is in the opposite direction, and the sign of the topological charge can be determined. This shows that the topological charge of a vortex beam can be obtained intuitively from the Young's double-slit interference.

5.2.4 *OAM Mode Separation Through Mach-Zehnder Interferometry*

In addition to the direct measurement of topological charge, interferometry can also be used to separate OAM components of a multiplexed vortex beam and then further probe the separated single-mode vortex beam. The most typical interferometric device for separating OAM components of a multiplexed vortex beam is the Mach-Zehnder interferometer (M-Z interferometer) with Dove prisms in two arms [12, 13], as shown in Fig. 5.15.

In Fig. 5.14, vortex beams propagate along the z-axis, with the x- and y-axes perpendicular and parallel to the paper surface, respectively. The two beam splitter S1 and S2 with a 1:1 beam splitting ratio and two reflectors R1 and R2 are employed. The Dove prism DP1 is rotated by an angle$\pi/2$ with respect to the Dove prism DP2, meaning that beams passing through DP2 will be flipped with respect to the

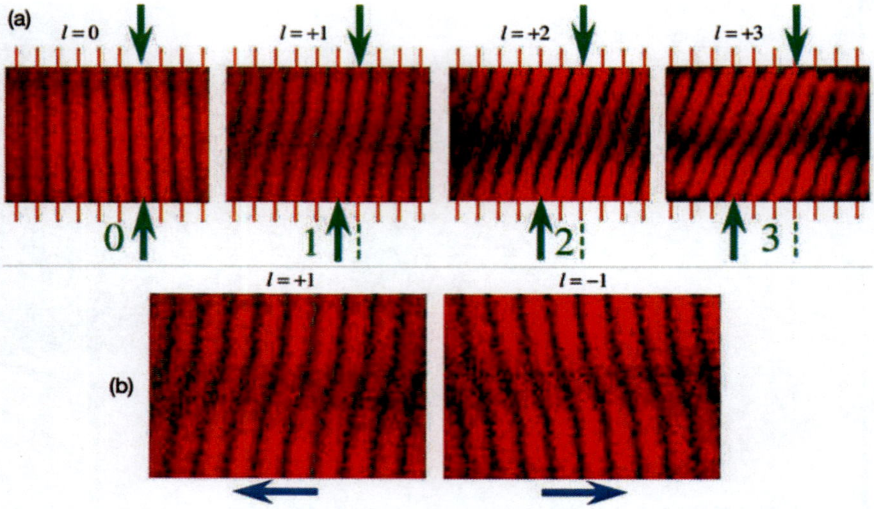

Fig. 5.13 Experimental results of vortex beam Young's double-slit interference. **a** The misalignment of the interference fringe is the same as the absolute value of the topological charge; **b** the misalignment direction shows the sign of the topological charge [11]

x-axis, while the same beam transmission through DP1 will be flipped and rotated by an angle π with respect to the x-axis. Since a vortex beam of topological charge l contains a phase factor $\exp(il\varphi)$, the helical phase term becomes $\exp[il(-\varphi)]$ when the vortex beam is transmitted through DP2, while the helical phase term becomes exp[il(π − φ)] when it is transmitted through the DP1. Obviously, a phase difference $l\pi$ is present when a vortex beam passing through DP1 and DP2 respectively.

The principle of separating the OAM of a vortex beam in an M-Z interferometer with two arms and a Dove prism can be understood by "interferometric phase extinction". Define two operators $S(\theta)$ and $R(\theta)$, when applied to the vortex beam $E(r, \varphi)$,which satisfy.

$$S(\theta)E(r, \varphi) = E(r, 2\theta - \varphi) \tag{5.2.18}$$

$$R(\theta)E(r, \varphi) = E(r, \theta + \varphi) \tag{5.2.19}$$

where $E(r, \varphi)$ can be expressed as:

$$E(r, \varphi) = E(r)\exp(il\varphi) \tag{5.2.20}$$

It is shown that $S(\theta)$ makes the light field symmetrical with respect to the line $\varphi = \theta$; while $R(\theta)$ a rotate the light field by an angle θ. Thus, the operator of the reflector can be expressed as $S(0)$, and the operator f of the Dove prism $D(\theta)$ can be expressed as

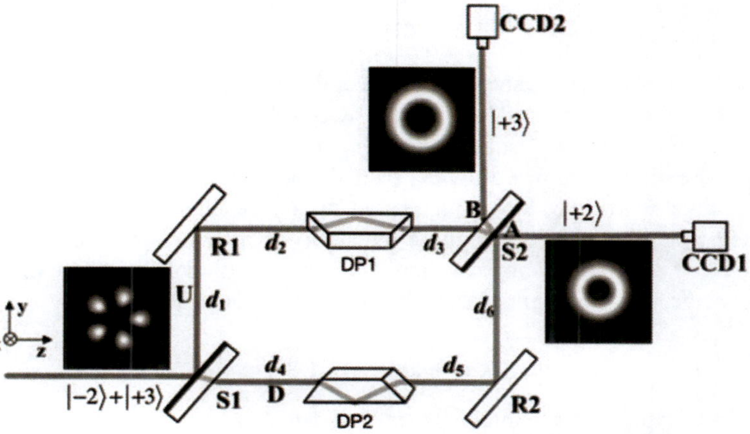

Fig. 5.14 System for separating OAM components of a multiplexed vortex beam by M-Z interferometry. S1 and S2: beam splitter; DP1 and DP2: Dove prisms; R1 and R2: reflector; CCD1 and CCD2: CCD cameras

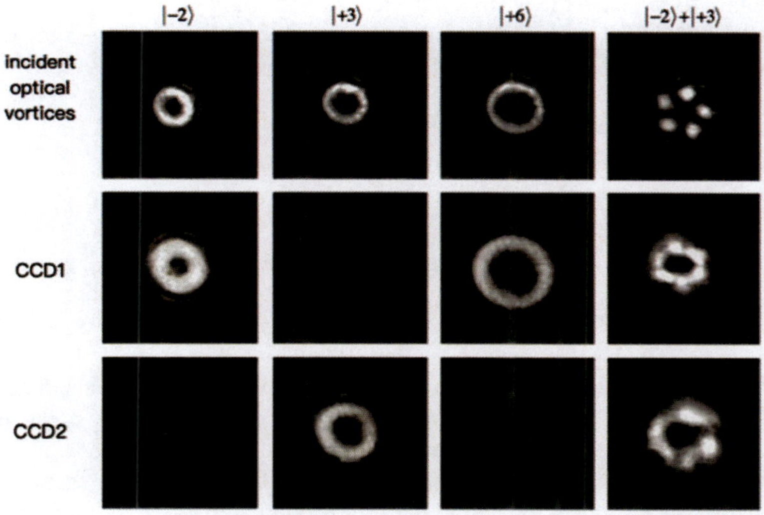

Fig. 5.15 Experimental results of separating OAM components of vortex beams using M-Z interferometry [13]

$$D(\theta)E(r,\varphi) = R(2\theta)S(0)E(r,\varphi) = E(r, 2\theta - \varphi) \tag{5.2.21}$$

In Fig. 5.14, d_1~d_6 indicate the distance between optical elements in the M-Z interferometer, by adjusting the optical path to satisfy:

$$d_1 + d_2 + d_3 = d_4 + d_5 + d_6 \tag{5.2.22}$$

If U and D are used to denote the upward optical path (S1 → R1 → DP1 → S2) and the downward optical path (S1 → DP2 → R2 → S2) of the M-Z interferometer, the two beams entering CCD1 undergo the following phase transition in their respective optical ranges. In the upward optical path U, the incident light is reflected by the beam splitter S1 to introduce a phase transition π result from the half wave loss, reflected by R1 to introduce a phase transition of π, transmitted through DP1 to introduce a fixed phase change σ_D and a phase delay $l\pi$, followed by a fixed phase change σ through the beam splitter S2, the process can be shown as:

$$\begin{aligned} E_1 &= \exp(i\sigma)\exp(ikd_3)\exp(i\sigma_D)D\left(\frac{\pi}{2}\right)\exp(ikd_2) \\ &\times \exp(i\pi)S(0)\exp(ikd_1)\exp(i\pi)S(0)E(r,\varphi) \\ &= E(r,\pi-\varphi)\exp\{i[2\pi+\sigma_D+\sigma+k(d_1+d_2+d_3)]\} \end{aligned} \tag{5.2.23}$$

In the downstream optical path D, the incident beam is transmitted through the beam splitter S1 to introduce a fixed phase change σ, transmitted through DP2 to introduce a fixed phase change σ_D, reflected by R2 to introduce a phase change of π, and reflected by the beam splitter S2 to introduce a phase change of π. Such process can be expressed as:

$$\begin{aligned} E_2 &= \exp(i\pi)S(0)\exp(ikd_6)\exp(i\pi)S(0)\exp(ikd_5) \\ &\times \exp(i\sigma_D)D(0)\exp(ikd_4)\exp(i\sigma)E(r,\varphi) \\ &= E(r,-\varphi)\exp\{i[2\pi+\sigma_D+\sigma+k(d_4+d_5+d_6)]\} \end{aligned} \tag{5.2.24}$$

Here, the field E_A at S2(A) is a superposition of E_1 and E_2, expressed as

$$E_A = E_1 + E_2 \tag{5.2.25}$$

From Eqs. (5.2.22)–(5.2.25), when the incident vortex beam has a helical phase $\exp(il\varphi)$, the phase difference of beams entering CCD1 between the upstream optical path U and the downstream optical path D is $l\pi$.

Similarly, for the two beams entering CCD2 the following phase changes occur in their respective optical ranges. In the upward optical path U, the incident beam is reflected by the beam splitter S1 to introduce a phase change of π, reflected by the R1 to introduce a phase change of π, transmitted through the DP1 to introduce a fixed phase change σ_D and a phase delay $l\pi$, transmitted through the beam splitter S2 to introduce a fixed phase change σ, reflected by the beam splitter S2 Reflected without a phase change (no half-wave loss if transmission from an optically dense medium to an optically sparse medium with), the second pass through beam splitter S2 introduces a fixed phase change σ.

$$
\begin{aligned}
E_3 &= \exp(i\sigma)S(0)\exp(i\sigma)\exp(ikd_3)\exp(i\sigma_D)D\left(\frac{\pi}{2}\right) \\
&\times \exp(ikd_2)\exp(i\pi)S(0)\exp(ikd_1)\exp(i\pi)S(0)E(r,\varphi) \\
&= E(r,\pi+\varphi)\exp\{i[2\pi+\sigma_D+2\sigma+k(d_1+d_2+d_3)]\}
\end{aligned} \tag{5.2.26}
$$

In the downstream optical path D, the incident beam transmits the beam splitter S1 to introduce a fixed phase change σ, transmits through DP2 to introduce a fixed phase change σ_D, reflected by R2 to introduce a phase change of π, and passes through the beam splitter S2 to introduce a fixed phase change σ.

$$
\begin{aligned}
E_4 &= \exp(i\sigma)\exp(ikd_6)\exp(i\pi)S(0)\exp(ikd_5) \\
&\times \exp(i\sigma_D)D(0)\exp(ikd_4)\exp(i\sigma)E(r,\varphi) \\
&= E(r,\varphi)\exp\{i[\pi+\sigma_D+2\sigma+k(d_4+d_5+d_6)]\}
\end{aligned} \tag{5.2.27}
$$

E_B at S2(B) is

$$
E_B = E_3 + E_4 \tag{5.2.28}
$$

For a vortex beam with a helical phase $\exp(il\varphi)$, the phase difference of beams entering CCD2 between the upstream optical path U and the downstream optical path D is exactly $(l+1)\pi$.

The above analysis shows that for a vortex beam with an even topological charge l, after passing through an M-Z interferometer with Dove prisms on both arms, at S2(B) the phase difference between the two arms is an odd multiple of π, producing an interferometric phase extinction, $E_B = 0$. Hence the incident beam will output from CCD1 end only. Similarly, for a vortex beam with an odd topological charge l, at S2(A) the phase difference between the two beams is an odd multiple of π, resulting in an interferometric phase extinction, $E_A = 0$. Hence the incident beam is output from CCD2 end only. Here it should also be noted that the beam transmission in both U and D paths from CCD1 end undergoes three times reflection, resulting in the opposite topological charge according to the mirroring property introduced in Sect. 2.2.4. While the beam transmission in the U and D paths from CCD2 end occurs four times and two times total reflection, respectively, leading to unchanged topological charge.

Figure 5.15 shows the experimental results of the vortex beam measurement using M-Z interferometer. When a two-fold multiplexed vortex beam whose topological charges is -2 and -6 is incident, the beam only output from CCD1 end. When a vortex beam with topological charge $+3$ is incident, the beam mainly output from CCD2 end. When the two-fold multiplexed vortex beam $(|-2\rangle + |+3\rangle)$ is incident, + 2 and $+3$ single-mode vortex beams are output at CCD1 and CCD2 ports respectively after M-Z interferometry. This confirms that the two-arm M-Z interferometer with Dove prisms separates vortex beams with odd and even topological charges.

There are some shortcomings of this scheme. It is invalid for multiplexed vortex beams whose topological charge is in same parity. In addition, this approach can only judge the order parity of single-mode vortex beams. The specific value of the topological charge can only be obtained by other mode diagnosing methods such as diffraction grating.

5.3 Diffractometry

When a vortex beam is incident in a specially designed diffraction grating, its far-field diffraction will show a special pattern associated with its topological charge. In other words, the topological charge of the incident vortex beam can be deduced from far-field diffraction. Common diffraction gratings for diagnosing vortex beams include triangular apertures, angular double slits, cylindrical lenses, periodic tape annular gratings, composite fork-shaped gratings, Dammann vortex gratings, etc.

5.3.1 *Triangular Aperture Diffraction*

Triangular aperture can be employed for OAM mode recognition, but it is only effective for single-mode vortex beams and invalid for multiplexed vortex beams [14, 15]. The triangular aperture is an amplitude grating with a triangular transmission region, as shown in Fig. 5.16, where the black part is opaque and the white part is transmissive.

The far-field diffraction of a vortex beam through a triangular aperture shows a special triangular spots array, the direction of which illustrate the sign of topological charge, and the number of spots in the array illustrate the absolute value of the topological charge. This means that the topological charge of the incident vortex beam can be determined by the number of spots in the far-field. The principle can be simply understood as the far-field diffraction of a triangular aperture can be regarded as the interference of the far-field diffraction of three sides of the triangular aperture [14], as shown in Fig. 5.17. Figure 5.17a–c show the far-field diffraction pattern when a +1st order vortex beam ($l = 1$) passing through the three sides of the triangular

Fig. 5.16 Triangular-aperture amplitude grating

aperture, respectively. Figure 5.17d shows the triangular spot array obtained after interference, which is actually the far-field diffraction when vortex beam ($l = 1$) passing through the triangular aperture. Figure 5.18 gives the theoretical simulation and experimental results of the far-field diffraction patterns when +1st, +2nd and + 3rd order vortex beams are incident in. When the +1st order vortex beam is diffracted through the triangular aperture, the far-field diffraction spot is also in triangular shape, and the number of spots on each side minus one is the absolute value of the topological charge. When vortex beams with topological charges of opposite sign are diffracted by a triangular aperture, the spots are found to be arranged in opposite directions, as shown in Fig. 5.19.

Similar to the triangular aperture, the triangular slit amplitude grating shown in Fig. 5.20 can achieve similar functions [16] and gives sharper diffraction results than the triangular aperture. In the triangular slit, a new parameter characterizing the slit

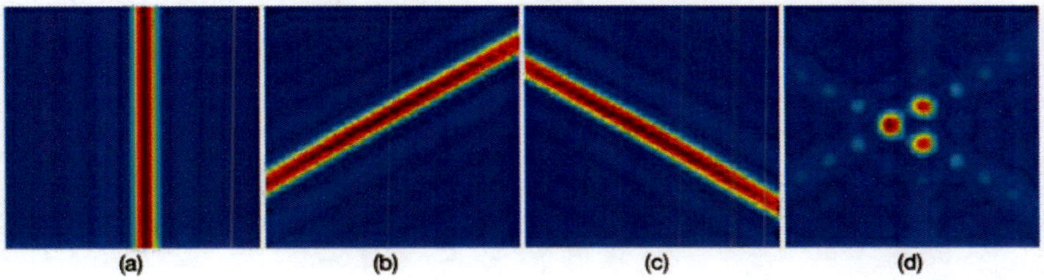

Fig. 5.17 Far-field diffraction patterns when a vortex beam ($l = -1$) passing through three sides of the triangular aperture: **a** bottom side; **b** right and **c** left. **d** is the triangular spot array interfered from the three fields of (a)(b)(c),namely, the far-field diffraction pattern a triangular aperture [14]

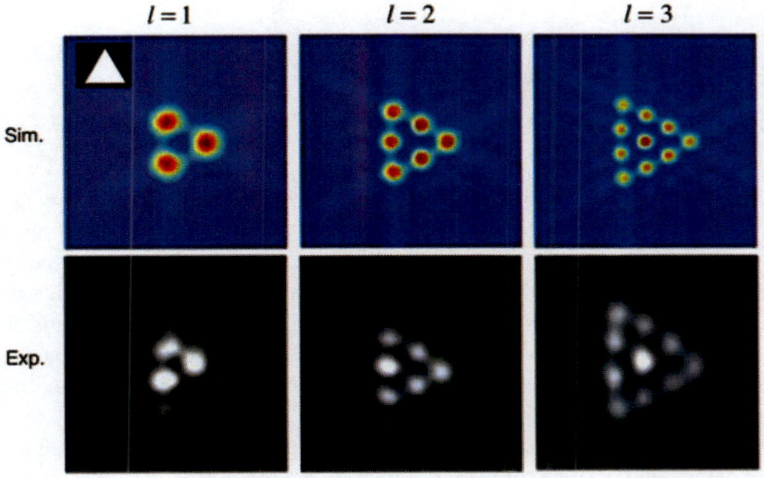

Fig. 5.18 Simulation and experimental results of the diffraction patterns when various vortex beams passing through a triangular aperture [14]

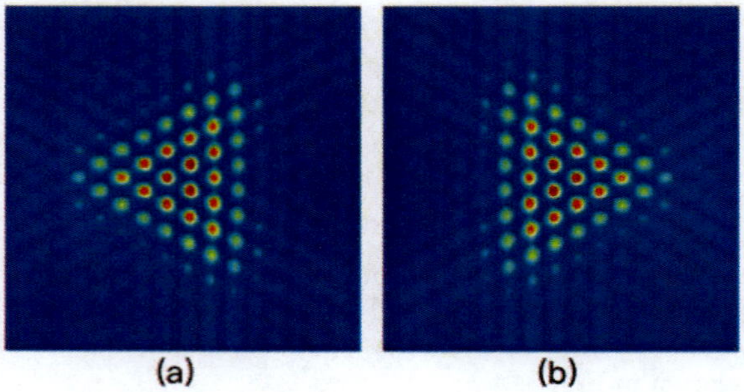

Fig. 5.19 Simulation results of diffraction patterns when vortex beams with opposite topological charges (|l| = 7) passing through a triangular aperture: **a** $l = -7$; **b** $l = 7$ [14]

Fig. 5.20 Triangular slit amplitude grating

width is introduced, namely the ratio η of the inner and outer edge lengths of the slit. When $\eta = 0$, the triangular slit degenerates to a triangular aperture. Studies in Ref. [16] have shown that the larger η is, the clearer the far-field diffraction, as shown in Fig. 5.21, where (a)–(c), $\eta = 0$, and (d)–(f), $\eta = 0.8$.

5.3.2 *Angular Double Slit Diffraction*

The double slits mentioned in Sect. 5.2.3 are precisely in the direction of the x-axis in Cartesian coordinates. By replacing the Cartesian coordinates with polar coordinates and introducing double slits in the φ direction, angular double slits are formed, as shown in Fig. 5.22, where black indicates the non-transmissive part and white the transmissive part. The angular double slit is also an amplitude grating, which only modulates the amplitude of the incident light, and its diffraction field can be interpreted as the interference field formed by the beams transmitted from the angular double slit.

Similar to Young's double slit, the angular double slit can also be used to detect vortex beams, which also exploits the principle that the vortex beams transmitted by the two slits have different phases [17]. As shown in Fig. 5.22, when a vortex beam with topological charge l is incident, the phase difference between the paths q_1p_1

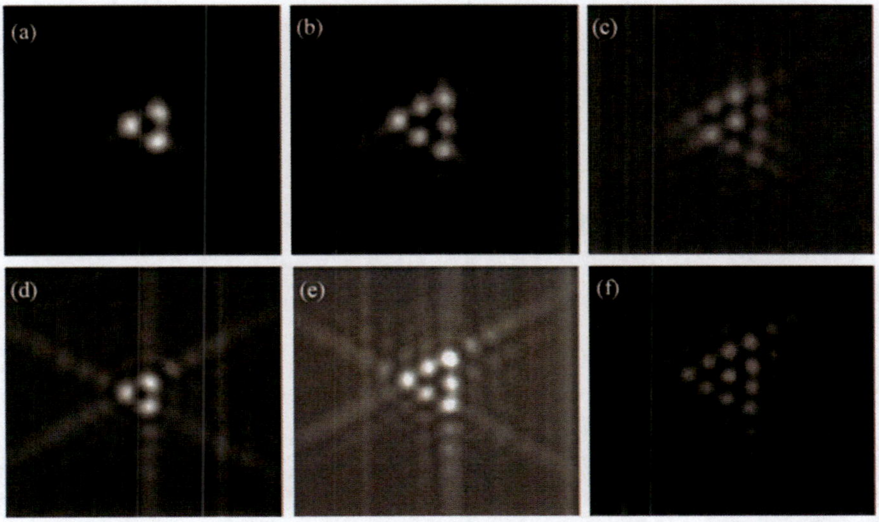

Fig. 5.21 Far-field diffraction patterns when vortex beams passing through triangular slits. **a–c** $\eta = 0$; **d–f** $\eta = 0.8$; **a**, **d**: $l = 1$; **b**, **e**: $l = 2$; **c**, **f**: $l = 3$ [16]

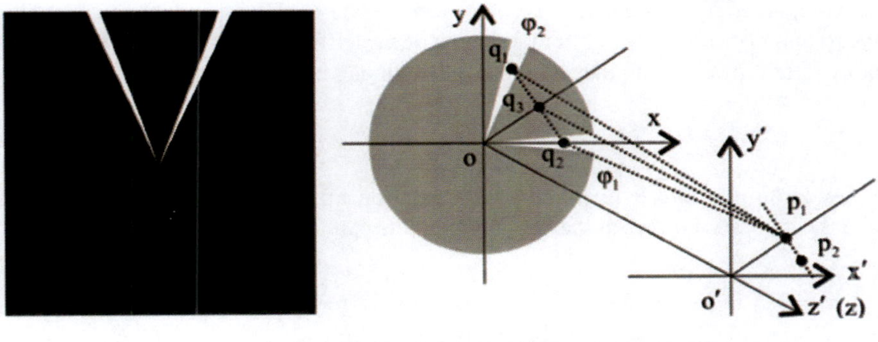

Fig. 5.22 Schematic diagram of the angular double slit and its diffraction [17]

and q_2p_1 can be expressed as:

$$\Delta\phi = l\Delta\varphi + 2\pi\frac{|q_1p_1| - |q_2p_1|}{\lambda} \tag{5.3.1}$$

where $\Delta\varphi = \varphi_2 - \varphi_1$ is the angle between the two slits and λ is the wavelength of the light. When $\Delta\phi = N\pi$ and N is even, a bright fringe appears on the receiving surface axis $o'p_1$. If N is odd, a dark fringe appears. Since $\Delta\phi$ is associated with the topological charge l, the OAM of the incident beam can be determined from the morphology of the fringe. Figure 5.23 shows the experimental results of measuring OAM through angular double slit, with the dashed line indicating the axis $o'p_1$.

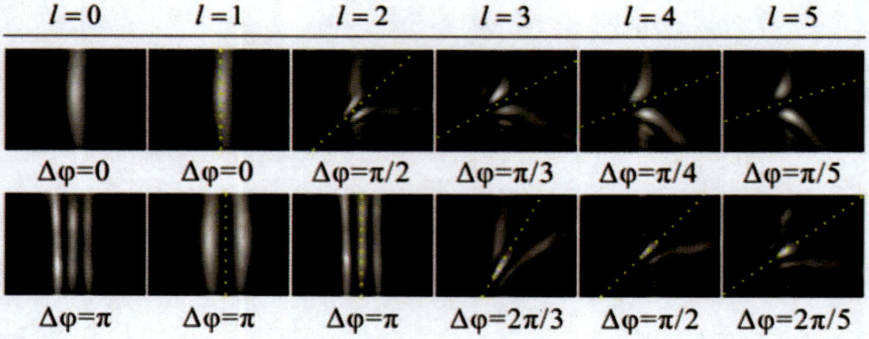

Fig. 5.23 Experimental results of vortex beam OAM detection using angular double slits [17]

The angular bisector of the angular slit, light and dark fringes appear in different directions. Hence, the topological charge of the incident beam can be determined from the movement of the fringes and the movement direction.

Diagnosing OAM mode of vortex beams by angular double slit diffraction requires the collection of a series of diffraction patterns and is inaccurate sufficiently. Improvement can be done on this basis to obtain more accurate results [18]. In this detection scheme, an additional phase σ is introduced to one of the slits while ensuring that $|q_1p_1|$ and $|q_2p_1|$ in Fig. 5.22 remain unchanged. Difference between the two beams at p_1 after diffraction by the angular double slit can be expressed as

$$\Delta\phi = l\Delta\varphi + \sigma \tag{5.3.2}$$

According to interference theory, the relationship between the intensity I at p_1 and the angle $\Delta\varphi$ between the two angular slits can be obtained as a function:

$$I = |E_1 + E_2|^2 \propto 2\cos^2\frac{\Delta\phi}{2} = (1 + \cos\Delta\phi) = \left\{1 + \cos\left[l\left(\Delta\varphi + \frac{\sigma}{l}\right)\right]\right\} \tag{5.3.3}$$

Equation (5.3.3) shows that when $\Delta\varphi \in [0, 2\pi]$, $|l|$ peaks appear in the I-$\Delta\varphi$ curve, so the absolute value of the topological charge l can be determined from the number of peaks. The $\Delta\varphi$-I curve will shift if the value of σ is changed, and the direction shift is associated with the sign of l. Therefore, the sign of the topological charge l can be determined by changing the direction of the observed curve shift.

5.3.3 Cylindrical Lens

All the previous introduced schemes in this section use amplitude gratings like aperture or slit. Actually, OAM mode diagnostic can also be achieved through common optical lenses.

A cylindrical lens is an aspherical lens which can effectively reduce spherical and chromatic aberrations and has a one-dimensional magnification with a phase distribution function of

$$\phi(x, y) = k\left(f - \sqrt{f^2 + x^2}\right) \tag{5.3.4}$$

with f the focal length.

In Chap. 3, Sect. 3.2.1 of this book, a mode converter consisting of two or three cylindrical lenses is described, which converts a Hermit Gauss beam into a vortex beam. Depending on the reversibility of the optical path, the mode converter must also be able to convert the vortex beam into a Hermit Gauss beam. The order of the vortex beam generated by the mode converter is associated with the values of the order m and n of the incident Hermit Gauss beam [Eq. (3.2.10)]. The intensity pattern of a Hermit Gauss beam is distributed in an ordered structure of multiple spots associated with its order. Therefore, the vortex beam can be transformed into a similar structure by means of a cylindrical lens. Its topological charge can then be determined directly from the diffraction distribution [19–22].

Figure 5.24 shows the Diffraction patterns when various vortex beams are focused by a cylindrical lens. The focused patterns have structures similar with that of Hermit Gauss beams, and its shape is determined by the topological charge of the incident vortex beam. The number of spots minus one is the absolute value of the topological charge $|l|$. And the spot arranged orientation illustrate the sign of the topological charge. Therefore, the vortex mode can be acquired directly by cylindrical lens focusing.

The principle of using a cylindrical lens to detect vortex beams is also illustrated as follows. Since the vortex beam has a helical wavefront, its Poynting vector direction is not parallel to the optical axis, but a function related to the radial coordinate r. When a vortex beam is incident in a cylindrical lens, the wavefront gradient or the Poynting vector at two points of central symmetry, is symmetry with respect to optical axis. These two points are focused at different positions on the focal plane. Thus forming a pattern similar with a Hermit-Gauss beam.

In addition to the detection rules described above, the topological charge of the incident beam can be calculated theoretically by analyzing the pattern at the back focal plane. This analysis is more complex and less straightforward than direct observation of the intensity distribution of the focused pattern, but it is of great importance for the detection of new structural vortex fields with non-integer topological charges [22].

According to the relationship between angular momentum and momentum, the angular momentum of a vortex beam propagating along optical axis (z-axis) reads

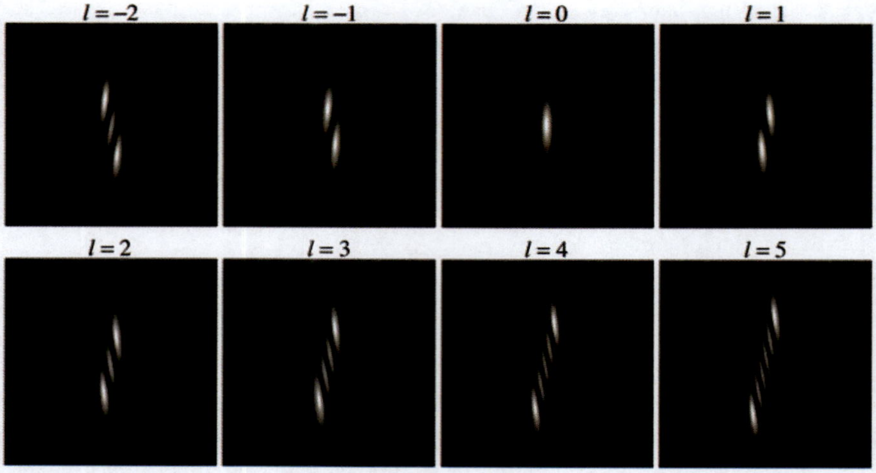

Fig. 5.24 Diffraction patterns when various vortex beams are focused by a cylindrical lens

$$L_z = \vec{r} \times \vec{p} \cdot \hat{z} = xp_y - yp_x = \hbar\left(xk_y - yk_x\right) \tag{5.3.5}$$

where $\hbar$ is the reduced Planck constant. p_x and p_y are the momentum components in the x and y-axis directions respectively. k_x and k_y are the optical wave numbers in the x and y-axis directions respectively. The initial diffraction plane cylindrical lens located is xOy, the back focal plane of the cylindrical lens is $x'Oy$. The focal length of the cylindrical lens be f. Then, from geometrical optics:

$$\tan\theta = \frac{x'}{f} = \frac{k_x}{k_z} \tag{5.3.6}$$

Thus,

$$k_x = \frac{k_z x'}{f} \tag{5.3.7}$$

where $k_z = k\cos\gamma$, γ is the angle between the Poynting vector and the optical axis. For a vortex beam, Eq. (1.2.26) shows that γ is negligible compared to the optical wave number k, so $\cos\gamma = 1$, $k_z = k$. Thereby, Eq. (5.3.7) reduces to

$$k_x = \frac{2\pi x'}{\lambda f} \tag{5.3.8}$$

Similarly,

$$k_y = \frac{2\pi y'}{\lambda f} \tag{5.3.9}$$

Then Eq. (5.3.5) can be written as

$$L_z = \frac{2\pi\hbar}{\lambda f}(xy' - yx') \tag{5.3.10}$$

Therefore, the average angular momentum of each photon in this vortex beam is

$$\langle L_z \rangle = \frac{2\pi\hbar}{\lambda f}(\langle xy' \rangle - \langle yx' \rangle) \tag{5.3.11}$$

where <ζ> represents an average of ζ. Since the cylindrical lens is focused in only one direction (in this case the x-axis direction), the light field on its back focal plane can be seen as a one-dimensional Fourier transformation of the incident beam. Therefore, $\langle xy' \rangle$ and $\langle yx' \rangle$ can be deduced from the diffraction distribution at the back focal plane, which in turn leads to $\langle L_z \rangle$. Consider the covariance $V_{x',y}$ of x' with y of intensity distribution at the back focal plane [22]:

$$V_{x',y} = -\langle yx' \rangle = \frac{\iint_\infty I(x', y)_l x' y dx' dy}{\iint_\infty I(x', y)_l dx' dy} \tag{5.3.12}$$

where $I(x', y)_l$ denotes the intensity distribution of a l-th order vortex beam with focused by a cylindrical lens. From the relationship between the topological charge l and the OAM of a single photon:

$$\langle L_z \rangle = l\hbar \tag{5.3.13}$$

one can get:

$$l = \frac{2\pi\hbar}{\lambda f}(V_{x',y} - V_{x,y'}) \tag{5.3.14}$$

The focused intensity distribution has a mirror symmetry about the central line (x = 0), thus $V_{x,y'} = -V_{x'y}$. Then substituting Eq. (5.3.12) into Eq. (5.3.14):

$$l = \frac{4\pi}{f\lambda} V_{x',y} = \frac{4\pi}{f\lambda} \frac{\iint_\infty I(x', y)_l x' y dx' dy}{\iint_\infty I(x', y)_l dx' dy} \tag{5.3.15}$$

Equation (5.3.15) shows that, the topological charge of a beam can be accurately calculated from the focused intensity distribution of a cylindrical lens.

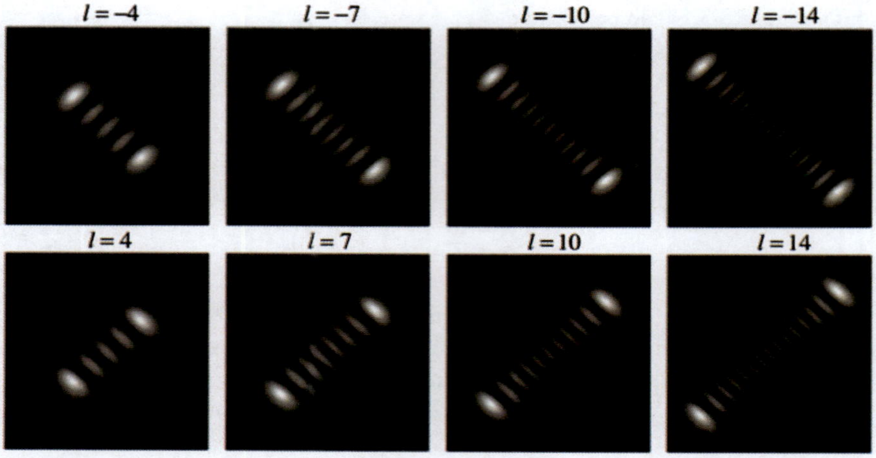

Fig. 5.25 Simulation results of OAM mode detection through a tilted lens [23]

5.3.4 Tilted Lens

Measuring OAM by a cylindrical lens can actually be understood as introducing different Fourier transformations to the optical field in both x and y directions. Similarly, qualitative detection of vortex beams can be achieved using a tilted placed convex lens [23]. Meanwhile, the tilted lens OAM measurement method is also applicable to the perfect vortex beam to be described in Chap. 8 [24].

A tilted convex lens has different equivalent focal lengths in the x and y directions, so that different Fourier transformations are accomplished for incident beams in x and y directions. Eventually, the incident vortex beam is diffracted into a field similar to Hermit Gauss mode, enabling the OAM detection of the vortex beam. Figure 5.25 gives the diffraction field when vortex beams passing through a tilted lens, which is very similar to that of a cylindrical lens. However, readers should be aware that here the diffraction field is not obtained at the back focal plane, but at another plane near the focal plane [23].

5.3.5 Gradually-Changing-Period Grating

Recently, researchers have proposed a method to measure the OAM state of a vortex beam using a gradually-changing-period grating (GCPG) [25]. As shown in Fig. 5.26, a GCPG is an amplitude grating that contains two types. Their transmittance functions are

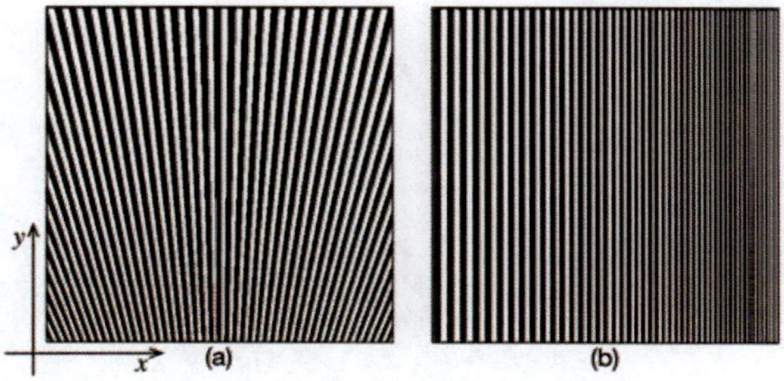

Fig. 5.26 **a** Type I and **b** type II gradually-changing-period grating

$$T_1(x,y)=\begin{cases}1, \text{ if } \cos\left[\frac{2\pi x}{d_0+ny}\right]\geq 0\\ 0, \text{ if } \cos\left[\frac{2\pi x}{d_0+ny}\right]<0\end{cases} \tag{5.3.16}$$

and

$$T_2(x,y)=\begin{cases}1, \text{ if } \cos\left[\frac{2\pi x}{d_0+nx}\right]\geq 0\\ 0, \text{ if } \cos\left[\frac{2\pi x}{d_0+nx}\right]<0\end{cases} \tag{5.3.17}$$

respectively. In Eqs. (5.3.16) and (5.3.17), d_0 is the grating constant at $x = 0$. n is the gradually changing factor, indicating the rate of the grating constant change. Gratings with the transmittance function shown in Eq. (5.3.16) have a period that varies asymptotically along y-axis and are known as Type I gradually-changing-period grating. Gratings with the transmittance function shown in Eq. (5.3.16), whose period varies asymptotically along the x-axis, are known as Type II gradually-changing-period grating. When a vortex beam is illuminated by a gradually-changing-period grating, the far field exhibits a diffraction field distribution that correlates with the topological charge of the incident vortex beam. Hence the topological charge can be acquired from the diffraction field.

According to scalar diffraction theory, the intensity distribution of the far-field diffraction is given in Fig. 5.27a, b when different vortex beams are diffracted by type I and type II gradually-changing-period gratings, respectively. The far-field diffraction can be seen as a number of subspots arranged in a particular position, with darker nodal lines between adjacent subspots whose number and direction are associated with the topological charge of the incident vortex beam. When the topological charge of the incident vortex beam is positive, the diffracted spots arrange in vertical at +1 diffraction order and horizontal at −1 diffraction order for the first type gradually-changing-period grating, and left down to right up at +1 diffraction order and left up to right down at −1 diffraction order for the second type gradually-changing-period grating. When the order of the incident vortex beam is negative, the diffracted spots arrange in horizontal at +1 level and vertical at −1 level for the first type of gradually-changing-period grating, and top-left to bottom-right at +1 level and bottom-left to top-right for the −1 level at second type of gradually-changing-period

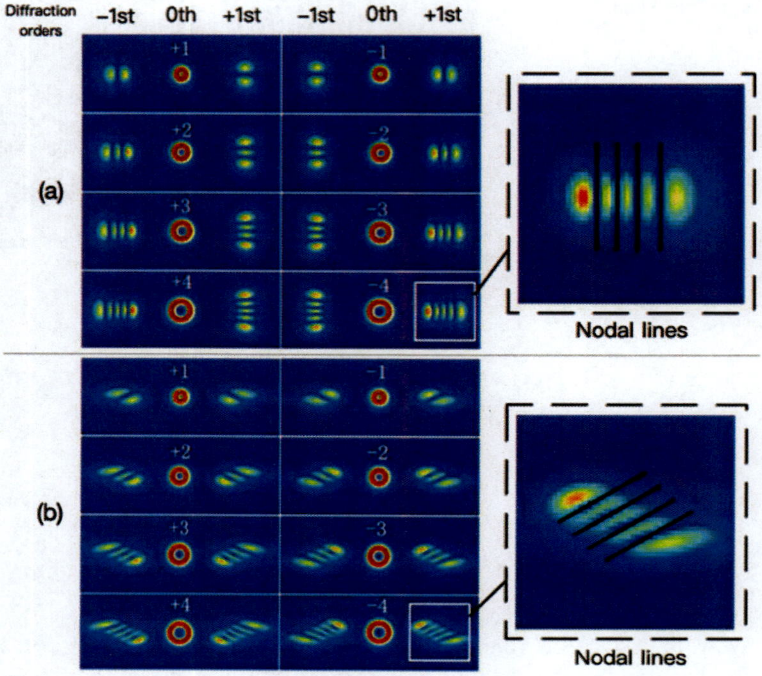

Fig. 5.27 Far-field diffraction patterns when different orders of vortex beams illuminate type I and type II gradually-changing-period grating, respectively [25]

grating. Secondly, the number of nodal lines in is associated with the absolute value of the topological charge. Therefore, for both Type I and Type II gradually-changing-period grating, the topological charge of the incident vortex beam can be measured by observing the +1 and −1 diffracted orders.

Since the gradually-changing-period grating is an amplitude grating, it can be produced in the same way as a fork grating, for example, laser etching. In experiments, gratings can also be obtained by printing directly onto hard cellophanes, as shown in Fig. 5.28a, which is easy and inexpensive to produce. Figure 5.28b shows the experimental far-field diffraction patterns when vortex beams incident a Type I gradually-changing-period grating printed on a hard cellophanes, which is in good agreement with the simulated results presented in Fig. 5.30.

Compared with other diffraction-based OAM detecting schemes, the main advantages of the gradually-changing-period grating are the lower optical path collimation requirements, thus ease of adjustment. The determination of OAM is still available if the incident vortex beam deviates from gradually-changing-period grating center, as shown in Fig. 5.29. Although easy to use, there are still limitations as the low diffraction efficiency and the inability to detect multi-mode multiplexed vortex beams.

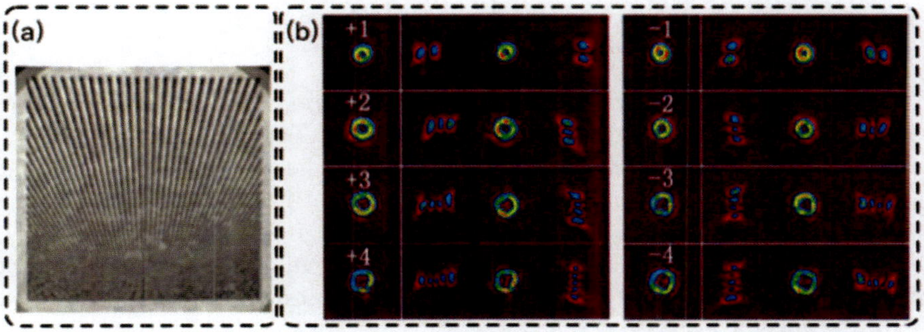

Fig. 5.28 Experimentally employed gradually-changing-period grating and its far-field diffraction [25]

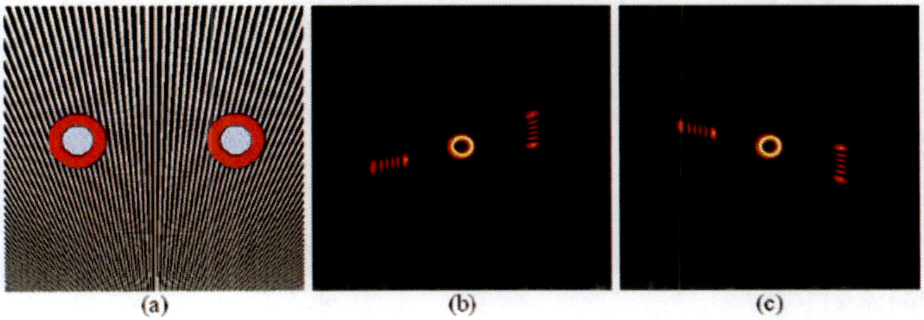

Fig. 5.29 Simulated far-field diffraction when a vortex beam is incidents in different positions of a gradually-changing-period grating. **a** Incident positions. **b** and **c** are the simulation results for both times

Fig. 5.30 Phase-only gradually-changing-period grating [26]

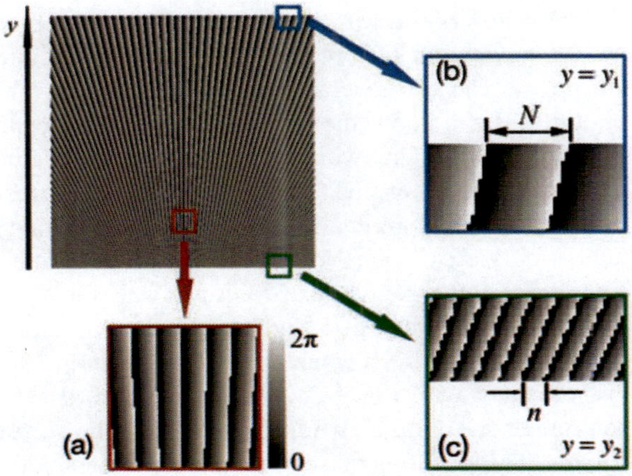

5.3.6 Phase-Only Gradually-Changing-Period Grating

When employing a **gradually-changing-period grating** to measure OAM, all we really need is +1st or −1st diffraction order in far-field. For an amplitude gradually-changing-period grating with a slit width ratio of 0.5, the intensity efficiency of the +1st or −1st diffraction order is only 10.13%. The diffraction efficiency is low. To overcome this issue, a phase-only gradually-changing-period grating [26, 27] is designed based on the gradually-changing-period grating, which diffracts all the incident beams into the first diffraction order, thus resulting in a significant increase in diffraction efficiency.

The phase distribution of a phase-only gradually-changing-period grating is [26]

$$\phi(x, y) = 2\pi \cdot frac(\frac{x}{a + by}) \tag{5.3.18}$$

where *frac*(ζ) denotes taking the fractional part of ζ. a and b are two parameters determining the gradient of the grating constant. The phase-only gradually-changing-period grating generated according to Eq. (5.3.18) is shown in Fig. 5.30.

How to determine values a and b of Eq. (5.3.18) is discussed below. In Fig. 5.30, the maximum value of y-axis is y_1 and the minimum value is y_2, and grating constants are N at $y = y_1$ and n at $y = y_2$, satisfying $N > n$. Since in this phase-only periodically asymptotic diffraction grating, the grating constants vary linearly with the y-axis, it follows:

$$\begin{cases} N = a + by_1 \\ n = a + by_2 \end{cases} \tag{5.3.19}$$

The values of the parameters a and b can be solved by means of Eq. (5.3.19).

The fabrication of a phase grating is very complex, the holographic grating of the device can be generated and encoded on a phase-only liquid crystal SLM. Figure 5.31a, b show the simulated and experimental results of far-field diffractions of vortex beams with positive and negative topological charges, respectively. Obviously, the experimental results are in good agreement with the simulation.

The phase-only gradually-changing-period grating can also be employed to detect multi-ring vortex beams (radial index $p \neq 0$) because of its higher diffraction efficiency. For multi-ring vortex beams, the radial index p should be determined as well as the topological charge l. Figure 5.32 gives the simulated and experimental results of far-field diffractions when multi-ring vortex beams are incident in.

Compared with Fig. 5.31, the distribution of sub-spots in far field arranged in one row or one column, while that of multi-ring vortex beams is multi-row or multi-column. This phenomenon can be directly understood as each ring of the vortex beam is diffracted into a row or column. Therefore, if the sub-spot of the diffraction field is $m \times n$, the number of concentric circles should be the same as the smaller ones of m and n. The number of concentric circles of a vortex beam equals $p + 1$, thus it is

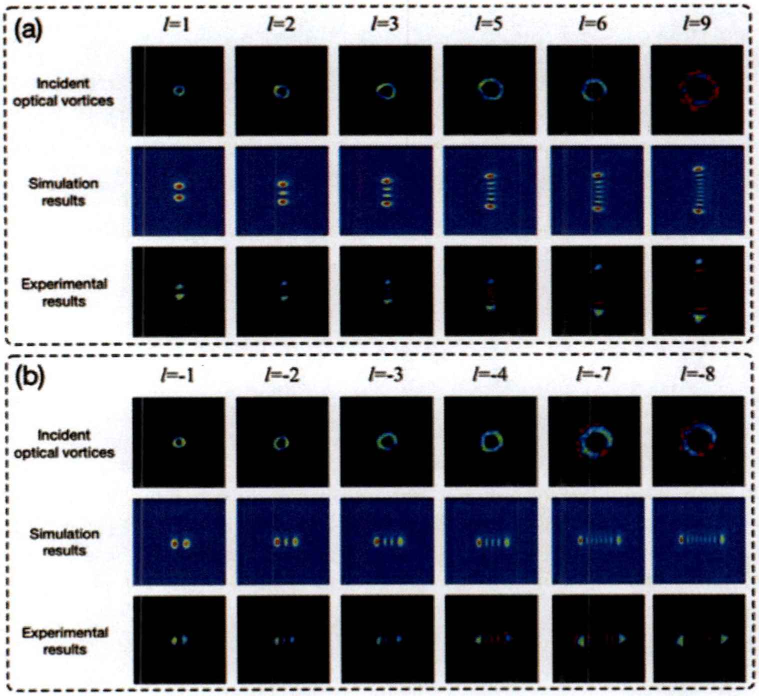

Fig. 5.31 Simulated and experimental results of far-field diffractions when vortex beams with positive and negative topological charges are incident in the phase-only gradually-changing-period grating given in Fig. 5.30 [26]

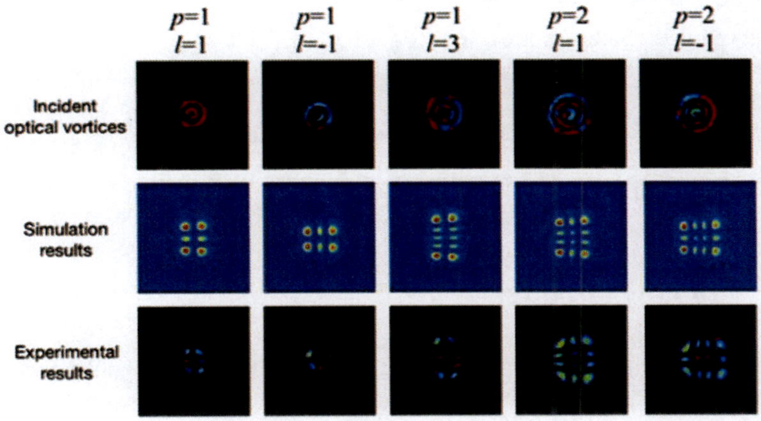

Fig. 5.32 Simulated and experimental results of far-field diffractions when multi-ring vortex beams are incident in the phase-only gradually-changing-period grating given in Fig. 5.30 [26]

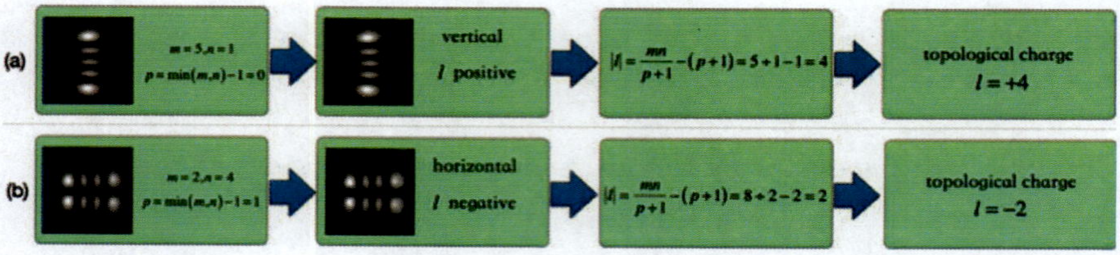

Fig. 5.33 Examples to recognize the topological charge from the far-field diffraction of a phase-only gradually-changing-period grating

easy to obtain:

$$p = \min(m, n) - 1 \tag{5.3.20}$$

where $\min(m, n)$ denotes taking smaller value.

Similar to gradually-changing-period grating, the sign of the topological charge l can be determined by observing the direction of the diffracted light spot arrangement, which is negative for horizontal arrangement and positive for vertical arrangement. The number of sub-spots in the diffracted field is associated with $|l|$ and the radial index p:

$$|l| = \frac{mn}{p+1} - (p+1) \tag{5.3.21}$$

In summary, a general method for deducing the topological charge l of the vortex beam from the far-field diffracted light field of a phase-only gradually-changing-period grating can be derived as follows:

Determine the radial index p from Eq. (5.3.20).
Determine the sign of the topological charge l from the orientation of the sun-spot arrangement.
Determine the absolute value of the topological charge $|l|$ according to Eq. (5.3.21).

Based on above steps, two examples of probing are given in Fig. 5.33 for reference.

5.3.7 *Annular Grating*

Annular grating has a ring structure [28], as shown in Fig. 5.34. Annular grating also has two types as amplitude and phase, and their transmittance functions are:

$$T_A(r) = \begin{cases} 1, \text{ if } \cos(2\pi r / \Lambda) \geq 0 \\ 0, \text{ if} \cos(2\pi r / \Lambda) < 0 \end{cases} \tag{5.3.22}$$

$$T_p(r) = \exp(i2\pi r / \Lambda) \tag{5.3.23}$$

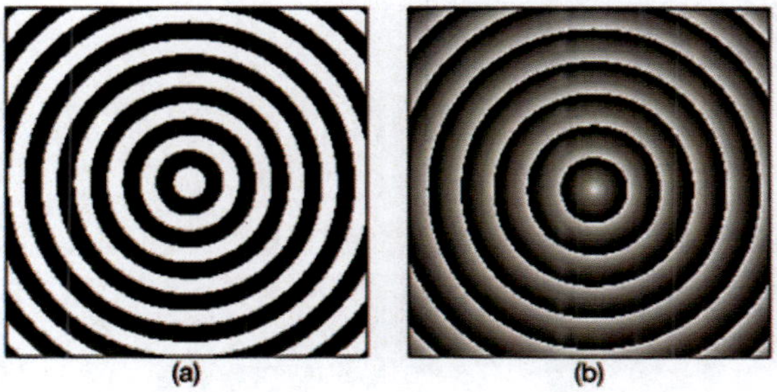

Fig. 5.34 Annular grating: **a** amplitude-type; **b** phase-only

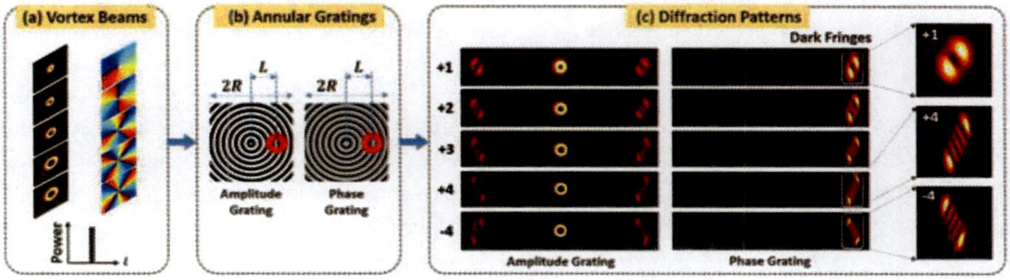

Fig. 5.35 Vortex beam mode recognition through an annular grating [28]

with r the radial coordinate and Λ the annular grating period.

Equation (5.3.23) is identical to Eq. (3.7.1) in Sect. 3.7.1, indicating that a phase-only annular grating has the same phase distribution function as an axicon. In other words, a phase-only annular grating is actually a 0th-order axicon. As mentioned in Chap. 3, a higher order BG beam can be generated when the vortex beam is diffracted by an axicon. If a vortex beam is incident off-axis but parallel to the optical axis, the diffraction pattern is associated with the topological charge, as shown in Fig. 5.35.

When a vortex beam passes off-axis but parallel to the optical axis through the annular grating, the diffraction patterns is very similar with the second type gradually-changing-period grating in Sect. 5.3.5 as shown in Fig. 5.36. The determination of the topological charge is also the same as that of the second type gradually-changing-period grating and will not repeated here.

5.3.8 Composite Fork-Shaped Grating Measurement

Previous discussed schemes are very effective for single-mode vortex beams, but they appear to be powerless for multiplexed vortex beams. The composite forked-shaped grating can solve this problem very well [29, 30].

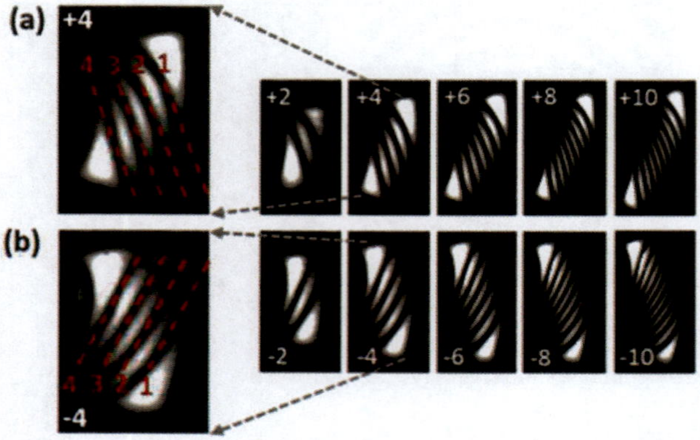

Fig. 5.36 Experimental results of vortex beam mode recognition a phase-only annular grating [28]

As described in Chap. 4, composite forked-shaped gratings can be used to generate 3×3 vortex beam arrays. The detection of vortex beams using a composite fork-shaped grating can actually be seen as the inverse of the generation process. From Eq. (4.1.8), the topological charges $l_{bx,\ by}$ of the vortex beam at the diffraction order (b_x, b_y) when a l-th order vortex beam are incident in reads:

$$l_{b_x,b_y} = l + b_x l_x + b_y l_y \tag{5.3.24}$$

where l_x and l_y are the forks number in x-axis and y-axis, respectively. In particular, if the central dislocation number of the composite forked-shaped grating is matched to the topological charge l of the incident light, i.e., if $l_{bx,by} = 0$ in Eq. (5.3.24), it must be possible to degenerate the vortex beam at the diffraction order (b_x, b_y) and obtain a fundamental mode beam with zero topological charge.

$$|l\rangle \cdot \exp\left[i\left(b_x l_x + b_y l_y\right)\varphi\right] \to |0\rangle \tag{5.3.25}$$

There is no longer a phase singularity at the beam center, so instead of having a ring, the beam presents a bright solid structure, as shown in Fig. 5.37. The topological charge of the incident vortex beam can be obtained according to the position of the bright spot in the far-field:

$$l = -b_x l_x - b_y l_y \tag{5.3.26}$$

Figure 5.38 gives the mode recognition results of vortex beams using a composite forked-shaped grating with the following parameters: $l_x = 1$, $l_y = 3$. In its far-field diffraction, the topological charge of the incident beam can be calculated by simply finding the location of the solid spot and combining the diffraction order (b_x, b_y) with Eq. (5.3.26).

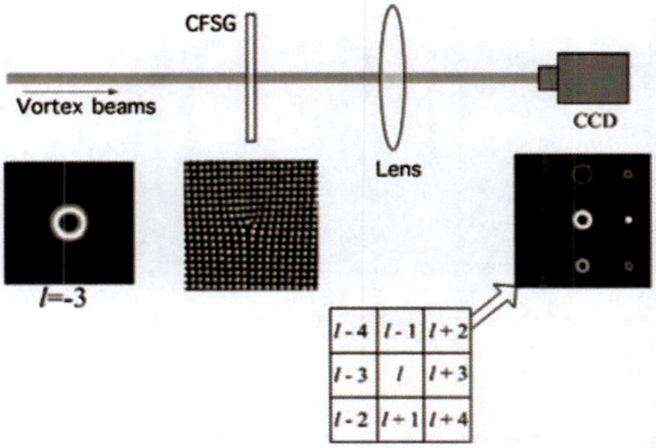

Fig. 5.37 Vortex beam mode recognition using a composite forked-shaped grating. CFSG, composite fork-shaped grating

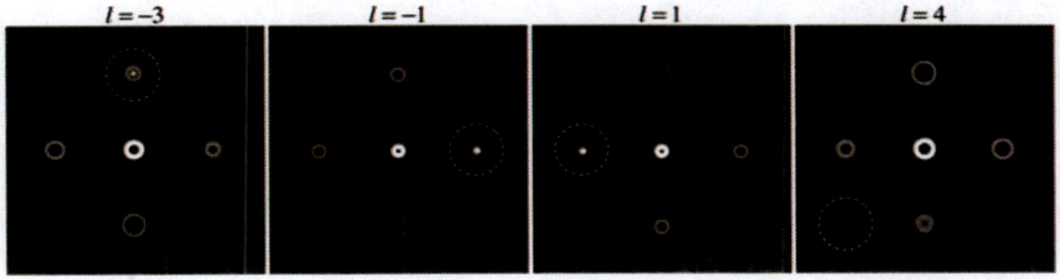

Fig. 5.38 Simulation far-field diffractions when single-mode vortex beams passing through a composite fork-shaped grating ($l_x = 1, l_y = 3$)

The composite fork-shaped grating is not only valid for single-ring vortex beams, but also available for multi-ring vortex beams with nonzero radial index p. A multi-ring vortex beam with 0 topological charge has multiple concentric circles over a solid central bright spot, so when a multi-ring vortex beam is diffracted by a composite fork-shaped grating, the topological charge can be determined in the same way as for a single ring, as shown in Fig. 5.39.

The nature of measuring OAM through a composite fork-shaped grating is to find where the bright spot present. When an N-fold multiplexed vortex beam is incident in, the diffraction field can be simply expressed as

$$\left(\sum_{a=1}^{N} |l_a\rangle\right) \cdot \exp\left[i\left(b_x l_x + b_y l_y\right)\varphi\right] \rightarrow \sum_{a=1}^{N} \left|l_a + b_x l_x + b_y l_y\right\rangle \tag{5.3.27}$$

If one of the terms on the right-hand side of Eq. (5.3.26) satisfies $l_a + b_x l_x + b_y l_y = 0$, the multiplexed vortex beam will contain a 0-th order, namely, a bright spot present as the beam center. This illustrates that composite forked-shaped grating are

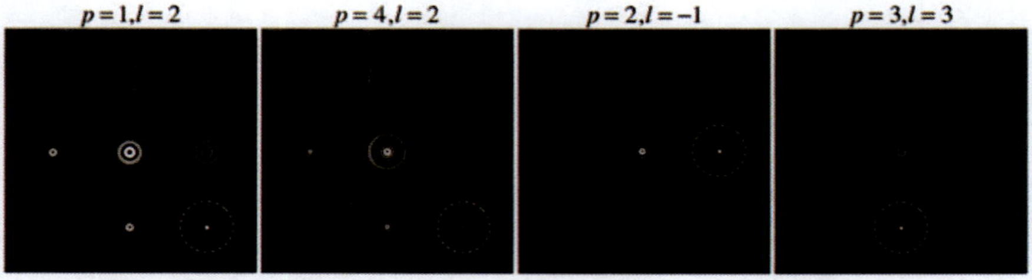

Fig. 5.39 Simulation far-field diffractions when multi-ring vortex beams passing through a composite fork-shaped grating ($l_x = 1, l_y = 3$)

also valid for multiplexed vortex beams. When a multiplexed vortex beam is incident, the far-field diffraction results in multiple solid spots. Depending on the position of the solid spots, the individual OAM components contained in the incident beam to be measured. Figure 5.40 shows the far-field diffraction distribution when a two-fold multiplexed vortex beam ($|{-4}\rangle + |2\rangle$) is incident in a composite forked-shaped grating. The presence of two solid spots in the diffraction field can be clearly seen in the figure.

The high diffraction orders of composite forked-shaped grating are uneasy to observe due to low diffraction efficiency. Moreover, the diffraction field contains only 9 clearly observable diffraction orders. So the continuous detectable topological charge range is only −4~+4. In practical applications of vortex beams, the orders used are not limited to −4~+4, but also contain higher orders, limiting the practical use of composite forked-shaped grating measurements.

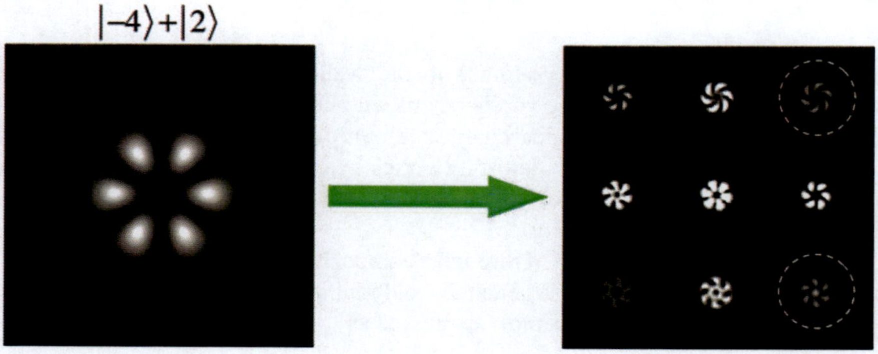

Fig. 5.40 Simulation far-field diffractions when two-fold multiplexed vortex beams passing through a composite fork-shaped grating ($l_x = 1, l_y = 3$)

5.3.9 Standard Dammann Vortex Grating Measurement

The $m \times m$ standard Dammann grating [31] introduced in Sect. 4.3.1 is based on Dammam vortex grating, which satisfies that m is odd, $l_x = 1, l_y =$ m, and the grating constants are the same in both orthogonal directions. When illuminated by a Gaussian beam, its far-field diffraction has m^2 vortex beams and their topological charges are $-(m^2 - 1)/2$~$(m^2 - 1)/2$ from the lower left to the upper right. Similar like composite forked-shaped grating s, when illuminated by a vortex beam, the topological charges $l_{bx,\,by}$ of the vortex beams located at the diffraction order (b_x, b_y) is:

$$l_{b_x,b_y} = l + b_x + mb_y \tag{5.3.28}$$

If $l_{bx,\,by} = 0$ exist in a certain diffraction order, then then a solid bright spot will appear at this diffraction order and the topological charge of the incident vortex beam can be deduced.

Standard Dammann vortex grating measurements are based on exactly the same principles as composite fork-shaped grating measurements. However, the significant advantage of the standard Dammann vortex grating is that the continuous detectable range of topological charges of the vortex beams can be greatly extended by setting a reasonable value of m. For example, if $m = 5$, the diffraction field has 25 vortex beams with topological charge distribution of −12~+12. If $m = 9$, the diffraction field has 81 vortex beams with topological charge distribution of −40~+40. In other words, the continuous detectable topological charge range can be −40~+40. Even though the continuous detectable range of the standard Dammann vortex beam is the same as that of the composite forked-shaped grating at $m = 3$, the intensity of the nine diffraction orders is equal and still has an advantage over the composite forked-shaped grating.

As the Dammann vortex grating is a $0 - \pi$ binarized phase-only grating, it is more complex to produce and is generally simulated by liquid crystal SLMs. As discussed previously, when performing grating design, the larger m is, the larger the topological charge detection range is. However, when m is relatively large, the high spatial frequency phase jump at the grating center makes it difficult to produce with the high accuracy. Furthermore, even if a liquid crystal SLM is employed, the resolution of existing SLMs cannot match such high spatial frequency phase jump. Therefore, m is usually set to 5 [31]. In this case, the continuous topological charge detectable range is −12~+12. Figure 5.41 gives the simulation results of using a 5 × 5 standard Dammann vortex grating to detect vortex beams, which is also valid for a multiplexed vortex beam.

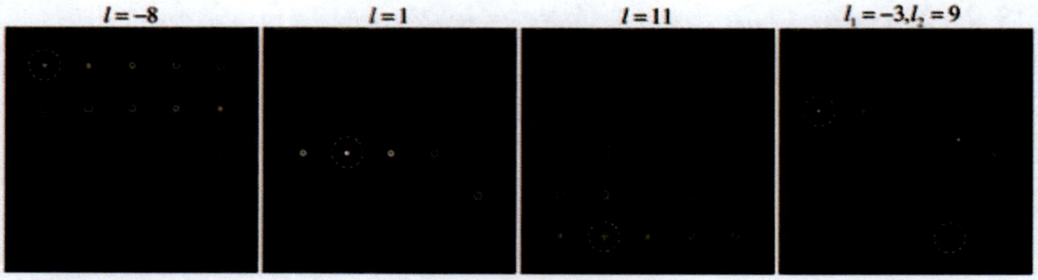

Fig. 5.41 Simulation results of vortex beam mode detection using a 5 × 5 standard Dammann vortex grating

5.3.10 Integrated Dammann Vortex Grating Measurement

The 5 × 5 standard Dammann vortex grating has a continuous detectable topological charge range of −12~+12. To further extend the detection range, $m \times m$ standard Dammann vortex gratings with larger m values can be proposed. The theory of this approach is simple, but is difficult to implement. As described in the previous subsection, such gratings require a high fabrication accuracy. In addition, the area of the far-field diffracted array is getting larger and larger as m increases. Nevertheless, the topological charge detection range of the vortex beam can be extended by integrating a Dammann vortex grating [32].

Section 4.3.1 has already introduced the integrated Dammann vortex grating, which is a phase-only grating consists of a standard $m \times m$ Dammann vortex grating and s vortex grating with $+(m^2 - 1)/2$ order or $-(m^2 - 1)/2$ order. When a Gaussian beam is incident in a integrated Dammann grating, the introduction of additional vortex phase leads to the topological charge shift of each diffraction order in far-field, resulting in a m^2-way vortex beam with topological charges ranging from $-(m^2 - 1)$~0 or 0~$(m^2 - 1)$. Figure 5.42 shows the generation of an integrated Darman vortex grating at $m = 5$, where the order of the superposed phase-only vortex grating is +12 or −12.

When a Gaussian beam is incident in the integrated Dammann vortex grating shown in Fig. 5.42d, the far-field diffraction pattern is a 5 × 5 vortex beams array. All the diffraction orders in the array have identical power, and the OAM states ranges from 0 to +24 from the lower left to the upper right, as shown in Fig. 5.43a. Similarly, if the integrated Dammann grating in Fig. 5.42e is used, the OAM states distribution in far-field is 0 to −24 from the upper right to the lower left, as shown in Fig. 5.43b.

The principle of detecting vortex beams with an integrated Dammann vortex grating is identical to that of the composite fork-shaped grating and standard Dammann vortex grating, as finding the location where the bright center spot emerge in far-field. It is easy to understand that the integrated Dammann vortex grating given in Fig. 5.42 extends the range of continuous topological charges detection to −24~+24. Since it consists of two parts, Fig. 5.42d, e, it is necessary to first split the incident vortex beam and then incident them to two integrated gratings and

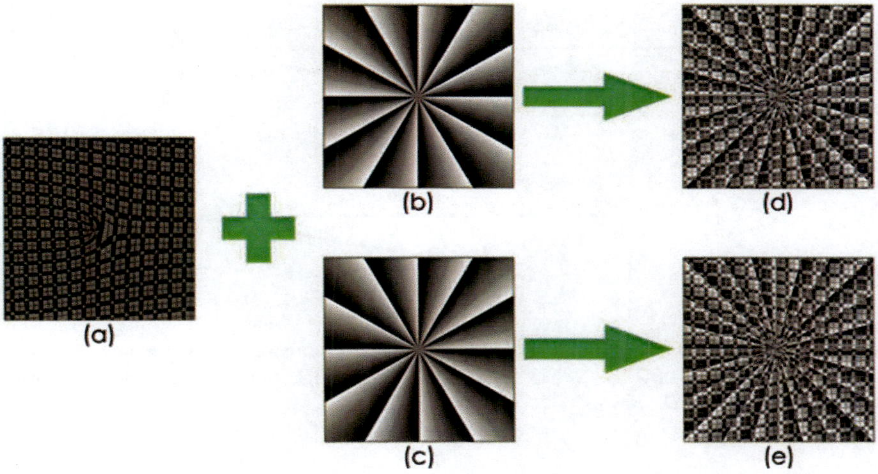

Fig. 5.42 Integrating Dammann vortex gratings. **a** 5 × 5 standard Dammann vortex grating; **b** + 12th order phase-only vortex grating; **c** −12th order phase-only vortex grating; **d** and **e** integrated Dammann vortex grating [32]

Fig. 5.43 Far-field topological charge distributions of integrated Dammann vortex grating shown in Fig. 5.42d, e, respectively, when a Gaussian beam is incident in

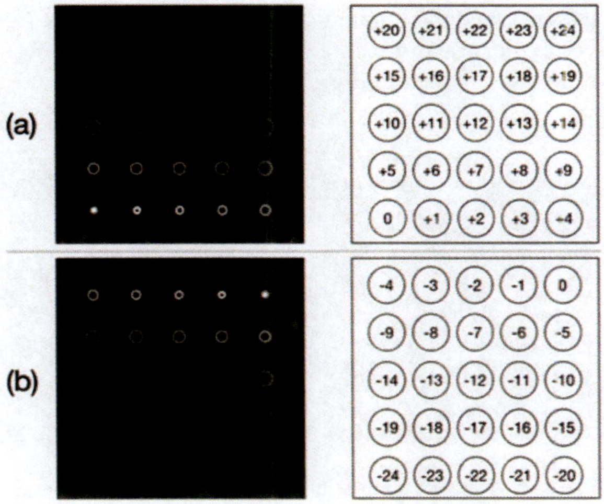

observe the far-field diffraction simultaneously. Figure 5.44 shows a typical setup for detecting a vortex beam using an integrated Dammann vortex grating, in which two liquid crystal SLMs simulate the two gratings given in Fig. 5.42d, e, respectively. The far-field diffraction is then observed by CCD1 and CCD2 to measure the topological charge distribution of the incident vortex beam. It is important to note that the reflector ® in Fig. 5.44 is essential, as the topological charge of the vortex beam in the reflected path of the beam splitter (BS) is opposite compared to the original

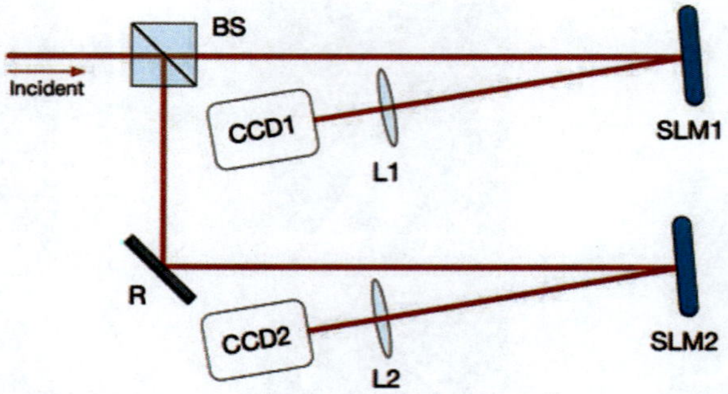

Fig. 5.44 Setup of OAM measurement using integrated Dammann vortex gratings. BS, beam splitter; R, reflector; SLM1 and SLM2, liquid crystal spatial light modulators; L1 and L2, lenses; CCD1 and CCD2, CCD cameras

due to the mirror reflection. Hence, an additional reflection must be introduced to compensate.

Figure 5.45 gives the simulation results of the detecting a three-fold multiplexed vortex beam ($|-7\rangle + |4\rangle + |18\rangle$) using integrated Dammann vortex gratings in Fig. 5.42. A total of three center bright spots appear in the two far-field diffraction patterns, and the topological charge distribution or OAM component of the incident vortex beam can be obtained directly by comparing it with the topological charge distribution given in Fig. 5.43.

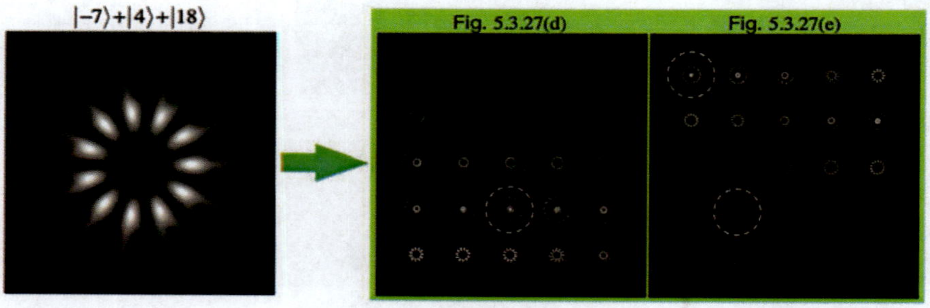

Fig. 5.45 Simulated far-field diffraction when a three-fold multiplexed vortex beam passing through integrated Dammann vortex gratings in Fig. 5.42

5.4 Polarimetry

5.4.1 Spatially Variable Half-Wave Plates Measurement

The spatially variable half-wave plates (SVHWP) introduced in Sect. 3.4.1 can be used not only for generating vortex beams but also for detecting vortex beams [33, 34]. Ideally, when the number of sub-half wave plates that make up the SVHWP $M \to \infty$, the Jones matrix of the SVHWP according to Eq. (3.4.3) can be written as

$$J_{SVHWP}(\varphi) = \begin{bmatrix} \cos m\varphi & \sin m\varphi \\ \sin m\varphi & -\cos m\varphi \end{bmatrix} \tag{5.4.1}$$

where φ is the azimuthal angle. For a 1st order SVHWP, when a horizontally linearly polarized vortex beam of topological charge l is incident in, the output reads:

$$\begin{aligned} E &= \begin{bmatrix} \cos m\varphi & \sin m\varphi \\ \sin m\varphi & -\cos m\varphi \end{bmatrix} \exp(il\varphi) \begin{bmatrix} 1 \\ 0 \end{bmatrix} = \exp(il\varphi) \begin{bmatrix} \cos m\varphi \\ \sin m\varphi \end{bmatrix} \\ &= \frac{1}{2}\left\{ \exp[i(l+m)\varphi] \begin{bmatrix} 1 \\ -i \end{bmatrix} + \exp[i(l-m)\varphi] \begin{bmatrix} 1 \\ i \end{bmatrix} \right\} \end{aligned} \tag{5.4.2}$$

Equation (5.4.2) shows that the output beam is a superposition of a right-hand circularly polarized vortex beam $|l + m\rangle$ and a left-hand circularly polarized vortex beam $|l - m\rangle$. When $l = \pm m$, one of the beams degenerates into a fundamental mode and its ring structure disappears, resulting in a solid bright spot.

Figure 5.46 gives the output intensity distributions when vortex beams of different orders is incident in SVHWPs of order 1 to 3 (sub-half wave plate number $M = 16$). The absolute value of the topological charge of the vortex beam $|l|$ can be determined by detecting the presence of a solid bright spot in the center of the output beam, but it is not possible to distinguish between positive and negative. Therefore, if a SVHWP is employed to detect the vortex beam, additional elements need to be introduced to separate the two orthogonal circular polarizations.

Two orthogonal circular polarizations can be converted into orthogonal linear polarization through a quarter wave plate [$J_{\lambda/4}(0°)$] with 0° arrangement fast axis. The converted linear polarizations are 45° and 135° respectively. A half wave plate [$J_{\lambda/4}(22.5°)$] with 22.5° arranged fast axis is placed behind to rotate the polarizations the two beams into horizontal and vertical, which are then separated by a Wollaston prism. The whole process can be expressed as follows.

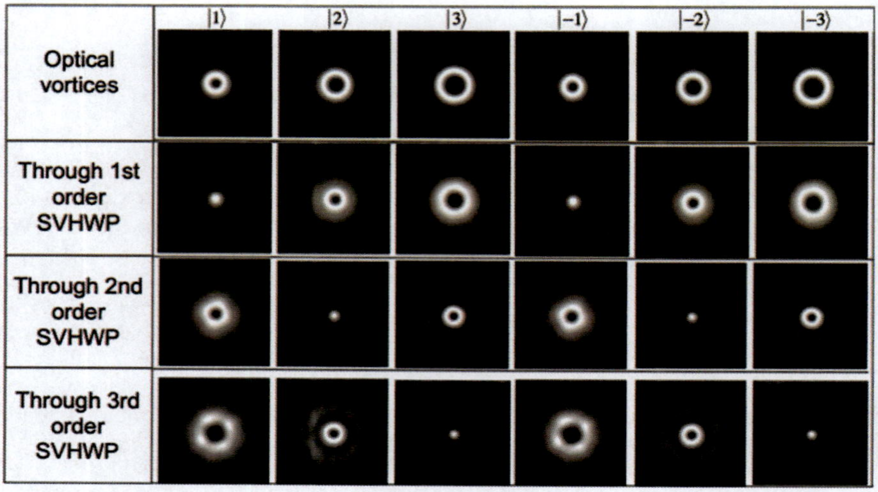

Fig. 5.46 Intensity distributions when various vortex beams passing through SVHWPs of orders [34]

$$
\begin{aligned}
E_{out} &= J_{\lambda/2}(22.5^\circ)\cdot J_{\lambda/4}(0^\circ)\cdot E \\
&= \frac{1}{2}J_{\lambda/2}(22.5^\circ)\cdot J_{\lambda/4}(0^\circ)\left\{\exp[i(l+m)\varphi]\begin{bmatrix}1\\-i\end{bmatrix}+\exp[i(l-m)\varphi]\begin{bmatrix}1\\i\end{bmatrix}\right\} \\
&= \frac{1}{2}J_{\lambda/2}(22.5^\circ)\left\{\exp[i(l+m)\varphi]\begin{bmatrix}1\\1\end{bmatrix}+\exp[i(l-m)\varphi]\begin{bmatrix}1\\-1\end{bmatrix}\right\} \\
&= \exp[i(l+m)\varphi]\begin{bmatrix}1\\0\end{bmatrix}+\exp[i(l-m)\varphi]\begin{bmatrix}0\\1\end{bmatrix}
\end{aligned}
\tag{5.4.3}
$$

Finally, the two separated beams are imaged on the CCD camera through a lens. The OAM detection can be achieved by observing the patterns of the two beams on the CCD camera. The whole detection system is shown in Fig. 5.47.

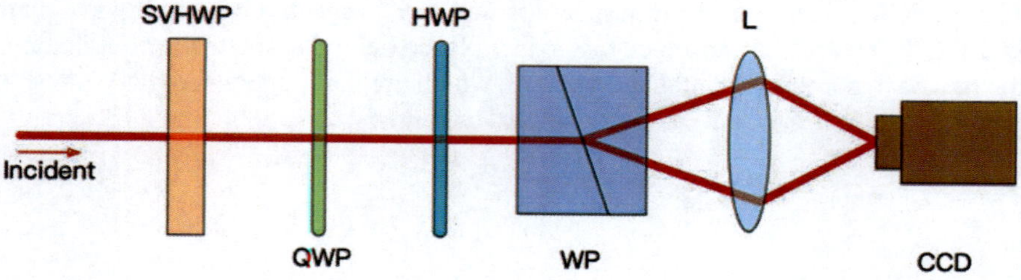

Fig. 5.47 Diagram for the measurement of topological charge by SVHWP. QWP, quarter wave plate; HWP, half wave plate; WP, Wollaston prism; L, lens; CCD, CCD camera

Simulation results and experimental results for the detecting vortex beams using first and second order SVHWP are presented in Figs. 5.48 and 5.49, respectively. Four vortex beams $|1\rangle$, $|-1\rangle$, $|2\rangle$, and $|-2\rangle$, are transformed into fundamental modes and vortex beams of other orders, after passing through the detection system. The position of the bright spot is associated with the positive or negative topological charge l of the incident beam. The bright spot is on the left for positive topological charge l, and the bright spot is on the right for negative topological charge l. In Figs. 5.48 and 5.49 the topological charges of the two beams after passing through the detection system have been marked with numbers.

The examples given in Figs. 5.48 and 5.49 can detect only four different orders of vortex beams, but this method can also be used to detect any order of vortex beams if higher order SVHWP are produced. Based on the above principle, a parallel detection system can be designed as shown in Fig. 5.50, thus enabling real-time detection of different vortex beams and multiplexed vortex beams.

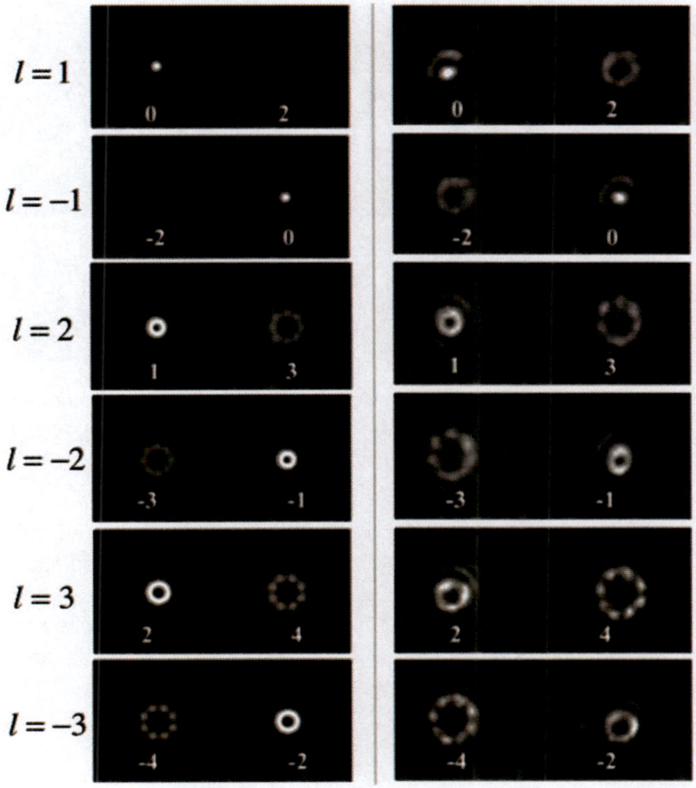

Fig. 5.48 Far-field patterns of the vortex beam after passing through a first-order SVHWP, where the left-hand side is the simulation result and the right-hand side is the experimental result [34]

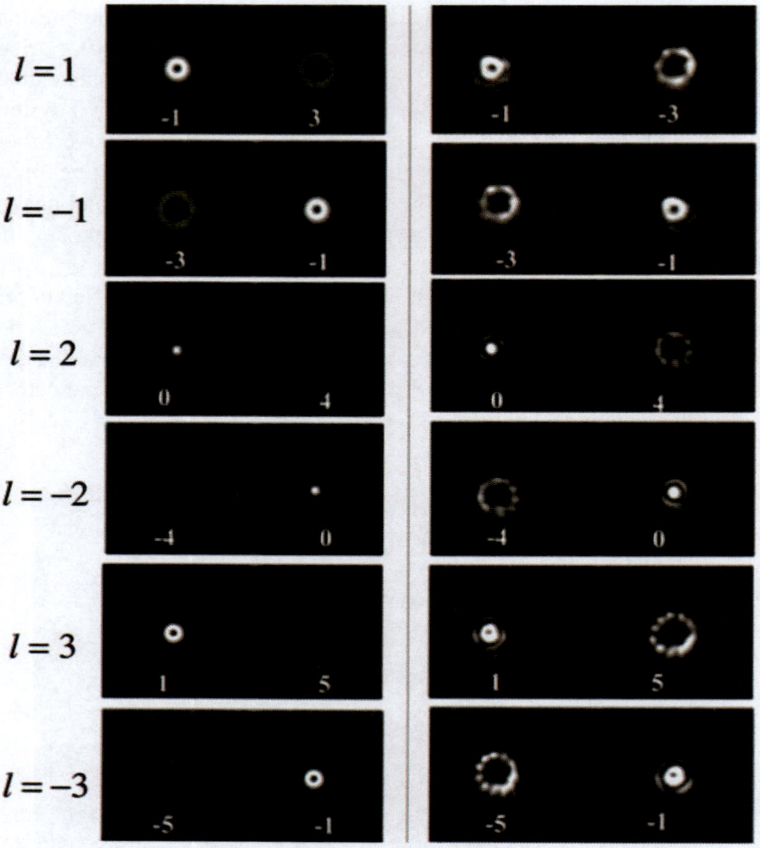

Fig. 5.49 Far-field patterns of the vortex beam after passing through a second-order SVHWP, where the left-hand side is the simulation result and the right-hand side is the experimental result [34]

In the detection system given in Fig. 5.50, the beam to be measured is first divided into four beams of equal energy without changing the polarization state, and then the four beams are allowed to pass first, second, third and fourth-order SVHWP respectively. After passing through the quarter wave plate, the circular polarization is converted into linear polarization, which is then split by a Wollaston prism and finally a photodiode with a small aperture diaphragm is used to detect recognize.

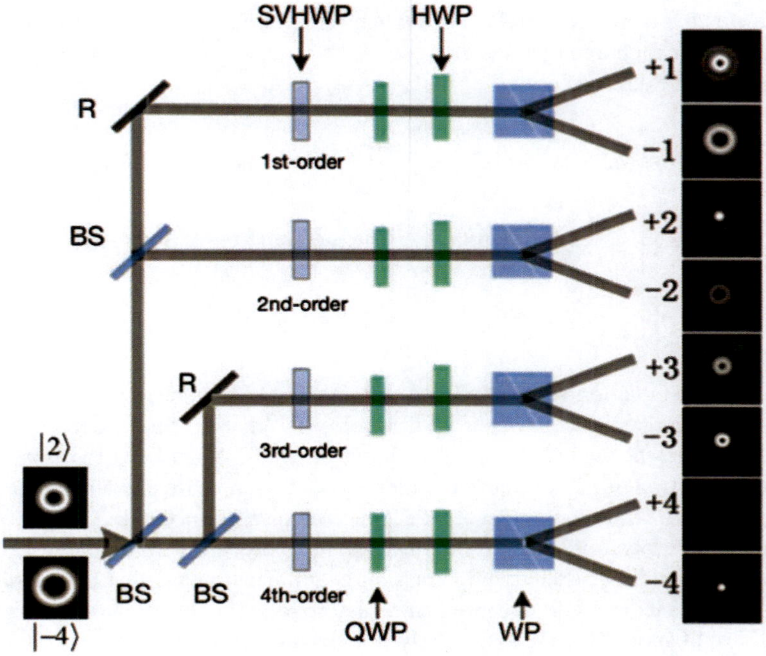

Fig. 5.50 Parallel vortex beam detection system based on a SVHWPs. BS, beam splitter; R, reflector; QWP, quarter wave plate; HWP, half wave plate; WP, Wollaston prism [34]

5.4.2 Spatially Variable Polarizers Measurement

Similar to the SVHWP, the spatially variable polarizer (SVP) is also a polarization element with azimuthal anisotropic direction of deviation. It consists of a combination of M sub-polarizers placed at different angles, each of which has a departure angle from the x-axis of

$$\theta_{SVP}(n) = \frac{2\pi m(n-1)}{M} + \theta_0 \tag{5.4.4}$$

where n is the serial number of the sub-polarizer, m is the order of the spatially variable polarizer, and θ_0 is the initial deviation direction.

The spatially variable polarizer can also be used to detect vortex beams, and when $M \to \infty$, the spatially variable polarizer is a continuous polarization element whose Jones matrix is [35]:

$$J_{SVP}(\varphi) = \begin{bmatrix} \cos^2(m\varphi + \theta_0) & \sin(m\varphi + \theta_0)\cos(m\varphi + \theta_0) \\ \sin(m\varphi + \theta_0)\cos(m\varphi + \theta_0) & \sin^2(m\varphi + \theta_0) \end{bmatrix} \tag{5.4.5}$$

where φ is the azimuthal angle. Then when a left circularly polarized vortex beam with topological charge l passes through a m-th order spatially variable polarizer, its output can be expressed as

$$\begin{aligned} E &= J_{SVP}(\varphi)\exp(il\varphi)\begin{bmatrix}1\\ i\end{bmatrix} = \frac{1}{\sqrt{2}}\begin{bmatrix}\cos m\varphi\\ \sin m\varphi\end{bmatrix}\exp[i(l+m)\varphi] \\ &= \frac{1}{2\sqrt{2}}\left\{\begin{bmatrix}1\\ i\end{bmatrix}\exp(-im\varphi) + \begin{bmatrix}1\\ -i\end{bmatrix}\exp(im\varphi)\right\}\exp[i(l+m)\varphi] \\ &= \frac{1}{2\sqrt{2}}\left\{\begin{bmatrix}1\\ i\end{bmatrix}\exp(il\varphi) + \begin{bmatrix}1\\ -i\end{bmatrix}\exp[i(2m+l)\varphi]\right\} \end{aligned} \tag{5.4.6}$$

The above equation shows that the output beam can be decomposed into a left circularly polarized vortex beam with topological charge l and a right circularly polarized vortex beam with topological charge $(2m + l)$. When $l = 0$, the output beam is a superposition of a left circularly polarized Gaussian beam and a right circularly polarized beam with topological charge 2 m. When $2m + l = 0$ and $l \neq 0$, the output beam is a superposition of a left circularly polarized beam with topological charge l and a right circularly polarized Gaussian beam. This means that the topological charge of the incident vortex beam can be determined by observing the presence of a solid bright spot of the right circularly polarized component, namely, $l = -2m$.

Figure 5.51 gives the simulated intensity distribution of the output right circularly polarized component when different left circularly polarized vortex beams are incident in various spatially variable polarizers (order 1/2, order 1, order 3/2 and order 2, $M = 18$). Only when the topological charge l and the order m of the spatially variable polarizers meet $l = -2m$, a bright solid spot presents at the output right circularly polarized component. This characteristic of the spatially variable polarizer can be used to measure the topological charge of a vortex beam.

5.5 OAM Spectra Measurement

OAM spectrum is defined as the ratio of the energy of each OAM component it carries, and can reflect properties of the beam, which is one of the most important parameters in evaluating vortex beams. For multiplexed vortex beams, the intensity distribution and wavefront distribution are completely different when the OAM components are the same but their proportion is different. Therefore, in addition to determining the topological charge or OAM component, the power ratio between each OAM components, namely, the OAM spectrum, should also be measured. The theoretical derivation of the OAM spectrum of a vortex beam has already been discussed in Sect. 1.3, on the basis of which several practical measurements of the OAM spectrum are presented in this section.

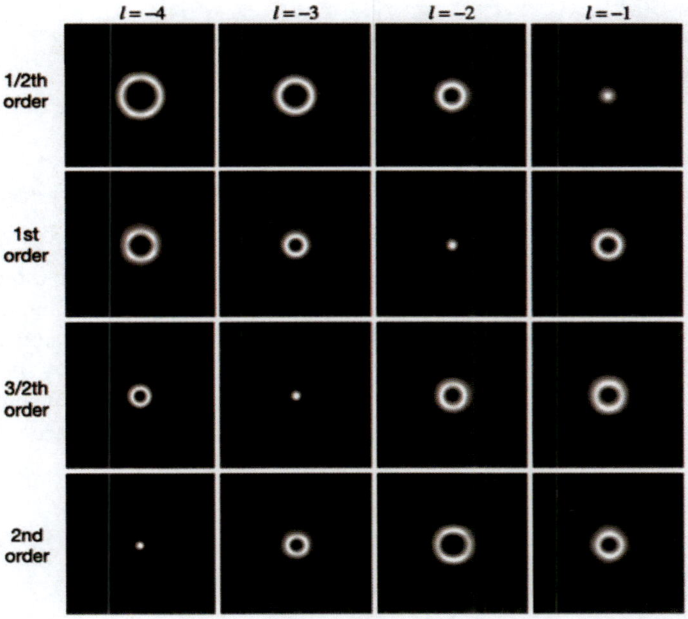

Fig. 5.51 Intensity patterns of output right circularly polarization when different vortex beams passing through spatially variable polarizers of various orders [34]

5.5.1 Complex Amplitude Derivation

From Sect. 2.3.1, the complex amplitude $E(x,y,z)$ of any beam can be expanded by the spiral harmonic $\exp(il\varphi)$ as

$$E(x, y, z) = \frac{1}{\sqrt{2\pi}} \sum_{l=-\infty}^{+\infty} a_l(r, z) \exp(il\varphi) \tag{5.5.1}$$

The intensity of each OAM component can be obtained by integrating the expansion coefficient a_l over the whole region [Eq. (2.3.3)]. Therefore, the expansion factor $\{a_l\}$ can be determined from Eq. (2.3.2) and the OAM spectrum can then be calculated using Eq. (2.3.3).

When calculating the expansion factor a_l through Eq. (2.3.2), the complex amplitude of the beam must be known, namely, both amplitude and phase distributions must be measured firstly. The measurement of the amplitude is relatively straightforward and can be obtained by measuring the intensity distribution through an array detector, for instance, a CCD camera, and then square rooting it. The measurement of the phase distribution is more complex and can be achieved by the interferometric method described in Sect. 5.2.2.

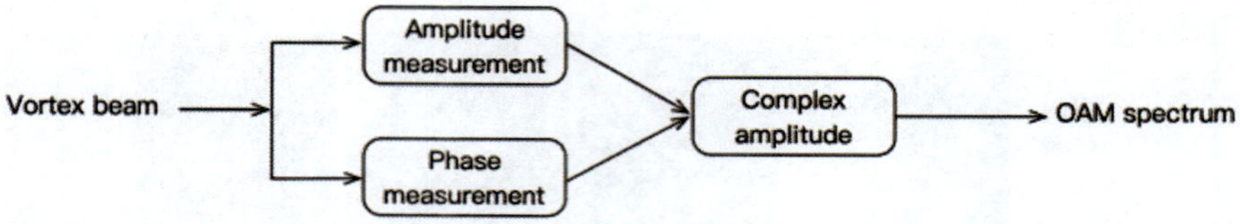

Fig. 5.52 Flow chart of measuring OAM spectrum through complex amplitude extrapolation

In summary, the amplitude distribution E_0 and the phase distribution σ of the beam should be measured separately. They can form the complex amplitude $E = E_0\exp(i\sigma)$, and then substitute E into Eq. (2.3.2) to obtain the expansion coefficient $\{a_l\}$, and finally substituted into Eq. (2.3.3) to calculate the OAM spectrum, as shown in Fig. 5.52. The principle of this method is very simple, but it is not easy to implement and requires the construction of a co-axial interference device to measure the phase, so it is rarely used in practice.

5.5.2 *Grey Scale Algorithm*

The grey-scale algorithm is a simple path for measuring OAM spectrum, and is effective for large mode intervals [36, 37]. It is based on a Damman vortex grating, and OAM spectrum is obtained by analyzing the intensity distribution of the far-field diffraction.

From Sect. 5.3.9, when a vortex beam is incident in a Dammann vortex grating, if a solid bright spot appears at the center of a diffraction order, the topological charge of the incident beam and the diffraction order satisfy $l = -b_x l_x - b_y l_y$, then l can be determined. From Eq. (5.3.27), the solid bright spot at the center of diffraction order (b_x, b_y) is entirely converted from the OAM component of the incident beam with topological charge $(-b_x l_x - b_y l_y)$, which indicates that if the intensity of the solid bright spot is measured, the proportion of the OAM whose topological charge is $(-b_x l_x - b_y l_y)$ can be obtained. By measuring the intensity of all components in turn, the OAM spectrum of the incident beam can be measured. Theoretically, the intensities of all the diffraction order in the far-field of a Dammann vortex grating are identical. Therefore, the OAM spectrum of the incident beam can be calculated by analyzing each diffraction order in turn as described above.

In summary, the key is the ability to efficiently and rapidly calculate the magnitude of the central spot intensity at different diffraction orders in far-field, which can be achieved by designing a new grey-scale algorithm for image processing. The diffraction field received by an array detector is represented as a grey scale (usually 8 bits, 0–255 grey values). The intensity of the received spot is proportional to the sum of the grey values of all the pixel points it contains, providing that the intensity threshold is unexceeded. Therefore, the relative intensity can be obtained by simply reading the sum of the grey values of each pixel contained by the central bright spot

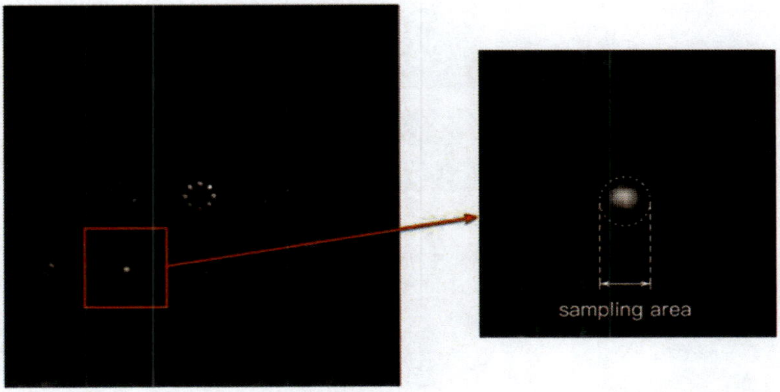

Fig. 5.53 The central bright spot sampling [36]

separately. Such relative intensity measurement uses no optical power meter, but is implemented as an image processing of calculating the grey scale, hence called the grey scale algorithm [36, 37]. The core of the grey-scale algorithm for measuring the OAM spectrum is that, for the captured far-field diffraction of a Dammann vortex grating, each diffraction order is firstly scanned in turn from end to end, and the presence or absence of a central bright spot in different diffraction orders is mastered. If not, the scan continues to the next spot, and if present, the intensity (i.e. the sum of the grey values) of the central bright spot is calculated. Note that when calculating the intensity of the central bright spot, the central bright spot sampling area should be selected to include exactly the entire central bright spot while excluding other side flaps, as shown in Fig. 5.53. Once all spots have been analyzed, the energy ratios of different modes, i.e. the OAM spectrum of the incident vortex beam, can be obtained. A flowchart of the grey-scale algorithm is given in Fig. 5.54, where taking diffraction fields of a 5×5 standard Darman vortex grating as an example.

However, the intensity of the beam at each diffraction order is not strictly the same due to various factors such as the practice operating environment. So the proportional power distribution of each diffraction order needs to be measured prior to the actual measurement, and the measured OAM spectrum needs to be compensated for. For the received diffraction pattern, the grey scale values of each pixel contained in each diffraction order are read in turn and summed before applying the grey scale algorithm. Then the intensity proportional distribution of each diffraction orders is measured. Figure 5.55 shows the intensity distribution between diffraction orders of a 5×5 standard Dammann vortex grating measured in the experiment. The actual intensity of each diffraction order is not equal due to the incident beam aperture, the resolution of the liquid crystal SLM and also other factors.

Figure 5.56 gives the measured OAM spectra of the three-, four- and six-fold multiplexed vortex beams in sequence [36]. Without the compensation of diffraction order intensity shown in Fig. 5.55, the experimental results differ significantly from

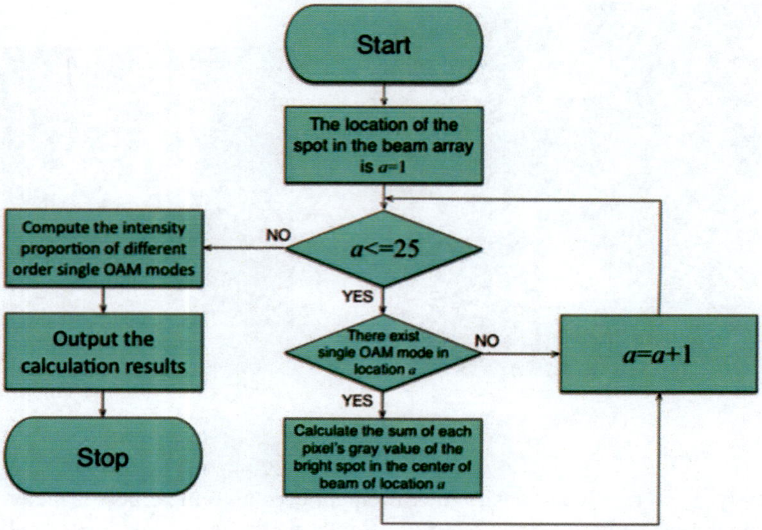

Fig. 5.54 Flow chart of the OAM spectrum measurement through the grey-order algorithm [36]

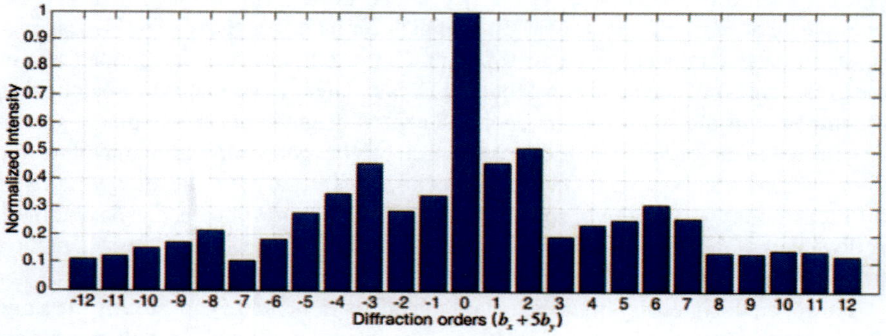

Fig. 5.55 Intensity distribution of diffraction orders of a 5 × 5 standard Darman vortex grating measured in the experiment [36]

the theoretical results. After compensation, the experimental values agree well with the theoretical values.

From Fig. 5.56c, the large difference between the experimental and theoretical results after compensation when measuring a six-fold multiplexed vortex beam, is due to the small mode spacing of different OAM modes contained in the incident multiplexed vortex beam (i.e. the difference in topological charges between adjacent modes) and the introduction of inter-mode crosstalk. As described in [36], this inter-mode crosstalk has very little effect on the measurement results when the adjacent

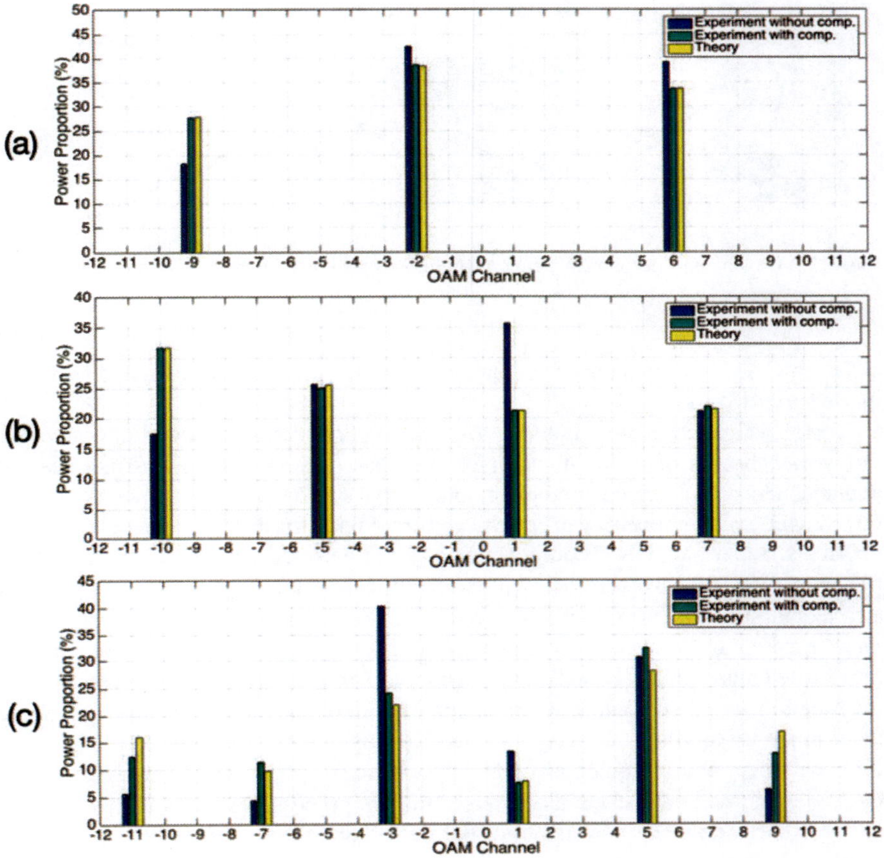

Fig. 5.56 Experimentally measured OAM spectrum of multiplexed vortex beams [36]

mode spacing is larger than or equal to 5. This suggests that the prerequisite for accurate measuring OAM spectrum using the grey-scale algorithm is that the topological charge difference between adjacent modes is as large as possible.

5.5.3 OAM Mode Sorter

When a beam is incident in a dispersive element such as a triple prism, different frequencies in the beam show different refractive indices and are separated. The spectrum of the incident beam can then be obtained by measuring the relative intensities of each frequency component with an array detector, which is actually the principle of a spectrometer. Similar to a spectrometer, the OAM spectrum of an incident beam can be measured by designing an optical element that separates the

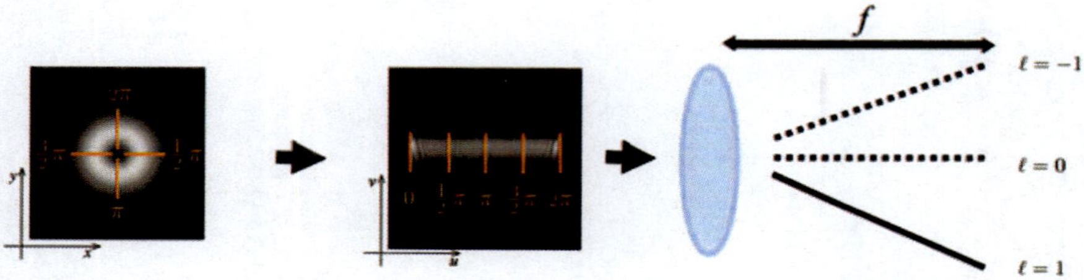

Fig. 5.57 Unwrapping the annular optical field of a vortex beam enable OAM modes separation [40]

different OAM components of the incident beam. Such element is called an OAM mode sorter [38–41].

Vortex beams have helical phases that varying $2l\pi$ along azimuthal in one circle. For vortex beams of different orders l, the phase change in azimuth is different and the helical wavefront gradient is also different. Therefore, the separation of different OAM components can be achieved employing the difference of helical phase gradients. Separation OAM components in azimuthal is difficult. Nevertheless, if the transverse annular structure of a vortex beam can be unwrapped and straightened to a rectangular structure, the helical phase gradient can be transformed into linear phase gradient in Cartesian coordinate ($2l\pi$). Hence, the helical phase is transformed into a tilted plane phase. Based on this principle, for different incident vortex beams, the transformed tilted plane wavefronts (phases) have various linear phase gradients determined by topological charges l. Thus, a convex lens can focus components of different linear phase gradients at different positions of the focal plane. And finally, OAM separation is achieved. The transformation process described above is given in Fig. 5.57, and the equiphasic points before and after vortex beam unwrapping are displayed.

The key to separate different OAM components is to unwrapped the annular optical field. When the angle-preserving mapping relationship is satisfied, the process of unwrapping can be achieved by a phase grating coupled with a Fourier transformation lens [38]. Under Cartesian coordinate where the input and output planes (Fourier planes) are xOy and uOv, respectively, then the unwrapping phase grating should achieve the mapping: $(x, y) \mapsto (u, v)$. Under the angle-preserving mapping, let:

$$v = a \arctan\left(\frac{y}{x}\right) \tag{5.5.2}$$

Then u should satisfy

$$u = -a \ln\left(\frac{\sqrt{x^2 + y^2}}{b}\right) \tag{5.5.3}$$

The phase distribution function of the unwrapping phase is [38]

$$\phi_1(x,y) = \frac{2\pi a}{\lambda f}\left[y\arctan\left(\frac{y}{x}\right) - x\ln\left(\frac{\sqrt{x^2+y^2}}{b}\right) + x\right] \tag{5.5.4}$$

In Eqs. (5.5.2)–(5.5.4), λ is the optical wavelength and f is the focal length of the lens behind the unwrapping phase. Parameter a scales the unwrapped field, and $a = d/2\pi$. d is the linearity of the transverse optical field after unwrapping (the length of the unwrapped rectangular field). Parameter b determines the position of the unwrapped field in the u-coordinate of the Fourier plane and is independent of parameter a.

The optical trajectories of different OAM components from the initial diffraction plane xOy to the Fourier plane uOv are different after unwrapping. Hence, there is a phase difference between them, and an additional phase needs to be introduced in the Fourier plane to compensate [38]:

$$\phi_2(u,v) = -\frac{2\pi ab}{\lambda f}\exp\left(-\frac{u}{a}\right)\cos\left(\frac{v}{a}\right) \tag{5.5.5}$$

The unwrapped field is compensated by the grating shown in Eq. (5.5.5) and then focused through another convex lens to separate the OAM components. By measuring the intensity of each component in turn, the OAM spectrum of the incident vortex beam can be obtained.

The unwrapped phase and correction phase generated according to Eqs. (5.5.4) and (5.5.5) are shown in Fig. 5.58. As shown in Fig. 5.59, when measuring OAM spectrum, such two phases can be simulated by two liquid crystal SLMs, and a lens L1 is placed in the middle of the two liquid crystal spatial light modulators. Their position relationship satisfies that the two SLMs are located at the front and back the lens L1. Another lens L2 is then employed to focus the output of the above element to separate various OAM components. Then the OAM spectrum of the incident beam can be measured.

In addition, the two phase gratings of a OAM mode sorter can also be fabricated through polymethyl methacrylate directly [39].

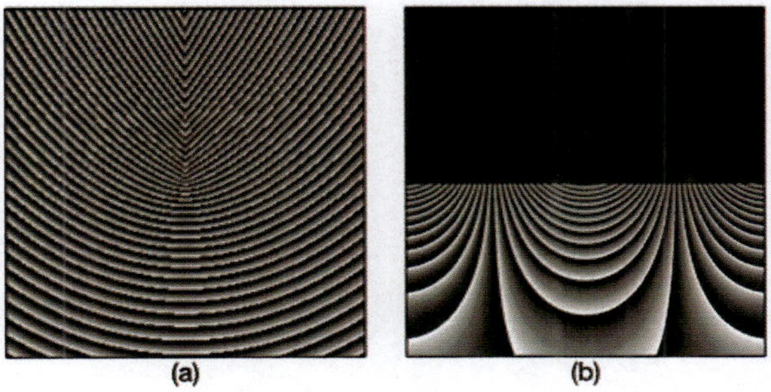

Fig. 5.58 **a** Unwrapped phase; **b** correction phase

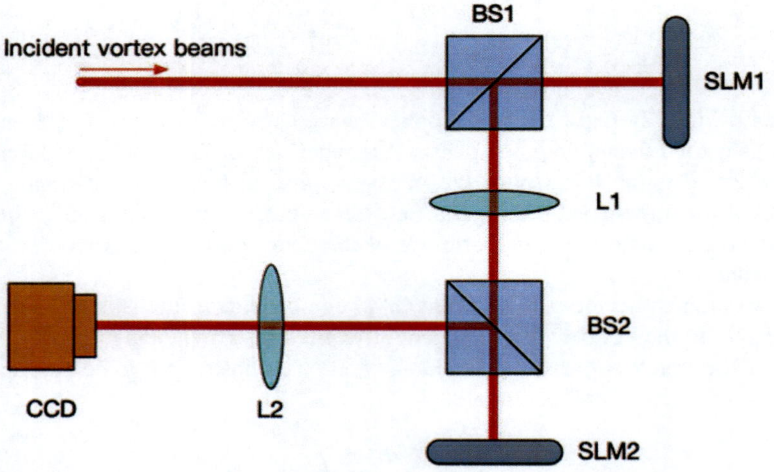

Fig. 5.59 Measuring OAM spectra through OAM mode sorter. BS1 and BS2, beam splitter; SLM1 and SLM2, liquid crystal spatial light modulators; L1 and L2, convex lenses; CCD, CCD camera

The experimental results of separating various OAM components are given in Fig. 5.60. After passing through the OAM mode sorter, vortex beams of different order l are focused at different positions on the back focal plane. When a multiplexed vortex beam is incident, the contained single OAM components can be separated from each other and the OAM spectrum can be obtained by measuring the intensity of each component. Compared to the grey-scale algorithm, the OAM mode sorter overcomes the problem of inaccurate measurement at small OAM mode intervals and shows good mode separating performance, becoming one of the most effective approaches for OAM spectrum measurement.

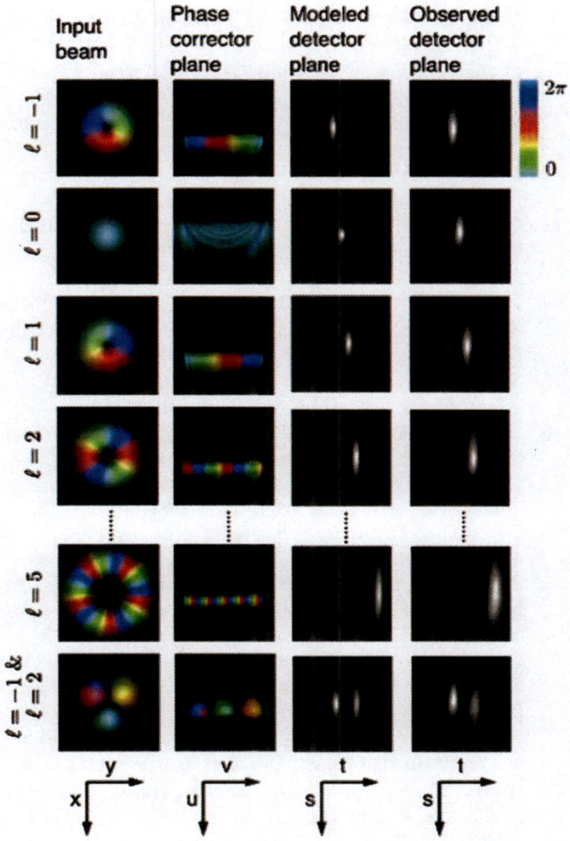

Fig. 5.60 Intensity patterns of various vortex beams before and after passing through the OAM mode sorter, where the phase distribution before and after unlooping is also indicated [38]

References

1. Beijersbergen MW. Measuring orbital angular momentum of light with a torsion pendulum. Proc SPIE - Int Soc Opt Eng. 2005;5736:111–25.
2. Volkesepúlveda K, Santillán AO, Boullosa RR. Transfer of angular momentum to matter from acoustical vortices in free space. Phys Rev Lett. 2008;100(2): 024302.
3. Gao C. Characterization and transformation of astigmatic laser beams. Wissenschaft und Technik Verlag; 1999.
4. Weber H. Propagation of higher-order intensity moments in quadratic-index media. Opt Quant Electron. 1992;24(9):S1027–49.
5. Lavery MPJ, Speirits FC, Barnett SM, et al. Detection of a spinning object using light's orbital angular momentum. Science. 2013;341:537–40.
6. Vasnetsov MV, Torres JP, Petrov DV, et al. Observation of the orbital angular momentum spectrum of a light beam. Opt Lett. 2003;28(23):2285–7.
7. Liu Y. Study on the orbital angular momentum of beams and their applications in data-transmission. Doctoral Dissertation of Beijing Institute of Technology; 2008 (in Chinese).
8. Zambrini R, Barnett SM. Quasi-intrinsic angular momentum and the measurement of its spectrum. Phys Rev Lett. 2006;96(11): 113901.

9. Huang H, Ren Y, Yan Y, et al. Phase-shift interference-based wavefront characterization for orbital angular momentum modes. Opt Lett. 2013;38(13):2348–50.
10. Sztul HI, Alfano RR. Double-slit interference with Laguerre-Gaussian beams. Opt Lett. 2006;31(7):999–1001.
11. Emile O, Emile J. Young's double-slit interference pattern from a twisted beam. Appl Phys B. 2014;117(1):487–91.
12. Leach J, Padgett MJ, Barnett SM, et al. Measuring the orbital angular momentum of a single photon. Phys Rev Lett. 2002;88(25): 257901.
13. Gao C, Qi X, Liu Y, et al. Sorting and detecting orbital angular momentum states by using a Dove prism embedded Mach-Zehnder interferometer and amplitude gratings. Opt Commun. 2011;284(1):48–51.
14. Hickmann JM, Fonseca EJ, Soares WC, et al. Unveiling a truncated optical lattice associated with a triangular aperture using light's orbital angular momentum. Phys Rev Lett. 2010;105(5): 053904.
15. Stahl C, Gbur G. Analytic calculation of vortex diffraction by a triangular aperture. J Opt Soc Am A: Opt Image Sci Vis. 2016;33(6):1175–80.
16. Liu Y, Tao H, Pu J, et al. Detecting the topological charge of vortex beams using an annular triangle aperture. Opt Laser Technol. 2011;43(7):1233–6.
17. Liu R, Long J, Wang F, et al. Characterizing the phase profile of a vortex beam with angular-double-slit interference. J Opt. 2013;15(12): 125712.
18. Fu D, Chen D, Liu R, et al. Probing the topological charge of a vortex beam with dynamic angular double slits. Opt Lett. 2015;40(5):788–91
19. Serna J, Encinas-Sanz F, Nemeş G. Complete spatial characterization of a pulsed doughnut-type beam by use of spherical optics and a cylindrical lens. J Opt Soc Am A Opt Image Sci Vis. 2001;18(7):1726–33.
20. Denisenko VG, Soskin MS, Vasnetsov MV. Transformation of Laguerre-Gaussian modes carrying optical vortices and their orbital angular momentum by cylindrical lens. Proc SPIE - Int Soc Opt Eng. 2002;4607:54–8.
21. Denisenko V, Shvedov V, Desyatnikov AS, et al. Determination of topological charges of polychromatic optical vortices. Opt Express. 2009;17(26):23374–9.
22. Alperin SN, Niederriter RD, Gopinath JT, et al. Quantitative measurement of the orbital angular momentum of light with a single, stationary lens. Opt Lett. 2016;41(21):5019–22.
23. Vaity P, Banerji J, Singh RP. Measuring the topological charge of an optical vortex by using a tilted convex lens. Phys Lett A. 2013;377(15):1154–6.
24. Chaitanya NA, Jabir MV, Samanta GK. Efficient nonlinear generation of high power, higher order, ultrafast "perfect" vortices in green. Opt Lett. 2016;41(7):1348–51.
25. Dai K, Gao C, Zhong L, et al. Measuring OAM states of light beams with gradually-changing-period gratings. Opt Lett. 2015;40(4):562–5.
26. Fu S, Wang T, Gao Y, et al. Diagnostics of the topological charge of optical vortex by a phase-diffractive element. Chin Opt Lett. 2016;14(8): 080501.
27. Li Y, Deng J, Li J, et al. Sensitive orbital angular momentum (OAM) monitoring by using gradually changing-period phase grating in OAM-multiplexing optical communication systems. IEEE Photonics J. 2016;8(2):1–6.
28. Zheng S, Wang J. Measuring orbital angular momentum (OAM) states of vortex beams with annular gratings. Sci Rep. 2017;7:40781.
29. Gibson G, Courtial J, Barnett S, et al. Increasing the data density of free-space optical communications using orbital angular momentum. Proc SPIE - Int Soc Opt Eng. 2004;5550:367–73.
30. Gibson G, Courtial J, Padgett M, et al. Free-space information transfer using light beams carrying orbital angular momentum. Opt Express. 2004;12(22):5448–56.
31. Zhang N, Yuan XC, Burge RE. Extending the detection range of optical vortices by Dammann vortex gratings. Opt Lett 2010;35(20):3495–7.
32. Fu S, Wang T, Zhang S, et al. Integrating 5×5 Dammann gratings to detect orbital angular momentum states of beams with the range of -24 to $+24$. Appl Opt. 2016;55(7):1514–7.

33. Xin J, Dai K, Zhong L, et al. Generation of optical vortices by using spiral phase plates made of polarization dependent devices. Opt Lett. 2014;39(7):1984–7.
34. Xin J. Study on the generation of vector beams and their applications. Doctoral Dissertation of Beijing Institute of Technology; 2013 (in Chinese).
35. Moh KJ, Yuan XC, Bu J, et al. Generating radial or azimuthal polarization by axial sampling of circularly polarized vortex beams. Appl Opt. 2007;46(30):7544–51.
36. Fu S, Zhang S, Wang T, et al. Measurement of orbital angular momentum spectra of multiplexing optical vortices. Opt Express. 2016;24(6):6240–8.
37. Gao C, Fu S, Zhang S. A setup for the measurement of beams orbital angular momentum spectra. Chinese Patent 201510867994.1, 2 Dec 2015 (in Chinese).
38. Berkhout GC, Lavery MPJ, Courtial J, et al. Efficient sorting of orbital angular momentum states of light. Phys Rev Lett. 2010;105(15): 153601.
39. Lavery MPJ, Robertson DJ, Berkhout GCG, et al. Refractive elements for the measurement of the orbital angular momentum of a single photon. Opt Express. 2012;20(3):2110–5.
40. Lavery MPJ, Berkhout GCG, Courtial J, et al. Measurement of the light orbital angular momentum spectrum using an optical geometric transformation. J Opt. 2011;13(13): 064006.
41. Mirhosseini M, Malik M, Shi Z, et al. Efficient separation of the orbital angular momentum eigenstates of light. Nat Commun. 2013;4(7):3781.

Chapter 6
Distortion Correction for Vortex Beams

Vortex beams propagating in free space will be disturbed by turbulent atmosphere, resulting in their intensity and phase distortions. It is already proved that atmospheric turbulence can disrupt the spiral phase of a vortex beam, leading to a broadening of its OAM spectrum and introducing crosstalk between different OAM modes. Therefore, it is essential to correct distorted vortex beams. Adaptive optics can be employed to correct distorted beams, where a wavefront sensor is encouraged to detect distorted wavefronts and a wavefront corrector is used to correct the distorted wavefront. However, vortex beams have complex spiral phases, which bring troubles to compensate using current adaptive correction techniques. Therefore, new distortion correction methods for distorted vortex beams must be developed. In this chapter, the effects of atmospheric turbulence on vortex beams and several adaptive distortion correction schemes for distorted vortex beams are presented.

6.1 Atmospheric Turbulence Model

6.1.1 Kolmogorov Theory of Atmospheric Turbulence

Atmospheric turbulence is a kind of random air motion, referring to the inhomogeneous distribution of the overall refractive index of the atmosphere from random changes of temperature, humidity and pressure in some regions of the atmosphere [1]. Atmospheric turbulence leads to small differences in density and temperature at different locations, and turbulence with different refractive indices is constantly created and extinguished as the wind moves. As a result, atmosphere in a flowing state has many swirls of constantly moving air, creating a random motion of atmospheric turbulence.

The refractive index of the turbulent atmosphere is a random variable for which accurate predictions cannot usually be made. In general, the refractive index $n(r,t,\lambda)$ of turbulent atmosphere varies with space r, time t and wavelength λ, and can be

S. Fu and C. Gao, *Optical Vortex Beams*, Advances in Optics and Optoelectronics,
https://doi.org/10.1007/978-981-99-1810-2_6

expressed as a function of space, time and wavelength as follows [2]:

$$n(r, t, \lambda) = n_0(r, t, \lambda) + n_1(r, t, \lambda) \tag{6.1.1}$$

where $n_0(r,t,\lambda)$ is the constant part of the atmospheric refractive index, approximated by 1 in general, and $n_1(r,t,\lambda)$ is the random rise and fall of the refractive index resulting from turbulent atmosphere motion.

The refractive index random variation of atmosphere results in a non-uniform distribution of the refractive index, and often called "turbulent vortices" [3]. Turbulences are appeared and disappeared as the atmosphere moves, so that the moving atmosphere has vortex cells with different pressures, wind speeds, temperatures and densities at the same time. As the atmosphere continues to move, the larger vortex cells become smaller and disappear, while the smaller vortices become larger and larger. Such succession of creation and disappearance gives rise to atmospheric turbulence.

Atmospheric turbulence can be considered to be composed of numerous large and small 'turbulent vortices'. The characteristic scale of large turbulent vortices is called the outer scale of turbulence and is denoted by L_0. For scales larger than L_0, turbulence is generally anisotropic. While turbulence with scales smaller than L_0 is isotropic, hence L_0 is also referred to as the maximum scale of turbulence when atmospheric turbulence is isotropic. In contrary, the inner scale of turbulence is denoted as l_0, where the kinetic and dissipative energies of the turbulent vortices cancel each other out, and all kinetic energy is converted into thermal energy. The much smaller 'turbulent vortices' cannot exist due to the lack of kinetic energy.

In the establishment of the turbulence theory, the mathematician Andrey Nikolaevich Kolmogorov from former Soviet Union made three assumptions [4].

① Both large and small eddies move randomly and are generally isotropic in character in atmospheric turbulence.

② Only internal frictional and inertial forces are present in isotropic atmospheric turbulence.

③ When the kinetic energy of turbulence is greater than the dissipative energy, the turbulent motion is affected only by inertial forces in the interval $l_0 \leq r \leq L_0$. Such range is called the inertial scale space.

Kolmogorov introduced a structure function to study the statistical characteristics of atmospheric turbulence that meets the three assumptions mentioned above, namely, the famous "2/3rd power" law. In the inertial scale space of turbulence, the structure constant between two positions is only related to the 2/3rd power of the distance Δr between the two positions, but not to the specific position of the two points or to their relative directions. The structure function $D_n(r)$ of the atmospheric refractive index can be expressed as:

$$D_n(r) = C_n^2 r^{2/3}, l_0 \ll r \ll L_0 \tag{6.1.2}$$

where C_n^2 is the structural constant of the atmospheric refractive index and is a function of space and time. It describes the strength of atmospheric turbulence and is one of the most important parameters for characterizing the strength of atmospheric turbulence. The strength of atmospheric turbulence can usually be classified according to the magnitude of C_n^2 [5]:

Strong turbulence: $C_n^2 > 2.5 \times 10^{-13}\text{m}^{-2/3}$;

Moderate turbulence:$6.4 \times 10^{-17}\text{m}^{-2/3} < C_n^2 < 2.5 \times 10^{-13}\text{m}^{-2/3}$;

Weak turbulence: $C_n^2 < 6.4 \times 10^{-17}\text{m}^{-2/3}$.

In the optical band, the atmospheric refractive index $n(r)$ can be approximated read:

$$n(r) = 1 + 77.6 \times 10^{-6}\left(1 + 7.52 \times 10^{-3} \cdot \lambda^{-2}\right)\frac{P(r)}{T(r)} \tag{6.1.3}$$

where λ is the wavelength at the micron scale; $P(r)$ and $T(r)$ are the pressure and the Thermodynamic temperature, respectively, and are functions of the space coordinate r. In the near-earth condition, the pressure change is small and can be neglected. Taking the derivative of r in Eq. (6.1.3), one can obtain:

$$dn(r) = 77.6 \times 10^{-6}\left(1 + 7.52 \times 10^{-3} \cdot \lambda^{-2}\frac{P}{T^2}\right)dT \tag{6.1.4}$$

This indicates that the change of near-earth refractive index is mainly result from temperature fluctuation. From Formula (6.1.4), the atmospheric refractive index structure constant C_n^2 can be expressed as:

$$C_n^2 = \frac{dn}{dT}C_T^2 = 77.6 \times 10^{-6}\left(1 + 7.52 \times 10^{-3} \cdot \lambda^{-2}\frac{P}{T^2}\right)C_T^2 \tag{6.1.5}$$

where C_T^2 is the structure constant of temperature fluctuation, a quantity related to the generation rate of temperature fluctuation and the dissipation rate of turbulent kinetic energy.

Equation (6.1.5) gives a direct path for the measurement of atmospheric refractive index structure constant C_n^2. Since the direct measurement of atmospheric refractive index is difficult and the measurement of temperature and temperature fluctuation is easy. It is usually to measure the temperature fluctuation when measuring C_n^2, and then calculate the structural constant of atmospheric refractive index through Eq. (6.1.5).

The atmospheric coherence length r_0 is also one of the important parameters for representing the strength of atmospheric turbulence, which can be used to express the spatial density and intensity of turbulence. It was firstly introduced by the American scientist Fried and is therefore also called the Fried parameter. In physics, this parameter represents the diffraction limit of light wave in turbulence [1]. Atmospheric coherence length r_0 can reflect comprehensive intensity of atmospheric turbulence, the optical turbulence effect of the beam on the atmospheric path, and the effect of spatial coherence. The smaller the turbulence, the better the atmospheric conditions, and the larger the corresponding r_0.

r_0 and C_n^2 satisfy [6]:

$$r_0 = \left[0.423k^2 \int_0^L C_n^2(z)dz\right]^{-3/5} \tag{6.1.6}$$

with k the wave number, L the total length of the path, z the integral variable along the path. Thus, atmospheric coherence length r_0 is a physical parameter related to the refractive index structure constant, the wavelength and the path of light beam. Under the same coherent length, the turbulence effect is different for different beam sizes. Therefore, the ratio of the beam cross-section diameter d to the atmospheric coherence length r_0 is often used to characterize the effect of turbulence on the beam. The larger d/r_0, the greater the influence of atmospheric turbulence on the beam is.

6.1.2 Power Spectral Density of Refractive Index

The power spectral density of the refractive index $\Phi_n(\kappa)$ is the physical quantity to describe the undulation of the refractive index of the atmosphere. In Kolmogorov turbulence theory it reads:

$$\Phi_n(\kappa) = 0.033C_n^2\kappa^{-11/3}, \frac{2\pi}{L_0} \ll \kappa \ll \frac{2\pi}{l_0} \tag{6.1.7}$$

with κ the space wave number.

From Eq. (6.1.7), $\Phi_n(\kappa)$ is a function of the spatial wave number κ and takes a range of values $2\pi / L_0 \ll \kappa \ll 2\pi / l_0$. In addition to the spectral distribution given in Eq. (6.1.7), there are also power spectrum models such as the Von Karman spectrum [7] and the Hill spectrum [8]. These models are more complex than the Kolmogorov power spectrum given in Eq. (6.1.7) and contain various parameter settings to ensure consistency between theory and experiment. Kolmogorov turbulence theory propose that, the power spectral density contains 3 regions.

① Input area ($\kappa < 2\pi / L_0$): the form of the power spectrum in this interval is determined by the generated turbulence. While at this point the turbulence is usually anisotropic and the form of the power spectrum cannot theoretically be obtained.

② Inertial area ($2\pi / L_0 < \kappa < 2\pi / l_0$): The refractive index power spectral density in this interval satisfies Eq. (6.1.7).

③ Dissipation area ($\kappa > 2\pi / l_0$): The volume dissipation in this interval is greater than the kinetic energy, and the energy dissipation is rapid. As a result, $\Phi_n(\kappa)$ also falls quickly.

On the basis of Eq. (6.1.7), Tatarskii gives a model shown in Eq. (6.1.8) to include dissipative region $\Phi_n(\kappa)$.

$$\Phi_n(\kappa) = 0.033C_n^2\kappa^{-11/3}\exp\left(-\frac{\kappa^2}{\kappa_m}\right) \tag{6.1.8}$$

where $\kappa_m = 5.92 / l_0$.

However, according to Eqs. (6.1.7) and (6.1.8), both power spectral density models have non-integrable singularities at the origin. There is a certain range in the earth's atmosphere. When $\kappa \to 0$ the spectrum cannot converge to infinity. To overcome these drawbacks, the Von Karman spectral model is often used [7]. Let $\kappa_0 = 2\pi / L_0$, Von Karman spectra can be expressed as

$$\Phi_n(\kappa) = \frac{0.033C_n^2}{\left(\kappa^2 + \kappa_0^2\right)^{11/6}}\exp\left(-\frac{\kappa^2}{\kappa_m}\right) \tag{6.1.9}$$

Apparently, When the outer scale of turbulence $L_0 \to \infty$, $\kappa_0 = 0$, Eq. (6.1.9) is identical to Eq. (6.1.8). But when $l_0 \to 0$, $\exp\left(-\kappa^2 / \kappa_m\right) \to 1$, Eq. (6.1.8) is again identical to Eq. (6.1.7). This suggests that Von Karman spectrum given in Eq. (6.1.9) is actually a modified Kolmogorov spectrum [Eq. (6.1.7)]. Hill proposed a relatively accurate numerical model based on experiments, called the Hill spectrum [8]:

$$\Phi_n(\kappa) = 0.033C_n^2\kappa^{-11/3}\left\{\exp\left(-1.2\kappa^2 l_0^2\right) + 1.45\exp\left[-0.97(\ln(\kappa l_0) - 0.452)^2\right]\right\} \tag{6.1.10}$$

Hill spectrum is more complex, making it difficult in research. Hence Von Karman spectrum is usually employed.

6.1.3 Power Spectral Density of Phase

All the power spectral models discussed above study the power spectral density of the refractive index. What we ultimately want is to obtain the effect of atmospheric turbulence on the phase undulation of the laser wavefront. Hence, we need to convert these models into the power spectral density of phase. The power spectral density of phase $\Phi_\phi(\kappa)$ and the power spectral density of refractive index $\Phi_n(\kappa)$ satisfy:

$$\Phi_\phi(\kappa) = 2\pi^2 k^2 \Delta z \Phi_n(\kappa) \tag{6.1.11}$$

where Δz is the propagation distance of beams in atmospheric turbulence.

Assuming that the atmospheric refractive index structure constant in Eq. (6.1.6) is a variable independent of the spatial position. When the laser propagates in turbulent flow, the atmospheric coherence length Δz reads:

$$r_0 = \left[0.423k^2 \Delta z C_n^2\right]^{-3/5} \tag{6.1.12}$$

Substituting Eq. (6.1.12) into Eq. (6.1.9) and then into Eq. (6.1.11), the Von Karman phase power spectral density can be obtained:

$$\Phi_\phi(f) = 0.023 r_0^{-5/3} \frac{\exp\left(-f^2 / f_m^2\right)}{\left(f^2 + f_0^2\right)^{11/6}} \tag{6.1.13}$$

where f is the space frequency, defined as $f = \kappa/(2\pi), f_0 = 1/L_0, f_m = 5.92/(2\pi l_0)$.

6.1.4 Establishment of an Atmospheric Turbulence Phase Screen

An atmospheric turbulence phase screen can be thought of as a phase grating that simulates atmospheric turbulence. When a beam is incident in the turbulence phase screen, the diffraction field is the distorted equivalent to passing through atmospheric turbulence. Atmospheric turbulence phase screens are important for simulating arbitrary realistic turbulence in the laboratory. This section will describe in detail how to constructed turbulence phase screens.

As the refractive index of the atmosphere changes as a stochastic process, the refractive index power spectrum model gives a statistical average. Hence, the key to generate an atmospheric turbulence phase screen is how to implement this stochastic process. The generation of an atmospheric turbulence phase screen, in other words, is transforming a set of computer-generated random numbers into a two-dimensional array of points representing phase values that have the same statistical properties as the phase fluctuations result from atmospheric turbulence.

Currently, there are two main methods of generating atmospheric turbulence phase screens:

① Summing Zernike's polynomial [9]

The Zernike polynomial method is now a common method for wavefront phase simulation and has wide application in adaptive optics. By exploring the Zernike polynomial coefficients under the statistical law of composite atmospheric turbulence, the turbulence phase screen is then obtained by summing the Nth order Zernike polynomials. The Zernike polynomial method can only be applied to the Kolmogorov spectrum given in Eq. (6.1.7), and the resulting phase screen is more accurate in the low spatial frequency region. In the high spatial frequency region, the simulation results are relatively different from the theoretical value. Although such issue can be

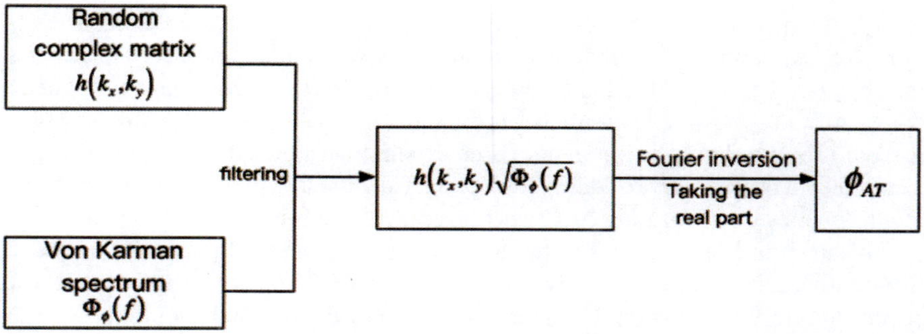

Fig. 6.1 The process of constructing an atmospheric turbulence phase screen through power spectrum inversion

addressed by increasing the order of the polynomial or changing the aperture of the system. It will cost more computational source to a certain extent.

② Power spectrum inversion

The power spectrum inversion method is suitable for a wide range of atmospheric turbulence models. The principle is to filter a complex Gaussian random matrix using the phase power spectrum of turbulence, and then perform an inverse Fourier transformation to obtain turbulence phase screen. Power spectrum inversion method is simple and convenient for computing a turbulence phase screen.

As shown in Fig. 6.1, the specific process of the power spectrum inversion is

① Generating a complex Gaussian random matrix $h(k_x,k_y)$ firstly, with mean 0 and variance 1 in the frequency domain.
② Obtaining the phase power spectral density matrix $\Phi_\phi(f)$ of the Von Karman spectrum from Eq. (6.1.13);
③ Using the power spectral density matrix $\Phi_\phi(f)$ in step ② to filter $h(k_x,k_y)$. Then we can obtain the matrix $h(k_x, k_y)\sqrt{\Phi_\phi(f)}$;
④ Inverse Fourier transforming is done on $h(k_x, k_y)\sqrt{\Phi_\phi(f)}$. Taking the real part to acquire the final atmospheric turbulence phase screen:

$$\phi_{AT} = \mathrm{Re}\left\{\mathcal{F}^{-1}\left[h(k_x, k_y)\sqrt{\Phi_\phi(f)}\right]\right\} \tag{6.1.14}$$

Constructing turbulent phase screens through power spectrum inversion has been widely adopted by researchers because of its simplicity and high-speed. The maximum and minimum spatial frequencies of the phase screens constructed here are:

$$\begin{cases} f_{\max} = \dfrac{N}{a} \\ f_{\min} = \dfrac{1}{a} \end{cases} \tag{6.1.15}$$

with N the resolution of the phase screen and a the size of the phase screen. Equation (6.1.15) shows that turbulence components whose the spatial frequency less than $f_{\min}$ are not represented in this phase screen. In other words, the low frequency component of atmospheric turbulence is missed. The low-frequency component plays a major role and should be ignored. In order to simulate atmospheric turbulence more accurately, it is necessary to adding low spatial frequencies components. Usually, this low frequency compensation can be achieved by the sub-harmonic compensation.

As shown in Fig. 6.2, the sub-harmonic compensation for low spatial frequency information is implemented by dividing the high spatial frequency part of the Fourier spectrum into 9 equal parts. Each part being 1/9 of the original sampling, with the sampling points distributed around 8 small regions in the center, thus forming a subharmonic network. P-level subharmonics ($P \geq 1$) correspond to a picking space of the size:

$$\Delta f_P = \frac{f_{\min}}{3^P} \tag{6.1.16}$$

The original high frequency sampling was then replaced by a 3^{2P} sampling component [10].

The phase distribution of the low spatial frequency term in the subharmonic compensation is

$$\phi_L(x, y) = \sum_{P=1}^{M} \sum_{n=-1}^{1} \sum_{m=-1}^{1} c_{n,m,P} \exp\left[i2\pi\left(f_{x_n}x + f_{y_m}y\right)\right] \tag{6.1.17}$$

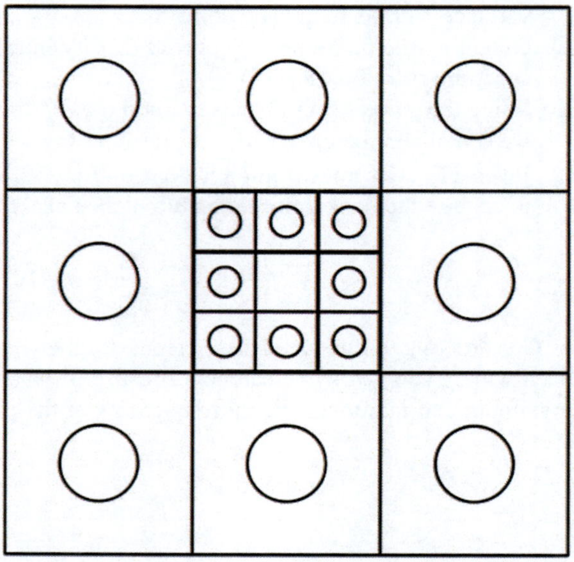

Fig. 6.2 Schematic diagram of sub-harmonic low spatial frequency compensation

where M is the highest order of the subharmonic grid; $c_{n,m,P}$ are the random Fourier coefficients, which can be obtained by filtering the subharmonic network by the Von Karman spectrum; and f_{xn} and f_{yn} are the spatial frequencies. The final atmospheric turbulence phase screen is obtained by superimposing Eq. (6.1.17) with Eq. (6.1.13).

$$\phi_T = \phi_{AT} + \phi_L \tag{6.1.18}$$

It can be clearly noticed that the minimum spatial sampling frequency becomes $f_{\min}/3^P$ after subharmonic compensation.

Note that although the lack of low frequency information in the power spectrum inversion scheme can be significantly improved by increasing the number of subharmonic stages P, the low spatial frequency deficit still exists. Figure 6.3 shows a phase screen of atmospheric turbulence generated employing Von Karman spectral model along with the power spectrum inversion and compensated by 3 levels of subharmonics, with resolution $N = 1000$ and phase screen size $a = 0.44$ m. From left to right, the structural constants of the atmospheric refractive index C_n^2 are 1×10^{-13} m$^{-2/3}$, 1×10^{-14} m$^{-2/3}$ and 1×10^{-15} m$^{-2/3}$, respectively. Obviously, the larger C_n^2, the more severe the phase fluctuations.

Since the refractive index or phase variation of atmospheric turbulence is a random process, complex Gaussian random matrices $h(k_x,k_y)$ with mean value 0 and variance 1 in the frequency domain are introduced in the construction of the turbulent phase screen. This means that the generated turbulent phase screen, even under the same turbulent strength, has different phase distributions due to the selection of the complex random matrix, as shown in Fig. 6.4.

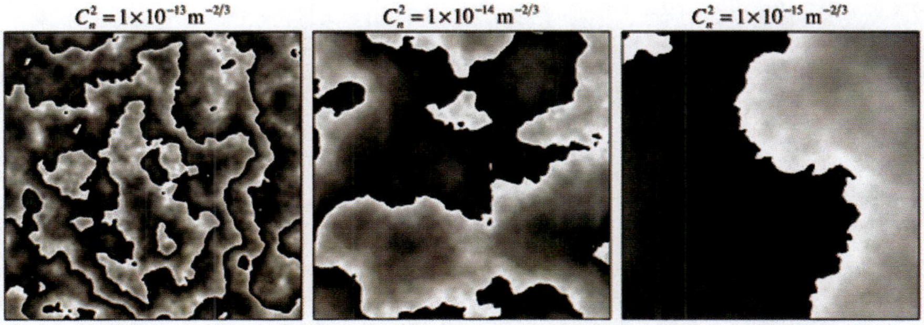

Fig. 6.3 Atmospheric turbulence phase screens generated through power spectrum inversion, C_n^2 from left to right is 1×10^{-13} m$^{-2/3}$,1×10^{-14} m$^{-2/3}$ and 1×10^{-15} m$^{-2/3}$ respectively

Fig. 6.4 Various atmospheric turbulence phase screens under identical turbulence strength ($C_n^2 = 1 \times 10^{-14} \text{m}^{-2/3}$)

6.2 Atmospheric Turbulence Effects on Vortex Beams

6.2.1 Beam Propagating in Atmospheric Turbulence

Beams propagating in atmospheric turbulence will suffer lots of effects such as intensity fluctuations, beam drift, jitter and beam expansion.

① Intensity undulation

Intensity fluctuation of a beam, also known as intensity flicker, is the random fluctuation of the intensity I around the mean value over time after propagating a certain distance. This variation result from the refractive index change caused by the temperature fluctuation.

In Kolmogorov weak turbulence, the logarithmic variance of the locally uniform, isotropic intensity is

$$\sigma_{\ln I}^2 = AC_n^2 \kappa^{7/6} \Delta z^{11/6} \tag{6.2.1}$$

where A is a constant. Usually $A = 0.496$ for spherical waves and $A = 1.23$ for plane waves [11]. Equation (6.2.1) shows that $\sigma_{\ln I}^2$ is proportional to the 11/6th power of the transmission distance Δz, the 7/6th power of the spatial wave number κ and C_n^2. It should be noted that Eq. (6.2.1) becomes more complicated when the transmission distance is long and the turbulence intensity is high, which is not discussed here.

② Beam drift

Beam drift is the most common beam distortion in atmospheric turbulence. After a certain distance of transmission, the position of the beam center in the transverse plane changes, as shown in Fig. 6.5. The beam drift effect has a serious impact on engineering applications such as LIDAR, laser ranging and optical communications, and is therefore widely taken into account.

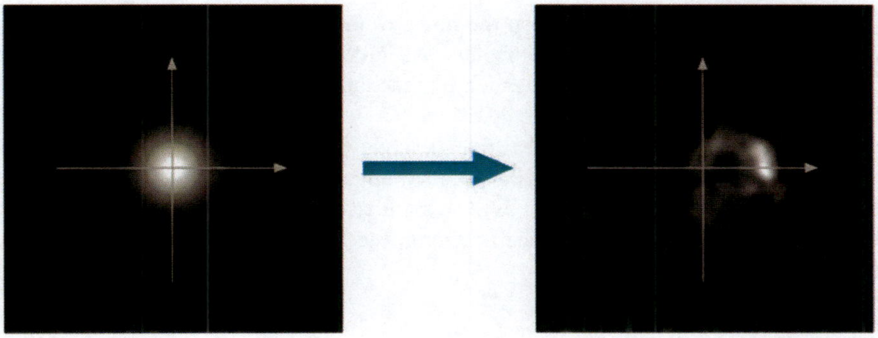

Fig. 6.5 Beam drift

For a beam, if the mean square deviation of the centroid drift in horizontal and vertical direction are σ_x^2 and σ_y^2, respectively, and they are statistically independent, the mean square deviation of the centroid drift of the beam can be expressed as:

$$\sigma_r^2 = \sigma_x^2 + \sigma_y^2 \tag{6.2.2}$$

For a plane wave and a collimated beam, when the atmospheric turbulence strength is uniform, that is, the atmospheric refractive index structure constant C_n^2 is a fixed value, the drift variance of the centroid of the facula is:

$$\sigma_r^2 = 2.03 C_n^2 D^{-1/3} \Delta z^3 \tag{6.2.3}$$

where D is the divergence aperture. Formula (6.2.3) shows that the drift of the pattern is independent of the wavelength, and the drift of the convergent beam is less than that of the collimated beam because of the smaller divergence aperture. Under strong turbulence, due to the serious beam distortion, the above discussion will no longer be meaningful.

③ Jitter

Atmospheric turbulence in the beam path results in the random change of refractive index, which affects the average angle of arrival of the beam and causes source-image jitter. When a beam propagates through the homogeneous medium, the far-field light field will have uniform wavefront. When the light wave propagates in atmospheric turbulence, the refractive index of the atmosphere is various in different parts of the beam, leading to the optical path difference and phase shift. These phase shifts cause changes in the shape of the isophase plane and fluctuations in received, i.e., the jitter of the image points.

④ Beam expansion

Beam expansion occurs when laser beams passing through turbulence of small scale. Beam expansion can be divided into short-term expansion and long-term expansion.

In a short period of time, a spot on the plane of the image will appear as a spot on the extended beam, which results from small turbulence. When the sampling time is long, the spot diameter will be much larger than that of the short-term expanded beam. This is because the large scale turbulence will lead to the refraction effect of the beam. The center position of the spot will change, so that the superposed image of the spot in different positions can be obtained under the condition of long exposure. And the imaging effect will appear as the expanded spot. In weak turbulence, the spot will disperse. When the turbulence is strong, the beam will be distorted seriously, and the spot will be broken.

6.2.2 Spiral Phase Distortion

In addition to the four transmission effects mentioned above, vortex beams have other more complex effects due to their spiral wave front and the OAM they carry.

When a vortex beam is transmitted through atmospheric turbulence, its spiral wavefront is distorted. Figure 6.6 shows the intensity and phase distribution of a vortex beam with Topological charges of +1, +3 and +5 after 1 km propagation in atmospheric turbulence with three different strengths. The specific parameters of the vortex beams are, wavelength $\lambda = 1550$ nm, and the radius of the fundamental waist $\omega_0 = 7$ cm. The atmospheric refractive structural constants of three atmospheric turbulences are 1×10^{-13} m$^{-2/3}$, 1×10^{-14} m$^{-2/3}$ and 1×10^{-15} m$^{-2/3}$, respectively. Their coherence lengths can be calculated from Eq. (6.1.6) as 1.97 cm, 7.85 cm and 31.24 cm, respectively. Since the beam size of the vortex beam and the probe Gaussian beam satisfy the relationship given in Eq. (2.4.10), namely, the spot radius of the vortex beam is different for different orders, d/r_0 is also different. As mentioned previously, the larger the d/r_0, the greater the effect of turbulence on the beam. It can therefore be predicted that, ceteris paribus, higher order vortex beams are somewhat more influenced by turbulence than lower orders.

From Fig. 6.6, vortex beams distort to some extent at different orders after atmospheric turbulence transport. Specifically, in terms of light intensity, the perfectly doughnut intensity structure begins to distort. The stronger the atmospheric turbulence, the more severe the pattern dispersion. When $C_n^2 = 1 \times 10^{-13}\text{m}^{-2/3}$, the doughnut intensity distribution is no longer visible. In terms of phase, the transverse spiral-shaped phase distribution distorted to varying degrees as the turbulence intensity increases. The stronger the turbulence, the more severe the distortion. When $C_n^2 = 1 \times 10^{-13}\text{m}^{-2/3}$ is present, the OAM state or topological charge can no longer be determined directly from the phase distribution.

Similar to single-mode vortex beams, multi-mode mixed vortex beams also produce phase distortions when transmitted through atmospheric turbulence. Figure 6.7 shows the intensity and phase distribution of a two-fold multiplexed vortex beam ($|2\rangle + |6\rangle$) with wavelength $\lambda = 1550$ nm and fundamental mode waist $\omega_0 =$ 7 cm after 1 km of atmospheric turbulence of different intensities. It is obvious from Fig. 6.7 that the spiral phase distortion becomes more and more severe as the

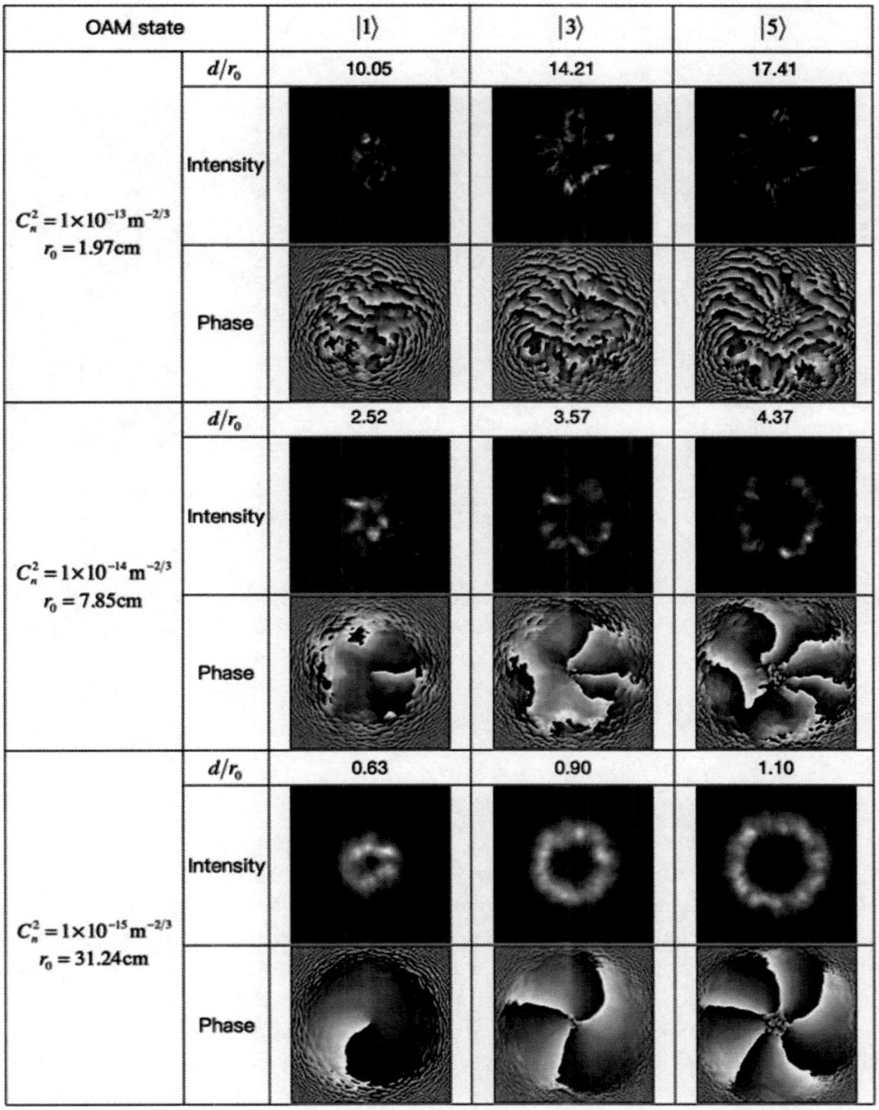

Fig. 6.6 Intensity and phase distribution of a vortex beam with topological charges +1, +3 and + 5 after 1 km propagation in atmospheric turbulence of three different strengths

turbulence intensity increases. And when $C_n^2 = 1 \times 10^{-13}\text{m}^{-2/3}$, the spiral phase is completely disordered.

This subsection gives a more intuitive picture of the atmospheric turbulence effect on the spiral phase of the vortex beam, through the analyzing intensity and phase distributions before and after atmospheric turbulence. However, the intensity and

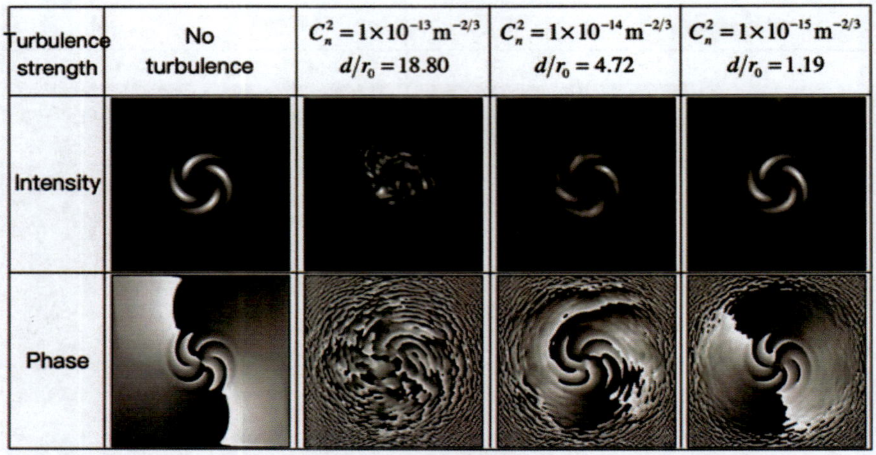

Fig. 6.7 Intensity and phase distributions of a two-fold multiplexed vortex beam ($|2\rangle + |6\rangle$) propagating through atmospheric turbulence of different strengths for 1 km

phase distributions cannot give the mode purity of distorted vortex beams. Hence a detailed analysis of the changes in the OAM spectrum before and after the turbulence is essential.

6.2.3 *OAM Spectrum Broadening*

Atmospheric turbulence contributes to spiral phase distortion, and such distortion will inevitably change the OAM spectrum. This change is manifested as the "leakage" of energy from the desired mode of transmission into adjacent modes, resulting in a decrease in the proportion of the final desired mode and a significant increase in the proportion of the irrelevant mode, as shown in Fig. 6.8, this phenomenon is known as the broadening of the OAM spectrum of the vortex beam [12].

OAM spectra of +1st, +3rd and +5th order vortex beams after 1 km propagation in three different atmospheric turbulence strength corresponding to Fig. 6.6 are given in Fig. 6.9, where all with the Topological charge in the horizontal coordinate and the relative intensity in the vertical coordinate. Here the OAM spectra are calculated by the complex amplitude derivation method introduced in Sect. 5.5.1.

From Fig. 6.9, the more intense the atmospheric turbulence, the more severe the dispersion of the OAM spectrum. When $C_n^2 = 1 \times 10^{-13}\text{m}^{-2/3}$, the Topological charge of the vortex beam can no longer be obtained directly from the OAM spectrum. The higher the order of the vortex beam and the larger the value of the parameter d/r_0, the greater the influence of turbulence on the vortex beam and the more severe the dispersion of the OAM spectra.

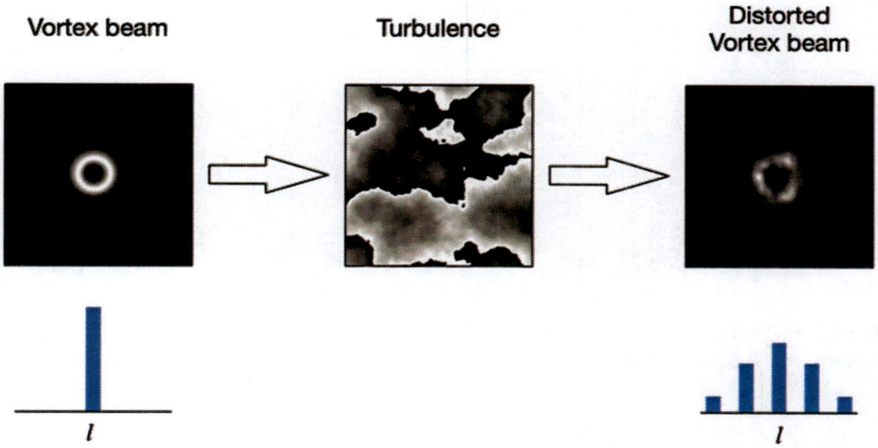

Fig. 6.8 OAM spectrum broadening after passing through atmospheric turbulence

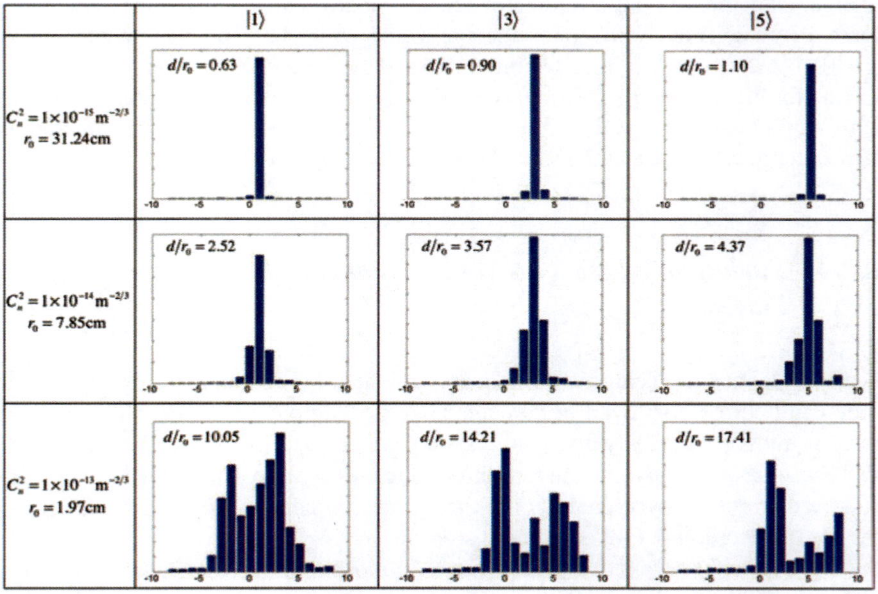

Fig. 6.9 OAM spectra of vortex beams after propagating in atmospheric turbulence of different strength for 1 km

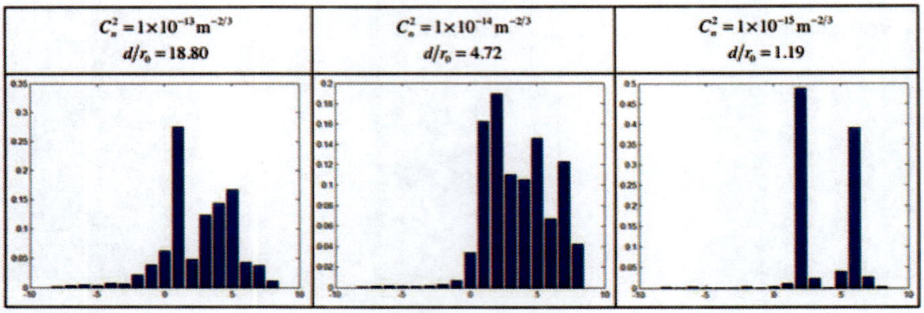

Fig. 6.10 OAM spectrum of a two-fold multiplexed vortex beam ($|2\rangle + |6\rangle$) after propagating in atmospheric turbulence of different strength for 1 km

Figure 6.10 shows the OAM spectra of the two-fold multiplexed vortex beam ($|2\rangle + |6\rangle$) after 1 km propagation in atmospheric turbulence of different strength corresponding to Fig. 6.7, where the horizontal coordinates are the Topological charges and the vertical coordinates are the relative intensities. Both modes are least affected in the weaker turbulence ($C_n^2 = 1 \times 10^{-15}\mathrm{m}^{-2/3}$). When $C_n^2 = 1 \times 10^{-14}\mathrm{m}^{-2/3}$, $l = 6$ is already obliterated and the $l = 2$ order component, although having the highest energy fraction, is not significantly different from adjacent orders. This again confirms that the higher order OAM components are more affected by atmospheric turbulence at the same turbulence strength.

6.2.4 Turbulence Effect on Vortex Beams with Different Patterns

Both single-ring LG beams and BG beams are the most common vortex beams, and they can both be employed in domains as optical communications. However, due to their transvers intensity distributions, they are necessarily affected or distorted to a different degree when transmitted through identical atmospheric turbulence. This sub-section will analyze the effect of turbulence on them in terms of the change in the OAM spectrum before and after distortion.

Since the refractive index distribution of atmospheric turbulence is a constantly changing quantity, the simulation can take the form of multiple turbulent phase screens to better simulate this stochastic process. For example, to simulate the transmission of a beam in turbulence with an atmospheric refractive index structure constant of C_n^2 at a distance of L, N C_n^2 turbulent phase screens with a transmission distance of L/N can be generated. The beam field after passing through these N phase screens can be calculated in sequence using scalar diffraction theory, as shown in Fig. 6.11. In other words, the initial complex amplitude is used to calculate fields after the first turbulent screen has transmitted L/N, and the calculated field is then used as the complex amplitude to calculate the light field after the second turbulent screen, until all turbulent screens have been calculated. The resulting diffracted field is the distorted field of the beam propagating in turbulence for distance L.

The OAM spectra of the third order ($l = 3$) single-ring LG beam and the BG beam after 1 km of identical turbulent transport are given in Fig. 6.12 [13]. Here, in order to control variate, the beam waist is set to 6 cm for both the central ring of the BG beam and LG beam. It can be predicted that the BG beam has a more complex phase structure than the LG beam, so its spiral phase distortion should be more severe after turbulence. Figure 6.12 shows that the OAM spectra of both two beams are broadened when the turbulence is weak ($C_n^2 = 1 \times 10^{-15}\text{m}^{-2/3}$), but their main intensities are still concentrated at the 3rd order position. In addition, the intensity of the BG beam is much smaller than that of the single-ring LG beam, indicating that the BG beam is somewhat more influenced by the turbulence at this time. As the turbulence intensity increases to $C_n^2 = 1 \times 10^{-14}\text{m}^{-2/3}$, the broadening of their OAM spectra is more severe, but the energy ratio of the non-transmitted mode of the BG beam is higher than that of the LG beam. When the turbulence intensity is further increased to $C_n^2 = 1 \times 10^{-13}\text{m}^{-2/3}$, their OAM spectra are completely disordered and the transmitted modes can no longer be distinguished at this point.

Figure 6.13 shows the OAM spectra of a four-fold multiplexed ($|-6\rangle + |-2\rangle + |3\rangle + |7\rangle$) single-ring LG beam and a BG beam before and after propagating 1 km through turbulence of the identical intensity [13]. Where the beam waist radius of the central ring of a BG beam of the same order are identical with that of a single-ring LG beam. The distribution of the OAM spectrum shown in Fig. 6.13 is similar to that in Fig. 6.12. When $C_n^2 = 1 \times 10^{-15}\text{m}^{-2/3}$, the main intensity of both beams is concentrated in the transmitted mode, but the intensity of the BG beam is much smaller than that of the single-ring LG beam. When the turbulence intensity is raised to $C_n^2 = 1 \times 10^{-13}\text{m}^{-2/3}$, their OAM spectra are both completely disordered and the transmitted modes annihilate in other unrelated modes.

The results in Figs. 6.12 and 6.13 show that when the beam waist radius of the central ring of the BG beam is equivalent to that of a single-ring LG of the same order, the wavefront distortion of the BG beam is worse when transmitted through the same atmospheric turbulence. The reason can be understood as, the BG beam has a paraflop structure and therefore its beam size is much larger than that of a single ring LG beam. That is, the BG beam has a larger d/r_0 value and is more influenced by turbulence. It also follows that increasing the number of rings of a LG beam (changing the value of the radial quantum number p), while keeping the central ring size constant, results in a larger wavefront distortion. Figure 6.14 shows a scatter

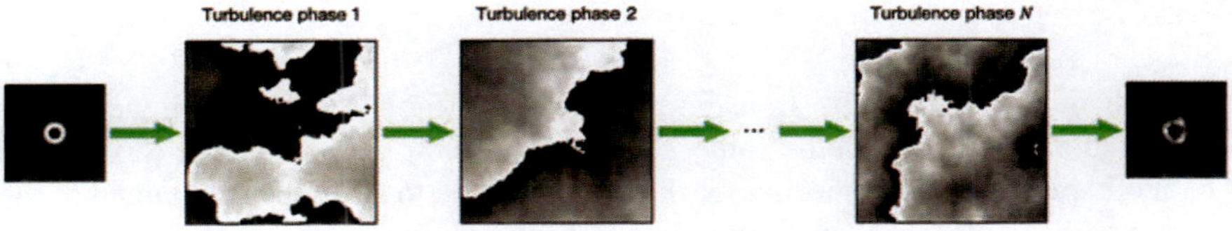

Fig. 6.11 Simulation of vortex beams propagation in atmospheric turbulence

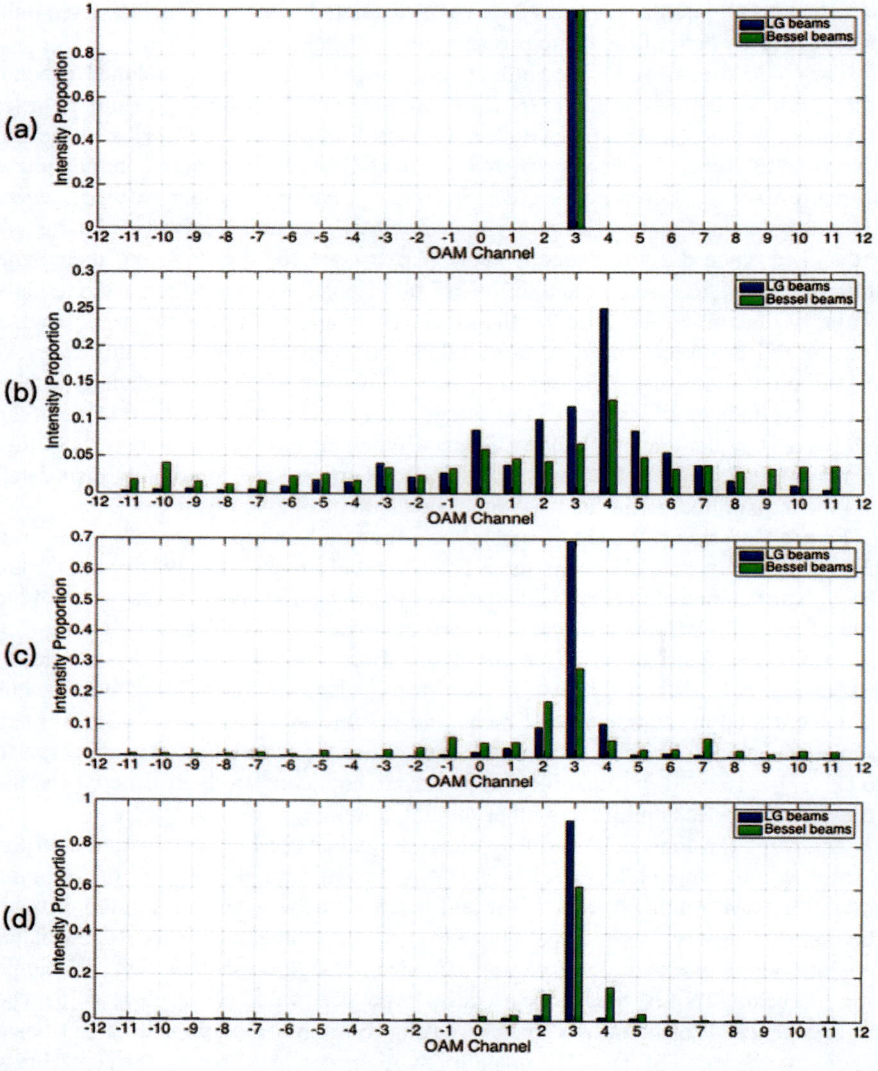

Fig. 6.12 OAM spectra of a single-ring LG beam and a BG beam when $l = 3$ before and after propagation through identical turbulence for 1 km. **a** Before propagation; **b–d** After propagation under C_n^2 1×10^{-13} $m^{-2/3}$, 1×10^{-14} $m^{-2/3}$ and 1×10^{-15} $m^{-2/3}$, respectively [13]

plot of the mode purity of the transmission mode as a function of radial indices p for a single mode ($l = 3$) with the same 1 km transmission in atmospheric turbulence. The mode purity is the proportion of the desired mode to all modes contained in the vortex beam. In Fig. 6.14, the dashed line indicates the mode purity of the BG beam under the same conditions. It can be seen that as p increases, the number of rings of

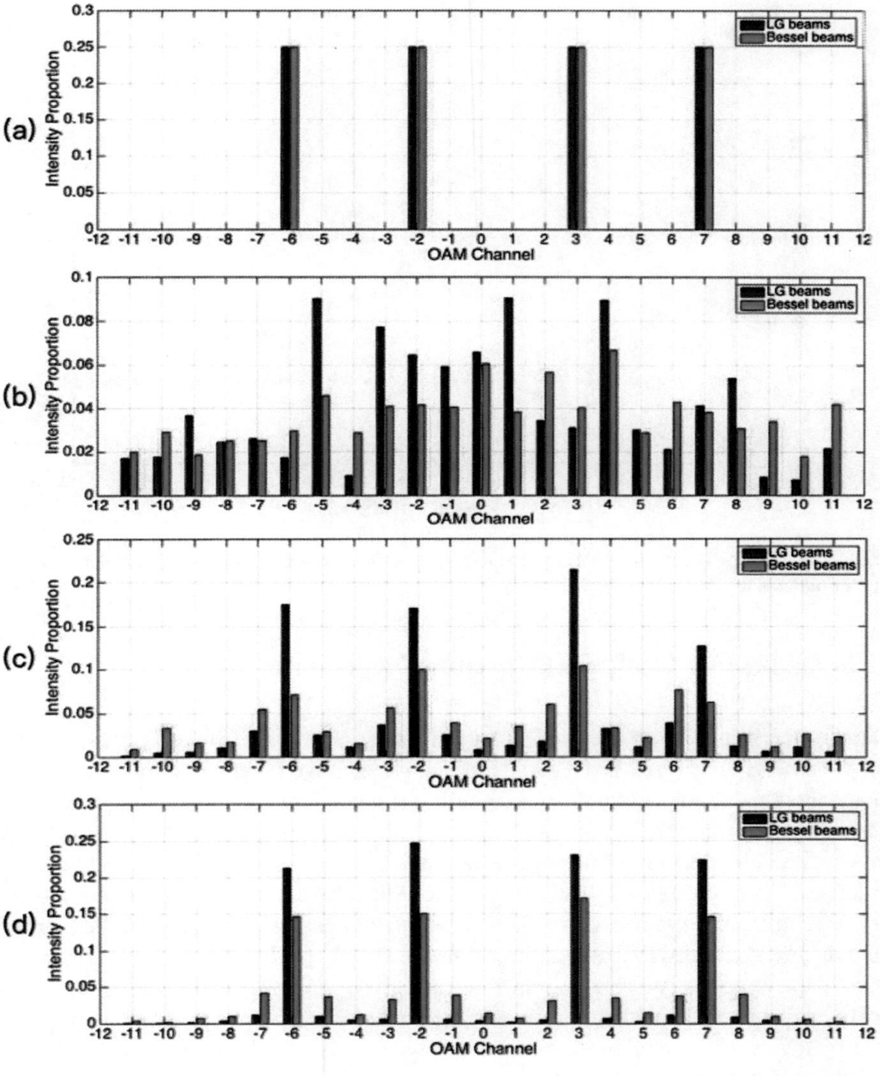

Fig. 6.13 OAM spectra of multiplexed ($|-6\rangle + |-2\rangle + |3\rangle + |7\rangle$) single-ring LG and BG beams before and after propagating through the same turbulence for 1 km. **a** Before propagation; **b–d** after propagation under C_n^2 1×10^{-13} m$^{-2/3}$, 1×10^{-14} m$^{-2/3}$ and 1×10^{-15} m$^{-2/3}$, respectively [13]

LG beams increases, the beam size becomes larger and the mode purity decreases. At $p = 9$, the mode purity is the same as for the BG beam and does not decrease significantly at $p = 10$. This is due to the fact that the beam sizes of the LG and BG beams are almost identical at this point, making their d/r_0 parameters almost identical and subject to the same turbulence.

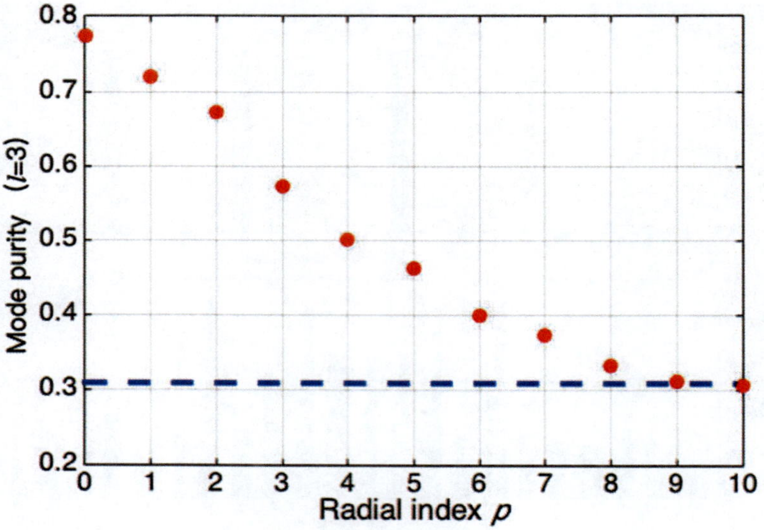

Fig. 6.14 The mode purity of the LG beam after passing through atmospheric turbulence versus radial indices p [13]

6.3 Adaptive Optical Distortion Correction

Due to the influence of atmospheric turbulence, spiral wavefront distortion present and OAM spectra are broadened, which go against practical applications. For instance, the broadened OAM spectra will introduce interchannel crosstalk in OAM based optical communications, and thus increase bit-error-rate. In addition, the OAM spectrum of beam emitted from celestial bodies in the universe can be used to determine their motion. Atmospheric turbulence at the Earth's surface will have a serious impact on its observations. Therefore, adaptive optics must be introduced to correct the distorted spiral wavefront and recover the OAM spectrum to its initial form.

Adaptive optics correction aims to recover the optical wavefront and to automatically measure, control and compensate the optical wavefront in real time to improve the quality of the optical field. As show in Fig. 6.15, a general adaptive optics system consists of three basic components: a wavefront detector, a wavefront controller and a wavefront corrector. The wavefront detector is used to capture information of beam's wavefront and to obtain the wavefront error in real time. The wavefront controller is used to obtain the wavefront error analyzed by the wavefront detector, which is then converted into a control signal for the wavefront corrector. The wavefront corrector is a phase control device that can quickly change the phase of the light wave, usually a deformation mirror or a phase-only liquid crystal spatial light modulator. The control signal from the wavefront controller controls the wavefront corrector to introduce a compensating phase for the distorted light field to correct the wavefront distortion.

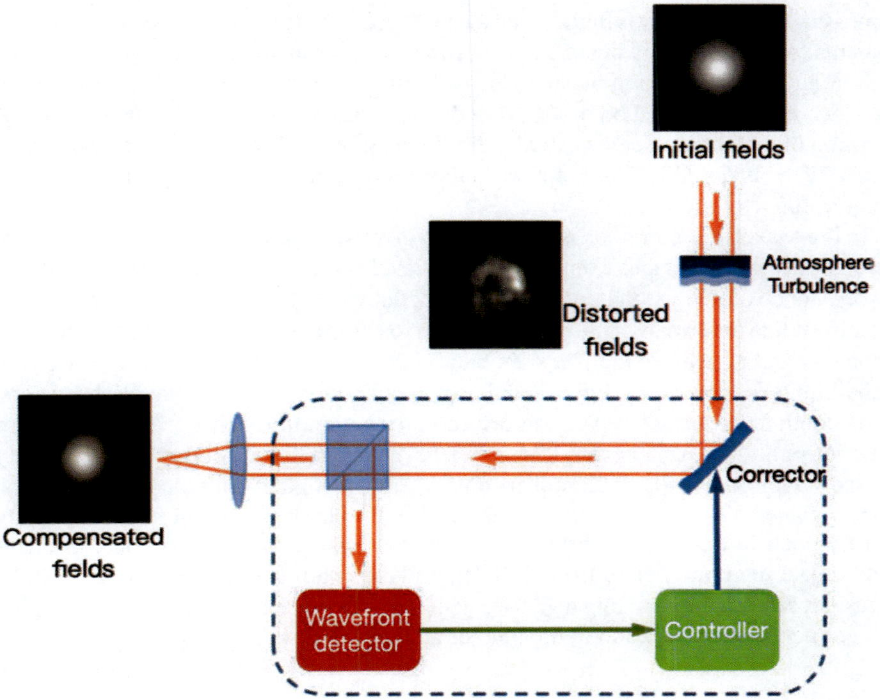

Fig. 6.15 Adaptive optics system

The method of obtaining information about the wavefront distortion of an optical field by means of a wavefront sensor called direct measurement of the wavefront phase. A variety of wavefront sensors are available, for instance, Shack-Hartmann wavefront sensors, curvature wavefront sensors, linear phase wavefront sensors and interferometric wavefront phase measurement sensors like Twyman Green interferometers. The use of Shack Hartmann wavefront sensors to detect distorted wavefronts has been widely used due to their relatively simple optical path design, low light source coherence requirements, large detection dynamic range and mature technology. The principle of the curvature wavefront sensor is to receive the light intensity of the front and rear out-of-focus surfaces of the system and obtain the wavefront phase and curvature information according to the corresponding algorithm, the disadvantage of which is that it is more difficult to detect the distortions of the medium and high frequency wavefronts. The linear phase wavefront sensor is calculated by measuring the distorted wavefront through the light intensity distribution of the far field after the system to obtain the phase information. Its disadvantage is that the application range is small and not suitable for detecting large scale distorted wavefronts. Interferometer wavefront sensors need to measure interference fringes, the measurement accuracy is higher than the previous wavefront sensors. But the requirements for optical path adjustment, experimental environment, etc. are very rigorous, limiting its application in the actual wavefront detection.

In addition, the method of inferring the phase information by the intensity distribution of the optical field is called indirect measurement of wavefront phase, which mainly includes GS phase retrieval algorithm, random parallel gradient descent algorithm, etc.

The wavefront corrector is the actuator of wavefront correction and is the core component of an adaptive optics system. Wavefront correctors usually work in two ways, one is to adjust the optical range by controlling the face shape of the optical element like deformation mirror, the other is to control the refractive index change of the element to achieve the phase modulation of the optical wavefront, for instance, the liquid crystal spatial light modulator. Among the various deformation mirrors, the continuous surface separation driven deformation mirror has the advantages of wavefront fitting, small error, high efficiency, etc., and is currently the most widely used wavefront corrector. Liquid crystals, as a very popular material in recent years, have gained a lot of applications in the research of laser beam tailoring. As already mentioned in Sect. 3.3.1, liquid crystal spatial light modulators are able to phase modulate the optical field from 0 to 2π, and the control of the optical field can be easily performed by loading a grey-scale image with high spatial resolution, making it an important application in the field of adaptive optics correction.

6.4 Shack Hartmann Compensation Method

6.4.1 Fundamentals of Shack Hartmann Wavefront Sensing

The Shack-Hartmann wavefront sensor consists of a microlens array and a CCD camera, and the image plane of the CCD camera coincides with the image-square focal plane of the microlens array, as shown in Fig. 6.16. When the light field passes through the microlens array, the subwave front incident on each microlens is focused on the image-square focal plane by the microlens. If the incident light field wave front is an ideal plane wave, the beam passing through each microlens is focused at the focal point and the CCD will receive an ideal dot array. If the incident beam has a wavefront distortion, i.e. a non-planar wavefront, the focused spot of the beam passing through the different microlenses will be out of focus, and the offset of the sub-focus point will vary from position to position. In this case, it is only necessary to measure the position information of each sub-spot to obtain the offset of the focal point of each molecular wavefront, and the reconstruction of the wavefront of the optical field can be carried out by means of a special adaptive optics algorithm.

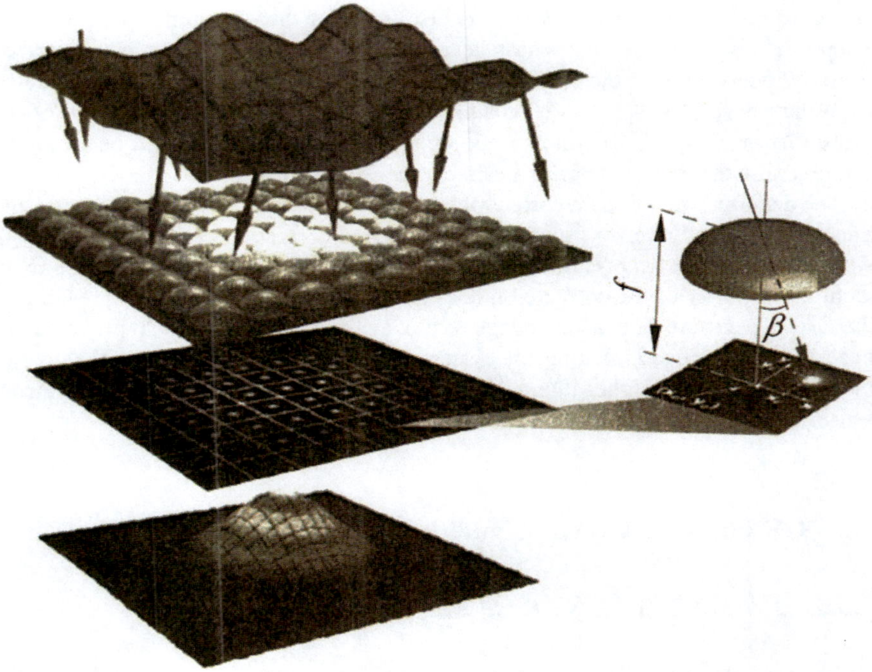

Fig. 6.16 Schematic diagram of the Shack Hartmann wavefront sensor [3]

6.4.2 Distorted Vortex Beam Compensation Through the Shack Hartmann Method

Shack Hartmann wavefront analyzers can measure the distortion of planar wavefronts very quickly and accurately, and are currently one of the main tools for wavefront detection. However, for vortex beams with spiral wavefronts and phase singularities, it is impossible to detect the optical wavefront through a Shack Hartmann wavefront sensor. And therefore correcting distortions of vortex beams using the conventional Shack Hartmann method is unavailable.

To address such issue, researchers have developed a solution that introducing a Gaussian beam probe, which effectively solves the problem of the wavefront sensor's inability to detect spiral wavefronts and enables adaptive distortion compensation of the vortex beam [14–16]. Firstly, an expanded Gaussian beam is used as the probe, which is required to have a larger beam waist radius than the vortex beam, and then the beam is coaxially combined with the vortex beam with orthogonal polarizations. When the combined beam is transmitted through atmospheric turbulence, the probe Gaussian beam and the vortex beam are subjected to the same turbulence and the same distortions are introduced. Therefore, as long as the wavefront of the probe Gaussian beam is measured by the Schack Hartmann wavefront sensor, the calculated corrected compensated phase screen is also valid for the transmitted vortex beam. Therefore, by placing a polarization beam splitting device, for instance, a polarized beam splitter or a Wollaston prism, at the receiver, the Gaussian beam can be separated from the transmitted beam as a probe. By detecting the distorted wavefront of the probe Gaussian beam, the compensation phase for the distorted vortex beam can be acquired and the distorted vortex beam can be well compensated.

The distorted vortex beam correction method based on Gaussian beam probes and Shack-Hartmann wavefront sensing achieves good correction performance for both single-mode vortex beams and multiplexed vortex beams, with significant improvement in the purity of the corrected modes and good suppression of the spreading of the OAM spectrum. The disadvantage of this method is that additional probe beams need to be introduced, making the system more complex. In addition, the cost of the whole system is higher due to the more expensive Shack Hartmann wavefront sensor.

6.5 GS Phase Retrieval Algorithm

6.5.1 Principles of GS Phase Retrieval Algorithm

GS algorithm has been introduced in Sect. 4.2.2 of this book. It calculates the required phase distribution of the optical field in the input plane from the known amplitude distribution of the optical field in the input plane and the required amplitude distribution in the output plane, which ultimately allows the optical system to modulate the incident beam according to our requirements.

The effect of atmospheric turbulence on the laser beam can be seen as a phase-only diffraction grating, namely, the modulating the beam by atmospheric turbulence phase screens. The distorted beam after atmospheric turbulence is in fact the diffraction field of the initial beam through the atmospheric turbulence phase screen. According to the properties of the GS algorithm, the phase distribution of the light field in the input plane can be calculated based on the intensity distribution of the initial light field and the distorted light field at the receiver, such that the distorted beam at the receiver can be generated after the initial beam has been sub-phase diffracted. This means that the phase distribution calculated by the GS algorithm plays the same role as the atmospheric turbulence phase screen, that is, the effect of atmospheric turbulence can be offset by introducing the inversion of the phase calculated by the GS algorithm before transmission via atmospheric turbulence. Thus the distorted beam can be pre-compensated.

Figure 6.17 gives a flow chart of the algorithm for calculating the pre-compensated phase $D(x,y)$ using the GS phase retrieval algorithm. The iterative process of the GS algorithm involves an iterative diffraction operation between the initial optical field and the distorted optical field located at the receiving plane. When the initial diffraction plane is at a finite distance from the receiving plane, this iterative process

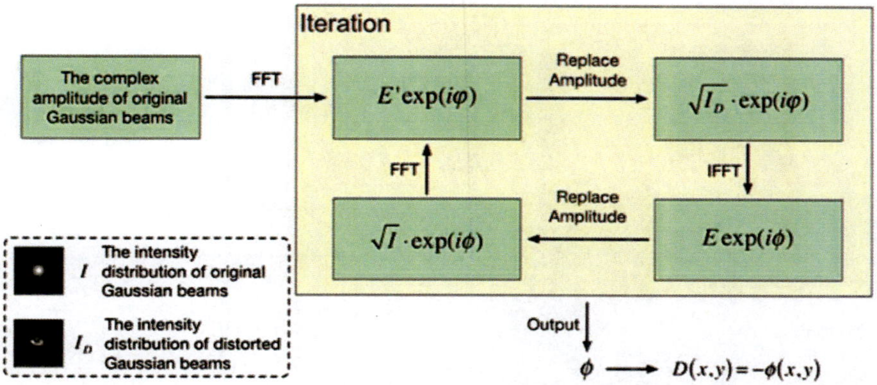

Fig. 6.17 Flow diagram of the GS algorithm based phase retrieval algorithm to calculate the pre-compensated phase [17]

can be realized by the angular spectrum in Sect. 1.4.3: from the initial light field $E_0(x, y)$ to the distorted light field $E_d(x, y)$, the angular spectrum operation is

$$E_d(x, y) = E_0(x, y) \otimes H_A\left(f_x, f_y\right) \tag{6.5.1}$$

where $H_A\left(f_x, f_y\right)$ is the transfer function of angular spectrum, $\otimes$ denotes the convolution operation. From the distorted light field $E_d(x, y)$ to the initial light field $E_0(x, y)$, require an inverse angular spectrum operation:

$$E_0(x, y) = E_d(x, y) \otimes H_A^*\left(f_x, f_y\right) \tag{6.5.2}$$

When the receiving plane is at infinity, the Fronhofer diffraction condition is satisfied and the diffraction iteration can be calculated simply by the Fourier transformation and inverse Fourier transformation.

Figure 6.18 shows the intensity distributions of the known distorted and initial Gauss Beams, the pre-compensation phase screen calculated by the GS phase retrieval algorithm, and the OAM spectra of the Gauss beams before and after compensation. The atmospheric refractive index structure constant $C_n^2 = 1 \times 10^{-13}\text{m}^{-2/3}$ and the parameter $d/r_0 = 1.79$. In Fig. 6.17, the broadening effect of the OAM spectrum of the probe Gaussian beam is obviously improved by the pre-compensated phase correction, and the main energy is concentrated at $l = 0$.

Unlike the Shack-Hartmann wavefront sensing, GS algorithm requires only the input of the initial and distorted light field intensity distribution. This can be achieved in a real adaptive system with a single CCD or CMOS camera, without a Shack-Hartmann wavefront sensor, which significantly reduce the cost. The GS phase retrieval algorithm is also known as adaptive optical correction without wavefront detection, as no wavefront detection is required to compensate for distorted beams.

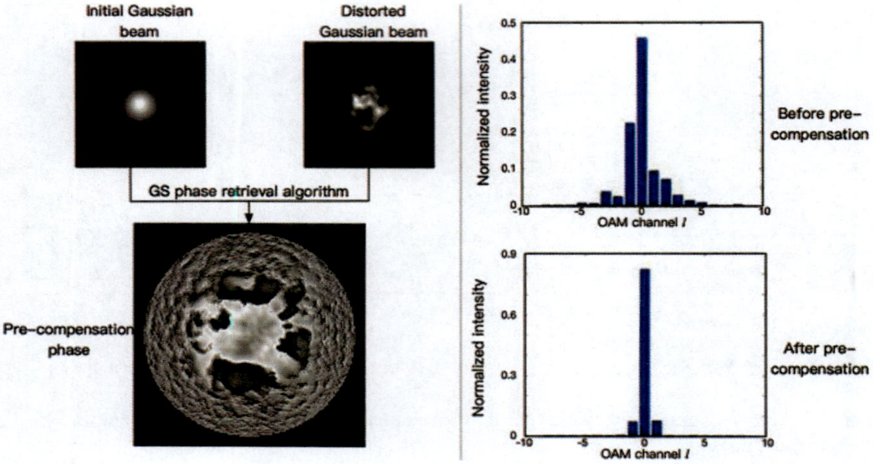

Fig. 6.18 Phase screen for distortion pre-compensation calculated by the GS phase retrieval algorithm, and also the OAM spectrum change of probe Gaussian beam before and after pre-compensation

6.5.2 *Probe Calibration Available*

The probe correction method is based on the GS phase retrieval algorithm, which is based on the same main idea as the Shack Hartmann method introduced in Sect. 6.4, as shown in Fig. 6.19. A probe Gaussian beam with a larger waist radius than the transmitted vortex beam is used as the probe, which is coaxially combined with the vortex beam in an orthogonal polarization through a device such as a polarized beam splitter. When the combined beam is transmitted through atmospheric turbulence, the fundamental mode Gaussian and vortex beams experience the same turbulence and are introduced with same distortions. A pre-compensated phase can then be calculated from the distortions and the intensity distribution of the initial Gaussian beam to pre-correct the vortex beam [17, 18].

Figure 6.20 gives the experimentally measured variation in mode purity for + 2nd and +4th order single-mode vortex beams before and after pre-correction by the GS phase retrieval algorithm for 100 iterations of the calculation. The turbulence intensity $r_0 = 1.4$ mm, parameter $d/r_0 = 2.47$ for +2nd order vortex beams and $d/r_0 = 3.19$ for +4th order vortex beams. It can be seen that the corrected mode purity has improved significantly, to more than 85% for all eight different turbulence scenarios [17]. Since only the intensity distribution of the initial and distorted Gaussian beams used as probes when calculating the pre-compensated phase, this method can also be used to correct distorted multiplexed vortex beams. Figure 6.21 gives the intensity ratio change of single OAM components of the two-fold multiplexed vortex beam ($|-4\rangle + |4\rangle$) before and after pre-compensation for 100 iterations, where $r_0 = 1.4$ mm and the parameter $d/r_0 = 3.19$. The intensity distribution of the uncorrelated modes

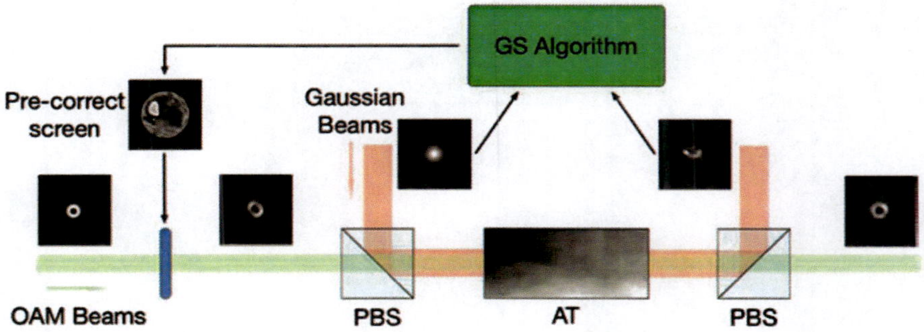

Fig. 6.19 Scheme for the pre-turbulence compensation based on a probe and the GS algorithm. PBS, polarized beam splitter; AT, atmospheric turbulence [17]

also appears in the absence of turbulence, due to the fact that the intensity of the single modes is measured here using the inverse mode conversion, that is, introducing an inverse vortex phase. The process of filtering out the central bright spot by a small aperture before measuring the power is obtained, a process very similar to the grey-scale algorithm introduced in Sect. 5.5.2, which inevitably introduces inter-mode crosstalk at small mode spacings (here 1). However, this crosstalk has no effect on the observed pre-correction effect, since as long as the intensity of the single modes after correction is essentially the same as in the absence of turbulence, the correction shows good performance.

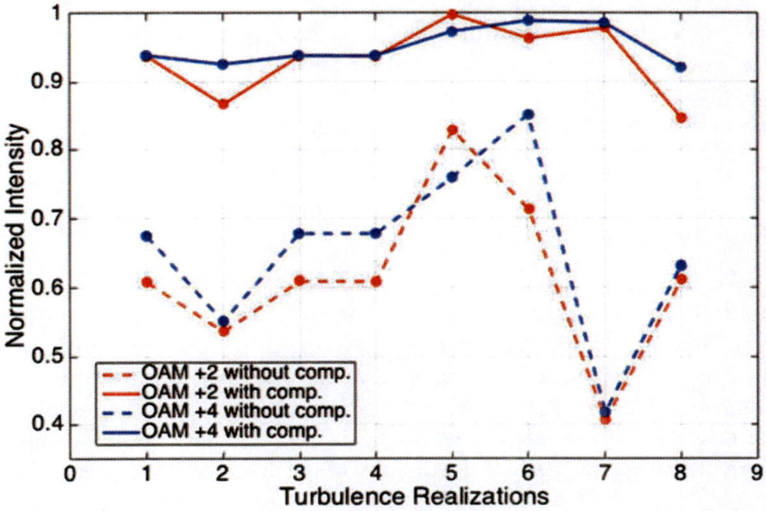

Fig. 6.20 Variation of mode purity before and after pre-correction of the single-mode vortex beam by GS phase retrieval algorithm [18]

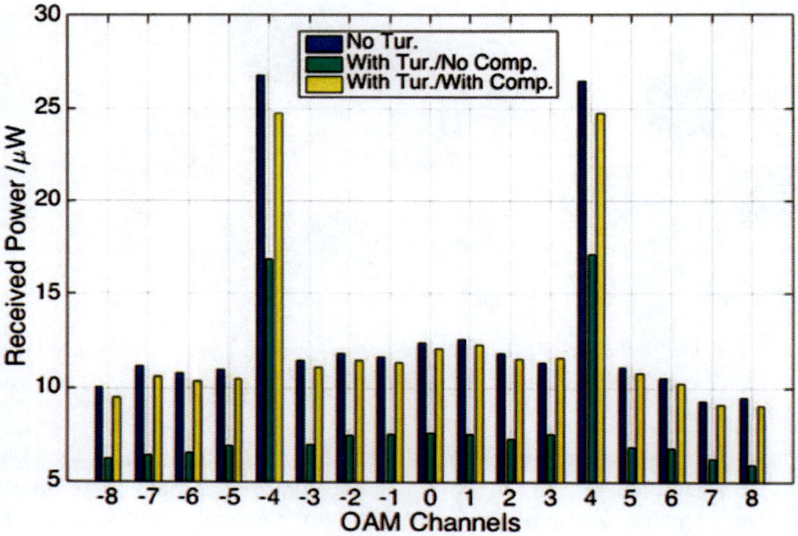

Fig. 6.21 Received power of various OAM components of two-fold multiplexed vortex beam before and after pre-correction [19]

The process of calculating the pre-compensated phase using the GS phase retrieval algorithm requires constant iterations. And the effect of pre-correction is related to the number of iterations. Figure 6.22 gives the experimentally measured variation of the mode purity of the distorted +2nd order vortex beam with the number of iterations, and also the intensity distribution of the vortex beam for a partial number of iterations. It can be seen that for single-mode vortex beams, a more satisfactory correction can be achieved at 70 iterations when calculating the pre-compensated phase, where the mode purity has been improved to 90%.

6.5.3 Non-probe Correction

In adaptive optical correction of distorted vortex beams using the Shack-Hartmann wavefront sensor, a probe beam must be introduced because the wavefront sensor cannot detect the distorted spiral wavefront directly. For the GS phase retrieval algorithm, in addition to introducing an additional probe beam, it is possible to operate directly on the distorted vortex beam and calculate the pre-compensated phase directly from the intensity distribution of the distorted vortex beam, as shown in Fig. 6.23 [18]. This probe free approach does not require the introduction of additional probe beams and greatly simplifies the complexity of the adaptive correction system. Figure 6.24 gives a flow chart of the GS phase retrieval algorithm that processing the distorted vortex beam directly. In non-probe scenario, the spiral phase

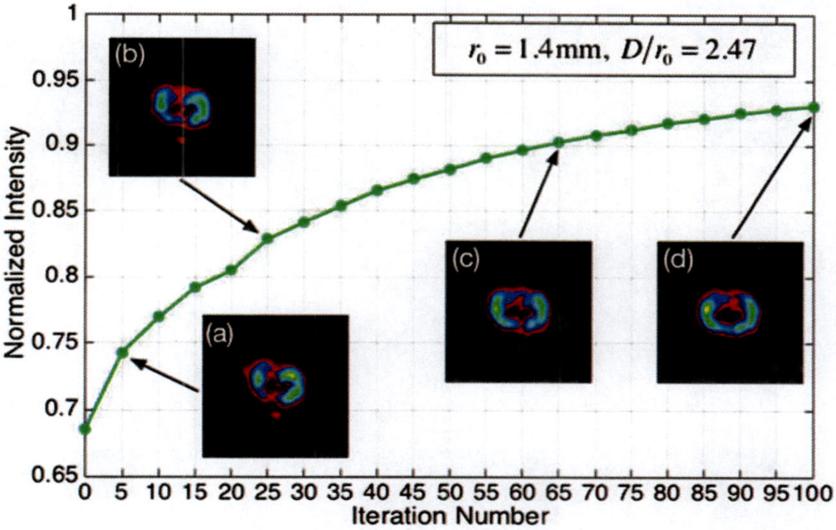

Fig. 6.22 Variation of mode purity with iteration numbers for distorted +2nd order vortex beams [17]

of the vortex beam must be considered during the iterative process. And finally the spiral phase of the transmitted vortex beam is subtracted from the phase output by the GS phase retrieval algorithm to obtain the pre-compensated phase $C(x,y)$.

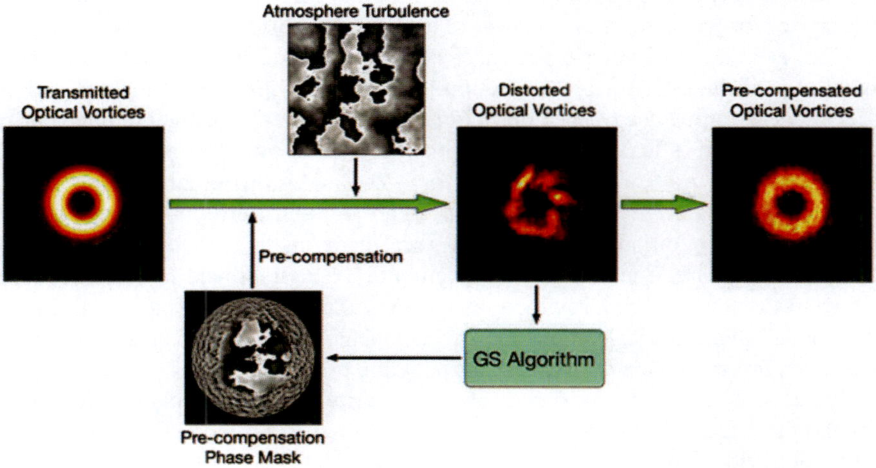

Fig. 6.23 Schematic diagram of non-probe correction of distorted vortex beam based on GS phase retrieval algorithm [18, 19]

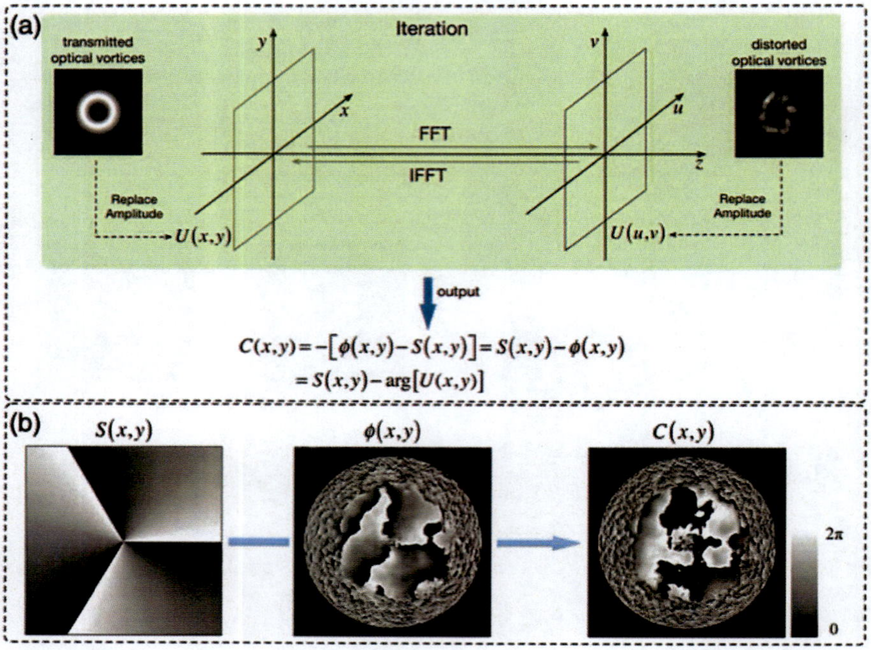

Fig. 6.24 Algorithm flow diagram of non-probe correction of distorted vortex beam based on GS phase retrieval algorithm [18]

The intensity distribution of the distorted vortex beam before and after non-probe GS phase retrieval algorithm is given in Fig. 6.25. The distortion of the vortex beam is more severe for stronger turbulence ($r_0 = 1$ mm) and the retrieval of the spot is more pronounced after pre-correction. The effect of pre-correction can also be seen at weak turbulence ($r_0 = 3$ mm). Figure 6.26 shows the change in mode purity of the +2nd and +3rd order vortex beams before and after pre-compensation. In this case, when $r_0 = 1$ mm, the d/r_0 for +2nd and +3rd order are 3.46 and 4 respectively. When $r_0 = 3$ mm, d/r_0 for +2nd and +3rd order are 1.15 and 1.33 respectively. In addition, 100 iterations were made in the calculation of the pre-compensated phase. The mode purity given in Fig. 6.26 improved significantly after compensation, with all improving to above 85% after pre-compensation, indicating that the GS phase retrieval algorithm can also achieve a more satisfactory adaptive distortion pre-correction without the introduction of a probe Gaussian beam.

Figure 6.27 shows the variation of mode purity with the number of iterations for the distorted +2nd order vortex beam measured using the GS algorithm pre-correction with no probe. When the turbulence is strong ($r_0 = 1$ mm), 100 iterations can improve the mode purity from 41.54 to 87.62%. When the turbulence is weak ($r_0 = 3$ mm), 50 iterations can increase the mode purity to over 90%.

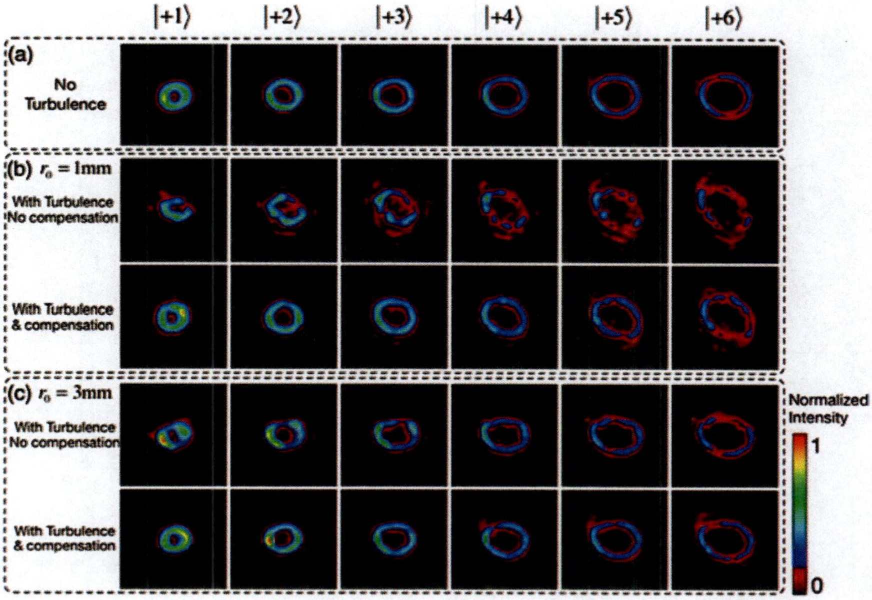

Fig. 6.25 Intensity distribution of the distortion of the vortex beam before and after non-probe correction under different turbulence strengths [18]

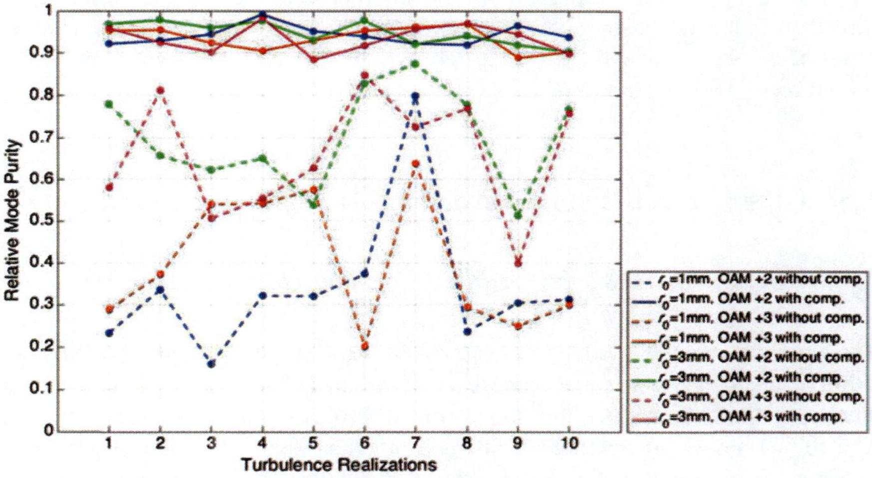

Fig. 6.26 Variation in mode purity of +2nd and +3rd order vortex beams before and after pre-correction under various turbulence realizations [18]

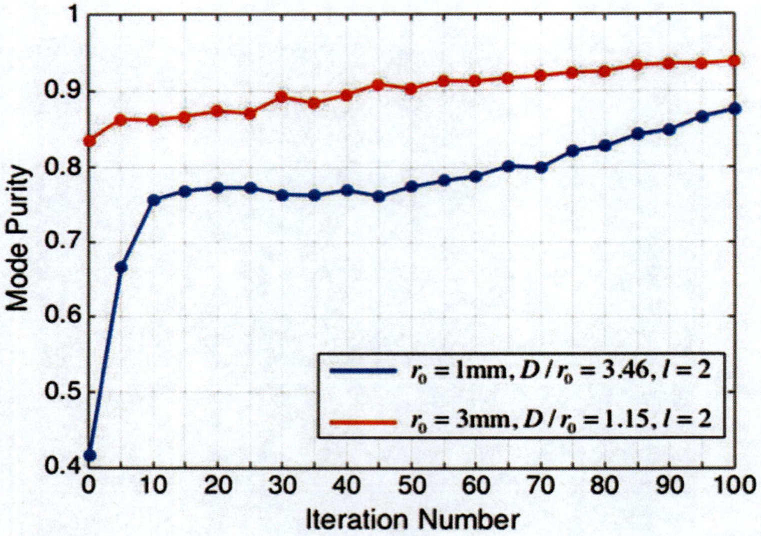

Fig. 6.27 Mode purity of 2nd order vortex beam versus iteration numbers in non-probe GS based pre-correction [18]

As already mentioned, spiral phases of vortex beams need to be taken into account when the GS phase retrieval algorithm is used to process the distorted vortex beam directly to calculate the pre-compensated phase, which requires us to be aware of the order of the transmitted vortex beam before pre-correction. Otherwise, probe free pre-correction will be unavailable.

6.6 Other Vortex Beam Correction Methods

6.6.1 Stochastic Parallel Gradient Descent Algorithm

The stochastic parallel gradient descent (SPGD) algorithm is an optimization algorithm based on the stochastic error descent algorithm and the simultaneous perturbed stochastic approximation algorithm for optical wavefront distortion correction [20]. The SPGD algorithm is efficient, fast and practical, and has a good background of application. Similar with GS algorithm, SPGD algorithm can also be employed for adaptive optical compensation without wavefront detection. Its key technologies include the performance evaluation function and the iterative method, where the performance evaluation function is to evaluate the good or bad correction effect. The traditional evaluation functions are Strehl Ratio (SR), Optical Transfer Function (OTF), image sharpness function, etc. [21].

The basic principle of the SPGD algorithm is described in detail as follows. Let $J(u)$ be the target performance evaluation function, where $u = (u_1, u_2, u_3, ..., u_N)$ is the control variable of the wavefront corrector. When the wavefront corrector is a device such as a deformation mirror, u is the driving voltage. When the wavefront corrector is a liquid crystal spatial optical modulator, u is the matrix of grey values loaded onto the modulator. J is unknown with respect to u and in order to calculate the corresponding gradient, it needs to be estimated from the measured values. Assuming a random perturbation Δu is applied to the control variable u, the amount of change in the objective function J is

$$\Delta J = J(u_1 + \Delta u_1, u_2 + \Delta u_2, u_3 + \Delta u_3, ..., u_N + \Delta u_N) - J(u_1, u_2, u_3, ..., u_N) \tag{6.6.1}$$

according to the Taylor series, expanding Eq. (6.6.1) at $(u_1, u_2, u_3, ..., u_N)$ gives

$$\Delta J = \sum_{j=1}^{N} \frac{\partial J}{\partial u_j}\Delta u_j + \frac{1}{2}\sum_{j,i=1}^{N} \frac{\partial^2 J}{\partial u_j \partial u_i}\Delta u_j \Delta u_i + ... \tag{6.6.2}$$

Multiply ΔJ with Δu_l,

$$\Delta J \Delta u_l = \frac{\partial J}{\partial u_i}(\Delta u_l)^2 + \chi_l \tag{6.6.3}$$

where

$$\chi_l = \sum_{j=1}^{N} \frac{\partial J}{\partial u_j}\Delta u_j \Delta u_l + \frac{1}{2}\sum_{j,i=1}^{N} \frac{\partial^2 J}{\partial u_j \partial u_i}\Delta u_j \Delta u_i \Delta u_l + ... \tag{6.6.4}$$

Averaging Eq. (6.6.3), we can obtain the expected value of $\Delta J \Delta u_l$

$$\langle \Delta J \Delta u_l \rangle = \frac{\partial J}{\partial u_j}\langle (\Delta u_l)^2 \rangle + \langle \chi_l \rangle \tag{6.6.5}$$

where

$$\langle \chi_l \rangle = \sum_{j=1}^{N} \frac{\partial J}{\partial u_j}\langle \Delta u_j \Delta u_l \rangle + \frac{1}{2}\sum_{j,i=1}^{N} \frac{\partial^2 J}{\partial u_j \partial u_i}\langle \Delta u_j \Delta u_i \Delta u_l \rangle + ... \tag{6.6.6}$$

Typically, the perturbation Δu_l is a statistically independent random variable that obeys the Bernoulli distribution. Its mean value is

$$\langle \Delta u_l \rangle = 0 \tag{6.6.7}$$

and its variance σ^2 satisfies.

$$\langle \Delta u_j \Delta u_l \rangle = \sigma^2 \delta_{jl} \tag{6.6.8}$$

where δ_{jl} is the Dirac function

$$\delta_{jl} = \begin{cases} 1, j = l \\ 0, j \neq l \end{cases} \tag{6.6.9}$$

Alternatively, for statistically independent random variables satisfying $\langle \Delta u_j \Delta u_i \Delta u_l \rangle = 0$, then Eq. (6.6.6) can be written as

$$\chi_l = o(\sigma^4) \tag{6.6.10}$$

where $o(\zeta)$ is a higher order infinitesimal of ζ. Therefore, Eq. (6.6.5) can be expressed as

$$\langle \Delta J \Delta u_l \rangle = \frac{\partial J}{\partial u_j} \sigma^2 + o(\sigma^4) \tag{6.6.11}$$

Dividing both sides of the equal sign of Eq. (6.6.11) by σ^2 simultaneously gives

$$\frac{\langle \Delta J \Delta u_l \rangle}{\sigma^2} = \frac{\partial J}{\partial u_j} + o(\sigma^2) \tag{6.6.12}$$

Equation (6.6.12) shows that $\langle \Delta J \Delta u_l \rangle / \sigma^2$ is an estimate of the gradient $\partial J / \partial u_j$ of the objective function J with an accuracy of $o(\sigma^2)$. Take the approximation

$$\frac{\partial J}{\partial u_j} \approx \frac{\Delta J \Delta u_j}{\sigma^2} \tag{6.6.13}$$

Let the number of iterations be k. Then consider the following iterations:

$$u_{k+1} = u_k - \gamma_k \left. \frac{\partial J}{\partial u} \right|_{u=u_k} \tag{6.6.14}$$

in Eq. (6.6.14), $\left. \frac{\partial J}{\partial u} \right|_{u=u_k}$ is the gradient of J at $u = u_k$. Substituting Eq. (6.6.13) into Eq. (6.6.14) gives

$$u_{k+1} = u_k - \mu \Delta J_k \Delta u_{l,k}, l = 1, 2, ..., N \tag{6.6.15}$$

where $\mu = \gamma / \sigma^2$ is defined as the search step size.

Equation (6.6.15) is the iterative formula of the SPGD algorithm. When $\mu > 0$, it corresponds to the minimum optimization process of the objective function. When $\mu < 0$, it corresponds to the maximum optimization process of the objective function.

Based on above principles, the specific process for calculating the compensated phase screen using the SPGD algorithm (Fig. 6.28) is as follows.

① Set the initial control variables $u = u_0$, $u_0 = (0,0,0,\ldots 0)$;

② Generating random perturbations Δu on the control variables obeying Bernoulli distribution;

③ Detecting the value of the change in performance $J^+ = J(u + \Delta u)$ after the perturbation;

④ Calculating the performance change $\Delta J = J^+ - J = J(u + \Delta u) - J(u)$;

⑤ Calculating the new control variable $u^+ = u - \mu \Delta J \Delta u$ according to Eq. (6.6.15) and update the variable $u = u^+$;

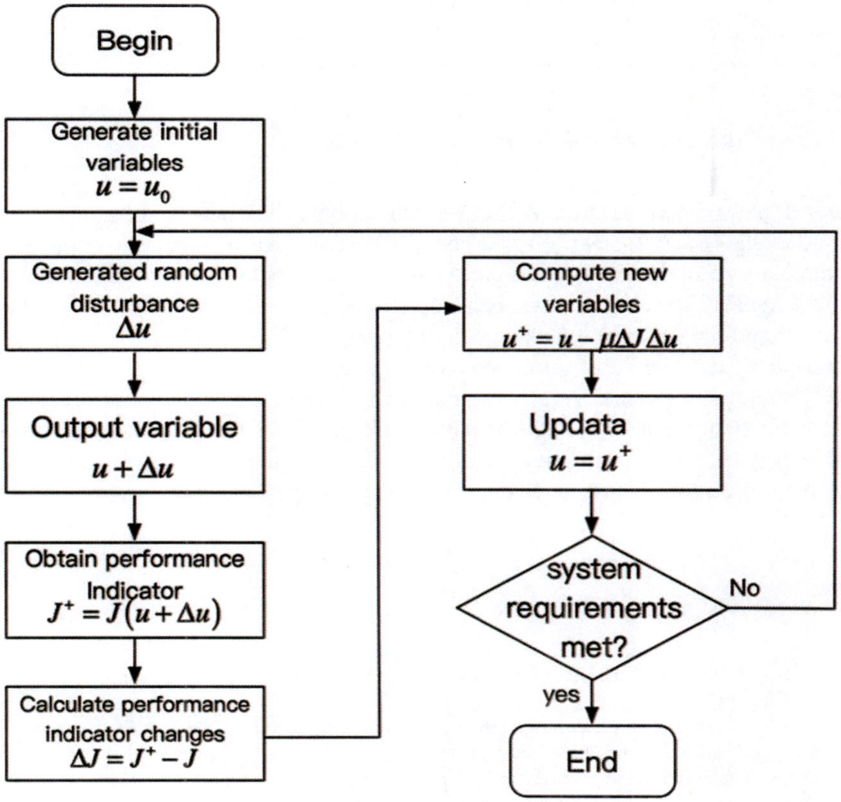

Fig. 6.28 SPGD algorithm calculation flow

⑥ Returning to step 2 and continue iterating until the system performance meets the requirements.

For vortex beams, the mode purity is related to the correlation coefficient of the corrected far-field intensity distribution, and satisfies the monotonic relationship, that is, the higher the correlation of the intensity distribution, the smaller the residual distortion and the higher the mode purity [22]. Therefore, the correlation coefficient of light intensity distribution can be used as the performance index of SPGD correction algorithm. First, the ideal intensity signal without atmospheric turbulence is collected by using CCD phase, and then the corresponding intensity signal of the distorted wavefront is collected by using CCD camera, according to the SPGD control algorithm, the system calculates the change of Performance Index, estimates the gradient, searches the control parameters in the gradient descent direction iteratively, and enters the next control cycle. The system controls the wavefront corrector in the above-mentioned iterative way, and corrects the distortion phase difference, finally obtains the distortion compensation near the diffraction limit more ideal vortex beam [23].

6.6.2 Digital Signal Processing Method

Current phase compensation methods for distorted vortex beams are usually based on adaptive optics. Recently, researchers have proposed a scheme to transfer the complexity of optical devices to electricity, interchannel crosstalk reduction using digital signal processing in vortex beam optical communication systems [24]. Single-channel and multiplexed vortex beams can generate crosstalk after passing through atmospheric turbulence. Multiple input multiple output (MIMO) equalizer can effectively suppress crosstalk, improve signal quality and reduce bit error rate, reducing the intensity of the irrelevant OAM pattern. However, in strong turbulence, most of the energy of vortex beam may transfer to other Topological charges, and MIMO equalizer is no longer helpful to improve the system performance.

References

1. Zhou RZ, Yan JX. Principles of adaptive optics. Beijing Institute of Technology Press; 1996 (in Chinese).
2. Han Y, Qiang X, Feng J, et al. Height distribution profiles and its application of atmosphere refractive index. Infrared Laser Eng. 2009;38(2):267–71 (in Chinese).
3. Dai K. The propagation property and adaptive distortion correction of OAM beams. Doctoral Dissertation of Beijing Institute of Technology; 2015 (in Chinese).
4. Kolmogorov A. The local structure of turbulence in an incompressible fluid at very high Reynolds number. Dokl Akad Nauk SSSR. 1941;30:299–303.

5. Davis JI. Consideration of atmospheric turbulence in laser systems design. Appl Opt. 1966;5(1):139–47.
6. Fried DL. Optical resolution through a randomly inhomogeneous medium for very long and very short exposures. J Opt Soc Am. 1966;56(10):1372–9.
7. Kármán TV. Progress in the statistical theory of turbulence. Proc Natl Acad Sci U S A. 1948;34(11):530.
8. Hill R. Models of the scalar spectrum for turbulent advection. J Fluid Mech. 1978;88:541–62.
9. Harding CM, Johnston RA, Lane RG. Fast simulation of a Kolmogorov phase screen. Appl Opt. 1999;38(11):2161–70.
10. Lane RG, Glindemann A, Dainty JC. Simulation of a Kolmogorov phase screen. Waves Random Media. 1992;2(3):209–24.
11. Zhang S. Adaptive distortion correction of OAM beams based on SPGD algorithm. Master's Thesis of Beijing Institute of Technology; 2017 (in Chinese).
12. Ren Y, Huang H, Xie G, et al. Atmospheric turbulence effects on the performance of a free space optical link employing OAM multiplexing. Opt Lett. 2013;38(20):4062–5.
13. Fu S, Gao C. Influences of atmospheric turbulence effects on the OAM spectra of vortex beams. 2016;4(5):B1–B4.
14. Ren Y, Xie G, Huang H, et al. Adaptive optics compensation of multiple OAM beams propagating through emulated atmospheric turbulence. Opt Lett. 2014;39(10):2845–8.
15. Ren Y, Xie G, Huang H, et al. Adaptive-optics-based simultaneous pre- and post-turbulence compensation of multiple orbital-angular-momentum beams in a bidirectional free-space optical link. Optica. 2014;1(6):376–82.
16. Li S, Wang J. Compensation of a distorted N-fold OAM multicasting link using adaptive optics. Opt Lett. 2016;41(7):1482–5.
17. Fu S, Zhang S, Wang T, et al. Pre-turbulence compensation of OAM beams based on a probe and the Gerchberg-Saxton algorithm. Opt Lett. 2016;41(14):3185–8.
18. Fu S, Wang T, Zhang S, et al. Non-probe compensation of optical vortices carrying OAM. Photonics Res. 2017;5(3):251–5.
19. Gao C, Fu S, Wang T, et al. Non-probe adaptive distortion correction of distorted vortex beams. Chinese Patent: ZL201610806984.1, 25 Jan 2017 (in Chinese).
20. Vorontsov MA, Sivokon VP. Stochastic parallel-gradient-descent technique for high-resolution wave-front phase-distortion correction. J Opt Soc Am A. 1998;15(10):2745–58.
21. Gao C, Zhang S, Fu S, et al. Adaptive optics wavefront correction techniques of vortex beams. Infrared Laser Eng. 2017;46(2):1–6 (in Chinese).
22. Huang H, Ren Y, Yan Y, et al. Phase-shift interference-based wavefront characterization for OAM modes. Opt Lett. 2013;38(13):2348–51.
23. Xie G, Ren Y, Huang H, et al. Phase correction for a distorted OAM beam using a Zernike polynomials-based stochastic-parallel-gradient-descent algorithm. Opt Lett. 2015;40(7):1197–200.
24. Huang H, Cao Y, Xie G, et al. Crosstalk mitigation in a free-space OAM multiplexed communication link using 4×4 MIMO equalization. Opt Lett. 2014;39(15):4360–3.

Chapter 7
Vector Beams and Vectorial Vortex Beams

Vector beams are a new class of beams whose polarization states are anisotropically distributed in transverse. Generally speaking, vector beams can take many forms. In this book, we only consider polarization vortices, which have a transverse inhomogenous polarization distribution and are the characteristic solutions of Maxwell's Eqs. in a cylindrical coordinate. To avoid ambiguity, beams with only polarization vortices, but no phase vortices, are usually called vector beams. Beams with both polarization vortices and phase vortices are called vectorial vortex beams. Vector beams carry no OAM, whereas vectorial vortex beams carry OAM due to their spiral phase structure. Vector beams and vectorial vortex beams have promising applications in laser processing, atomic cooling and surface plasma excitation.

7.1 Overview of Vector Beams

A vector beam with a transverse vortex polarization distribution can be represented by the Jones vector as:

$$\vec{E}(r,\varphi) = A(r)\begin{bmatrix} \cos(p\varphi + \theta_0) \\ \sin(p\varphi + \theta_0) \end{bmatrix} \tag{7.1.1}$$

where p is the polarization order or polarization topological charge, indicating the number of periods of polarization state changes around the transverse plane for one round. θ_0 is the initial polarization direction at $\varphi = 0$. $A(r)$ describes the amplitude distribution of the beam, commonly known as LG, BG, etc. Several common vector beams and their polarization distributions are listed in Fig. 7.1. Since vector beams have anisotropic transverse polarization distributions, their polarization states can be analyzed through a rotational polarizer, as shown in Fig. 7.1.

First-order ($p = 1$) vector beams are also known as cylindrical vector beams [1]. Depending on the initial point of polarization (θ_0), first-order vector beams can be

S. Fu and C. Gao, *Optical Vortex Beams*, Advances in Optics and Optoelectronics,
https://doi.org/10.1007/978-981-99-1810-2_7

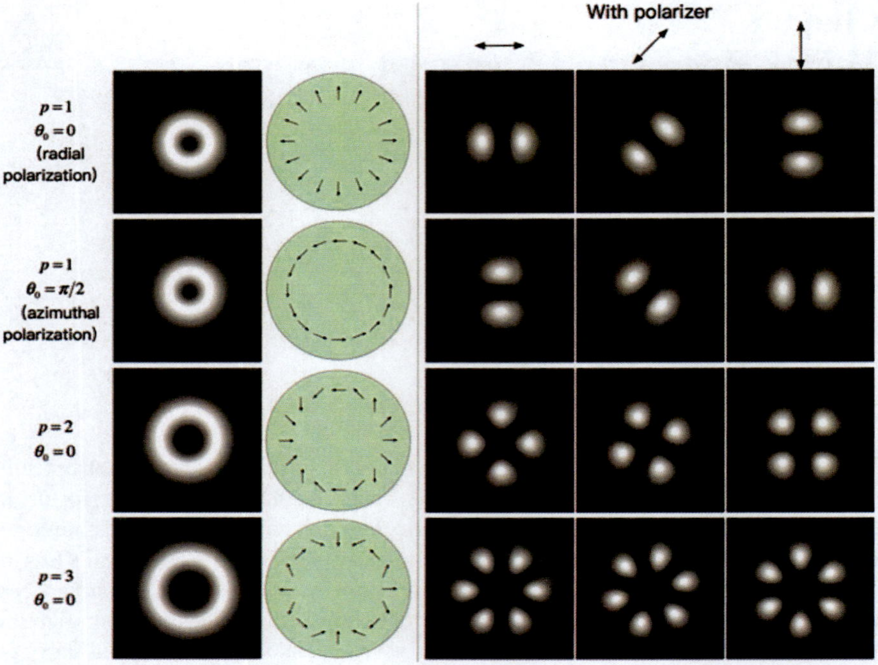

Fig. 7.1 Several common types of vector beams

divided into radially polarized beams ($\theta_0 = 0$) and azimuthally polarized beams ($\theta_0 = \pi/2$). Radially polarized beams and azimuthally polarized beams are the earliest researched and most widely employed vector beams. Normally, vector beams with polarization orders larger than 1 is regarded as higher order vector beams.

7.1.1 Bessel Solutions of Vector Beams

The Hermit Gauss and Laguerre Gauss modes are solutions of the scalar Helmholtz equation under paraxial conditions. They are often used to describe a component of a linearly polarized beam or a vector beam, whereas the vector wave Equation can describe the transvers electric field distribution under paraxial conditions.

Studying the vector wave equation requires some physical concepts drawn from the scalar fluctuation equation. Therefore, a brief description of the process of solving the scalar Helmholtz equation is given firstly. The scalar Helmholtz equation under paraxial condition is:

$$(\nabla^2 + k^2)E = 0 \tag{7.1.2}$$

The fundamental Gaussian beam is one of the its most common solutions. In a cylindrical coordinate, a beam propagating along z-axis reads:

$$E = f(r, \varphi, z)\exp[i(kz - \omega t)] \tag{7.1.3}$$

where k and ω are the wave number and angular frequency, respectively. And $\exp(-i\omega t)$ is the time term. Under the paraxial approximation, $\partial^2 f/\partial z^2 \ll \partial f/\partial z$, $\partial^2 f/\partial z^2$ can therefore be neglected. Thus $f(r, \varphi, z)$ is a solution of the scalar wave equation under the paraxial condition.

$$\frac{1}{r}\frac{\partial}{\partial r}\left(r\frac{\partial f}{\partial r}\right) + \frac{1}{r^2}\frac{\partial^2 f}{\partial \varphi^2} + 2ik\frac{\partial f}{\partial z} = 0 \tag{7.1.4}$$

The fundamental Gaussian mode solution is independent of the azimuthal angle φ and can therefore be expressed by the following equation.

$$f(r, z) = \frac{\omega_0}{\omega(z)}\exp[-i\Phi(z)]\exp\left(-\frac{r^2}{\omega_0^2}\cdot\frac{1}{1 + iz/z_R}\right) \tag{7.1.5}$$

In Eq. (7.1.5), z_R is the Rayleigh length $z_R = k\omega_0^2/2$, $\omega(z) = \omega_0\left[1 + (z/L)^2\right]^{0.5}$, $\Phi(z) = \arctan(z/z_R)$ and ω_0 is the waist radius.

On the basis of the scalar Helmholtz equation, defining the electric field into vector form, then vector wave equation of beam propagation is obtained as:

$$\nabla \times \nabla \times \vec{E} - k^2\vec{E} = 0 \tag{7.1.6}$$

To facilitate mathematical modelling and solution, considering a special case, namely, azimuthal polarization, $\vec{E}$ can be expressed as

$$\vec{E}(r, z) = \vec{F}(r, \varphi, z)\exp[i(kz - \omega t)] \tag{7.1.7}$$

where $\partial^2\vec{E}/\partial z^2$ is omitted under the paraxial approximation. Assuming that the transverse intensity of the beam meets with Gaussian distribution, it can be written in the form of the product of two factors:

$$\vec{F} = \vec{\varepsilon}(r, \varphi, z)f(r, z) \tag{7.1.8}$$

where $f(r, z)$ is the Gaussian function given in Eq. (7.1.5) and $\vec{\varepsilon}(r, \varphi, z) = \left(\varepsilon_r, \varepsilon_\varphi\right)$ is the unit vector in the radial and azimuthal directions, representing the electric field distribution in transverse. Substituting Eq. (7.1.8) into Eq. (7.1.6) after paraxial approximation, two differential Eqs. can be obtained as:

$$\frac{1}{r}\frac{\partial}{\partial r}\left(r\frac{\partial \varepsilon_r}{\partial r}\right) - \frac{4r/\omega_0^2}{1 + iz/L}\frac{\partial \varepsilon_r}{\partial r} + \frac{1}{r^2}\frac{\partial^2 \varepsilon_r}{\partial \varphi^2} - \frac{\varepsilon_r}{r^2} - \frac{2}{r^2}\frac{\partial \varepsilon_\varphi}{\partial \varphi} + 2ik\frac{\partial \varepsilon_r}{\partial z} = 0 \tag{7.1.9}$$

$$\frac{1}{r}\frac{\partial}{\partial r}\left(r\frac{\partial \varepsilon_\varphi}{\partial r}\right)-\frac{4r/\omega_0^2}{1+iz/L}\frac{\partial \varepsilon_\varphi}{\partial r}+\frac{1}{r^2}\frac{\partial^2 \varepsilon_\varphi}{\partial \varphi^2}-\frac{\varepsilon_\varphi}{r^2}+\frac{2}{r^2}\frac{\partial \varepsilon_r}{\partial \varphi}+2ik\frac{\partial \varepsilon_\varphi}{\partial z}=0 \quad (7.1.10)$$

The solutions of Eqs. (7.1.9) and (7.1.10) are [2],

$$\varepsilon_{r,m}(r,\varphi,z)=Q(z)\left[a_m J_{m-1}(u)+b_m J_{m+1}(u)\right]\begin{cases}\cos m\varphi\\ \sin m\varphi\end{cases} \quad (7.1.11)$$

$$\varepsilon_{\varphi,m}(r,\varphi,z)=Q(z)\left[a_m J_{m-1}(u)-b_m J_{m+1}(u)\right]\begin{cases}-\sin m\varphi\\ \cos m\varphi\end{cases} \quad (7.1.12)$$

where $J_\tau(\zeta)$ is τ-th order Bessel function of the first kind

$$Q(z)=\exp\left(-\frac{i\beta^2 z/2k}{1+iz/z_R}\right) \quad (7.1.13)$$

$$u=\frac{\beta r}{1+iz/z_R} \quad (7.1.14)$$

β is a parameter determining the spot size. m is the topological charge. a_m and b_m are constants. When $a_m=b_m$, Eqs. (7.1.11) and (7.1.12) can be simplified to

$$\varepsilon_{r,m}(r,\varphi,z)=a_m Q(z)\left[J_{m-1}(u)+J_{m+1}(u)\right]\begin{cases}\cos m\varphi\\ \sin m\varphi\end{cases} \quad (7.1.15)$$

$$\varepsilon_{\varphi,m}(r,\varphi,z)=a_m Q(z)\left[J_{m-1}(u)-J_{m+1}(u)\right]\begin{cases}-\sin m\varphi\\ \cos m\varphi\end{cases} \quad (7.1.16)$$

Assuming the coefficient $a_m=1$, Eqs. (7.1.15) and (7.1.16) can be expressed uniformly as

$$\vec{E}(r,\varphi,z)=\begin{pmatrix}\left[J_{m-1}(u)-J_{m+1}(u)\right]\begin{cases}-\sin m\varphi\\ \cos m\varphi\end{cases}\cdot\vec{e}_\varphi\\ +\left[J_{m-1}(u)+J_{m+1}(u)\right]\begin{cases}\cos m\varphi\\ \sin m\varphi\end{cases}\cdot\vec{e}_r\end{pmatrix}Q(z)f(r,z) \quad (7.1.17)$$

where $\vec{e}_r$ and $\vec{e}_\varphi$ are the unit direction vectors in polar coordinates. If the unit direction vectors in Cartesian coordinates are $\vec{e}_x$ and $\vec{e}_y$, the coordinate transformation relationship between them and polar coordinates reads

$$\begin{bmatrix}\vec{e}_r\\ \vec{e}_\varphi\end{bmatrix}=\begin{bmatrix}\cos\varphi & \sin\varphi\\ -\sin\varphi & \cos\varphi\end{bmatrix}\begin{bmatrix}\vec{e}_x\\ \vec{e}_y\end{bmatrix} \quad (7.1.18)$$

Equation (7.1.17) can be expressed in the form of the Jones vector in a Cartesian coordinate as:

$$\begin{bmatrix} E_x \\ E_y \end{bmatrix} = \begin{Bmatrix} [J_{m-1}(u) - J_{m+1}(u)]\alpha_\iota \begin{bmatrix} -\sin\varphi \\ \cos\varphi \end{bmatrix} \\ +[J_{m-1}(u) + J_{m+1}(u)]\beta_\xi \begin{bmatrix} \cos\varphi \\ \sin\varphi \end{bmatrix} \end{Bmatrix} Q(z)f(r,z) \tag{7.1.19}$$

where

$$\alpha_\iota = \begin{cases} -\sin(m\varphi), \iota = 0 \\ \cos(m\varphi), \iota = 1 \end{cases} \tag{7.1.20}$$

$$\beta_\xi = \begin{cases} \cos(m\varphi), \xi = 0 \\ \sin(m\varphi), \xi = 1 \end{cases} \tag{7.1.21}$$

Figure 7.2 shows the intensity distribution of the vector beam calculated according to Eq. (7.1.19). When $m = 0$, $J_{-1}(u) + J_1(u) = 0$ and the light field distribution is independent of β_ξ. When $\iota = 0$, $\alpha_\iota = 0$, the mathematical expression for the light field is constantly equal to zero; when $\iota = 1$, $\alpha_\iota = 1$, the distribution of the light field can be expressed as

$$\begin{bmatrix} E_x \\ E_y \end{bmatrix} = 2Q(z)f(r,z)J_{-1}(u) \begin{bmatrix} -\sin\varphi \\ \cos\varphi \end{bmatrix} \tag{7.1.22}$$

Equation (7.1.22) is the azimuthally polarized beam with a BG type amplitude distribution. When $m \neq 0$, the beam has a complex intensity and polarization distribution. As the topological charge m increases, the number of flaps of the intensity distribution increases and the number of main flaps is equal to $2m$.

7.1.2 Laguerre–Gauss Solutions of Vector Beams

On the basis of derivation of the BG solution for vector beams proposed by Hall et al. [2], Tovar derived the LG solution of vector beams in 1998 [4]:

$$\vec{E}(r,\varphi,z) = E_0 \exp\left[-i\int_0^z k_0(z')dz'\right] \exp\left\{-i\left[\frac{k_0}{q(z)}\frac{r^2}{2} + P(z)\right]\right\} \cdot \\ \left[\frac{2r^2}{W^2(z)}\right]^{\frac{l\pm 1}{2}} L_p^{|l|}\left[\frac{2r^2}{W^2(z)}\right] \begin{Bmatrix} \cos(l\varphi)\vec{e}_\varphi \mp \sin(l\varphi)\vec{e}_r \\ \pm\sin(l\varphi)\vec{e}_\varphi + \cos(l\varphi)\vec{e}_r \end{Bmatrix} \tag{7.1.23}$$

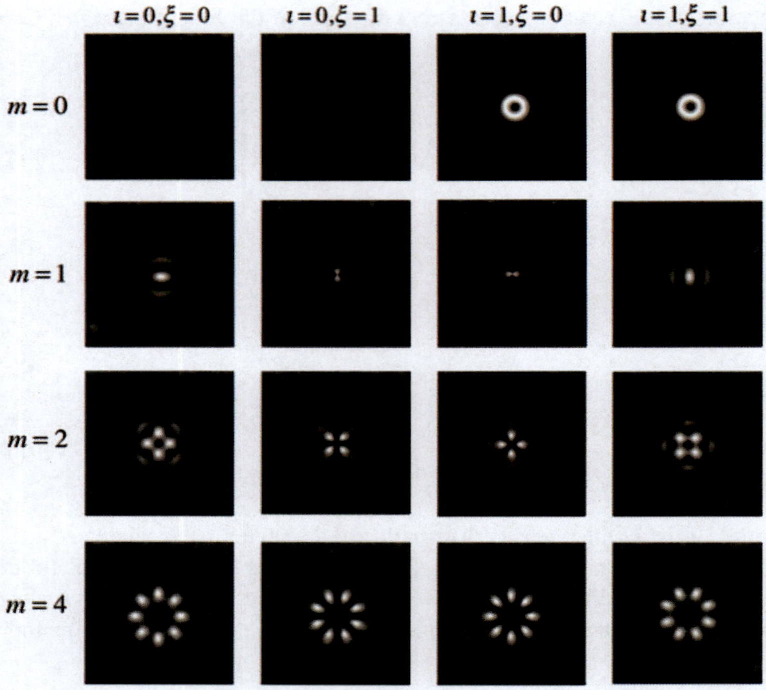

Fig. 7.2 Intensity distribution of various vector beams represented by Eq. (7.1.19) [3]

Consider the LG_{q0} mode with topological charge 0 and radial index of q. Letting $q = m \pm 1$, Eq. (7.1.23) can be written as

$$\begin{aligned} E_{LG}^{(v)}(r, \varphi, z) = E_0 \frac{\omega_0}{\omega(z)} \left[\frac{\sqrt{2}r}{\omega(z)} \right]^{|m\pm 1|} & \exp[-ikz + i(|m \pm 1| + 1)\psi(z)] \\ \cdot \exp\left\{ -r^2 \left[\frac{1}{\omega^2(z)} + \frac{ik}{2R(z)} \right] \right\} \cdot & \begin{cases} \cos(m\varphi)\vec{e}_\varphi \mp \sin(m\varphi)\vec{e}_r \\ \pm \sin(m\varphi)\vec{e}_\varphi + \cos(m\varphi)\vec{e}_r \end{cases} \end{aligned} \tag{7.1.24}$$

As the expression describing the polarization distribution in Eq. (7.1.24) (the last term) is complex in form and not easy to see its physical connotation, it is transformed by the coordinate transformation given by Eq. (7.1.18), so that the last term in Eq. (7.1.24) can be expressed as

$$\begin{cases} \mp \sin[(m \pm 1)\varphi]\vec{e}_x + \cos[(m \pm 1)\varphi]\vec{e}_y \\ \cos[(m \pm 1)\varphi]\vec{e}_x \pm \sin[(m \pm 1)\varphi]\vec{e}_y \end{cases} \tag{7.1.25}$$

Writing Eq. (7.1.25) in the form of the Jones vector in four Cartesian coordinates:

$$\vec{E}_{\text{up}} = \begin{bmatrix} \mp\sin(m\pm 1)\varphi \\ \cos(m\pm 1)\varphi \end{bmatrix} = \begin{bmatrix} \mp\sin p\varphi \\ \cos p\varphi \end{bmatrix} \tag{7.1.26}$$

$$\vec{E}_{\text{down}} = \begin{bmatrix} \cos(m\pm 1)\varphi \\ \pm\sin(m\pm 1)\varphi \end{bmatrix} = \begin{bmatrix} \cos p\varphi \\ \pm\sin p\varphi \end{bmatrix} \tag{7.1.27}$$

Equations (7.1.26) and (7.1.27) describe four different modes of polarization distribution. Substituting them into Eq. (7.1.24) yields four different field distributions. It is easy to see that the intensity distribution given by Eq. (7.1.24) is identical to that of the LG beam and has a one-to-one correspondence with the parameter q. Alternatively, identical parameter q means identical intensity distributions. However, vector beams with identical parameter q have four different polarization distributions given by Eqs. (7.1.26) and (7.1.27).

7.1.3 Jones Vector Representation of Vector Beams

The Jones vector and the Jones matrix have a very simple mathematical form and thus become widely used mathematical tools when analyzing polarization and polarization optical components. In the Jones vector, the electric field component of a beam propagating along the z-axis can be decomposed into x and y components and represented by a column vector. In addition, the polarization optical element can be represented as a 2×2 Jones matrix.

Equation (7.1.1) has already given the Jones vector representation of the vector beam:

$$\vec{J}(\varphi) = \begin{bmatrix} \cos(p\varphi + \theta_0) \\ \sin(p\varphi + \theta_0) \end{bmatrix} \tag{7.1.28}$$

When the polarization order $p = 0$, the above Jones vector no longer represents a vector beam, but a linearly polarized beam with the polarization direction at an angle θ_0 to the x-axis

$$\vec{J}(\varphi) = \begin{bmatrix} \cos\theta_0 \\ \sin\theta_0 \end{bmatrix} \tag{7.1.29}$$

When the polarization order $p = 1$ and the initial direction of polarization $\theta_0 = 0$, denote radially polarized beam,

$$\vec{J}(\varphi) = \begin{bmatrix} \cos\varphi \\ \sin\varphi \end{bmatrix} \tag{7.1.30}$$

When the polarization order $p = 1$ and the initial direction of polarization $\theta_0 = \pi/2$ indicate azimuthally polarized beam,

$$\vec{J}(\varphi) = \begin{bmatrix} -\sin\varphi \\ \cos\varphi \end{bmatrix} \tag{7.1.31}$$

Equations (7.1.28)–(7.1.31) are the Jones vectors representing the polarization characteristics of the beam in a Cartesian coordinate. For rotationally symmetric systems, the Jones vector and Jones matrix can also be described in the polar coordinate. Such representation in polar coordinates is also known as the polar Jones vector and the polar Jones matrix, denoted as $\widehat{J}$ and $\widehat{M}$, respectively. The polar Jones vector divides the electric field into radial and azimuthal components, i.e.

$$\widehat{J}(\varphi) = \begin{bmatrix} E_r(\varphi) \\ E_\varphi(\varphi) \end{bmatrix} \tag{7.1.32}$$

According to the coordinate transformation, the *p*-order vector beam given by Eq. (7.1.28) can be represented by the polar Jones vector as

$$\widehat{J}(\varphi) = \begin{bmatrix} \cos\left[(p-1)\varphi + \theta_0\right] \\ \sin\left[(p-1)\varphi + \theta_0\right] \end{bmatrix} \tag{7.1.33}$$

Setting $p = 0$, we obtain the polar Jones vector of linearly polarized beam

$$\widehat{J}(\varphi) = \begin{bmatrix} \cos[-\varphi + \theta_0] \\ \sin[-\varphi + \theta_0] \end{bmatrix} \tag{7.1.34}$$

Letting $p = 1$, $\theta_0 = 0$, we get the polar Jones vector of radially polarized beam

$$\widehat{J}(\varphi) = \begin{bmatrix} 1 \\ 0 \end{bmatrix} \tag{7.1.35}$$

Letting $p = 1$, $\theta_0 = \pi/2$, the polar Jones vector of azimuthally polarized beam can be described as:

$$\widehat{J}(\varphi) = \begin{bmatrix} 0 \\ 1 \end{bmatrix} \tag{7.1.36}$$

It is easy to see that polar Jones vectors have a more concise expression for describing the radially and azimuthally polarized vector beam than Jones vectors in Cartesian coordinate.

Table 7.1 gives the Jones vectors of several common beams in Cartesian coordinates and polar coordinates. And Table 7.2 gives the Jones matrices of several common polarization optical elements in Cartesian coordinate system and polar coordinates.

Table 7.1 Jones vectors of common beams in Cartesian and polar coordinates

Beam	Jones vector	Polar Jones vector	Polarization distribution
Vector beam	$\begin{bmatrix} \cos(p\varphi + \theta_0) \\ \sin(p\varphi + \theta_0) \end{bmatrix}$	$\begin{bmatrix} \cos\left[(p-1)\varphi + \theta_0\right] \\ \sin\left[(p-1)\varphi + \theta_0\right] \end{bmatrix}$	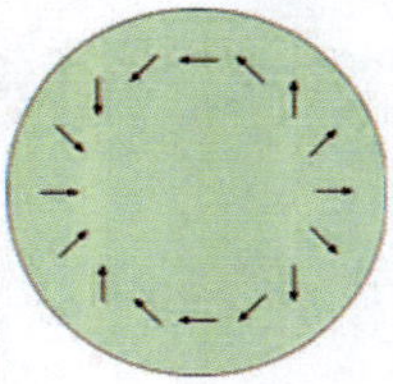
Radial polarization	$\begin{bmatrix} \cos\varphi \\ \sin\varphi \end{bmatrix}$	$\begin{bmatrix} 1 \\ 0 \end{bmatrix}$	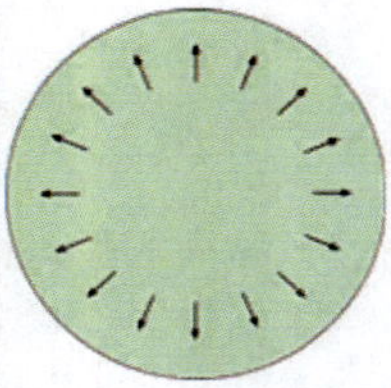
Azimuthal polarization	$\begin{bmatrix} -\sin\varphi \\ \cos\varphi \end{bmatrix}$	$\begin{bmatrix} 0 \\ 1 \end{bmatrix}$	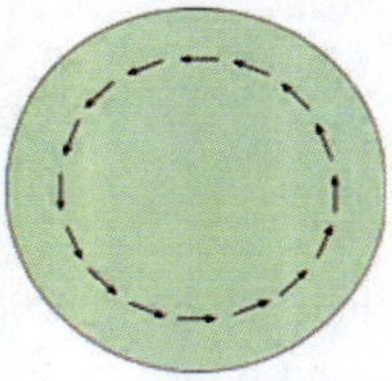
Linear polarization	$\begin{bmatrix} \cos\theta_0 \\ \sin\theta_0 \end{bmatrix}$	$\begin{bmatrix} \cos[-\varphi + \theta_0] \\ \sin[-\varphi + \theta_0] \end{bmatrix}$	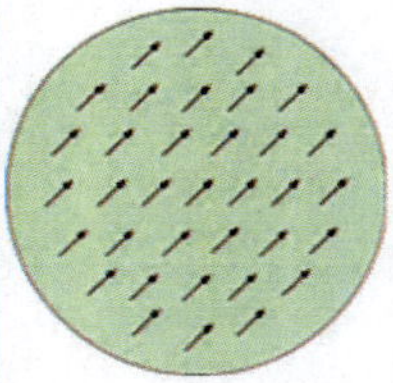
Right circular polarization	$\frac{1}{\sqrt{2}}\begin{bmatrix} 1 \\ i \end{bmatrix}$	$\frac{1}{\sqrt{2}}\begin{bmatrix} \cos\varphi + i\sin\varphi \\ -\sin\varphi + i\cos\varphi \end{bmatrix}$	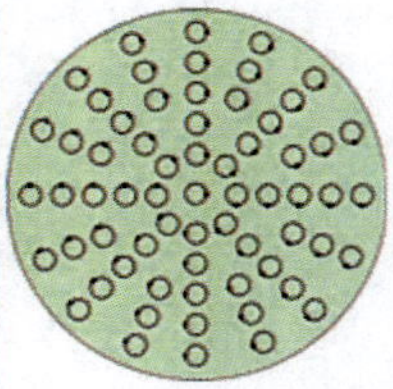

(continued)

Table 7.1 (continued)

Beam	Jones vector	Polar Jones vector	Polarization distribution
Left circular polarization	$\frac{1}{\sqrt{2}}\begin{bmatrix} 1 \\ -i \end{bmatrix}$	$\frac{1}{\sqrt{2}}\begin{bmatrix} \cos\varphi - i\sin\varphi \\ -\sin\varphi - i\cos\varphi \end{bmatrix}$	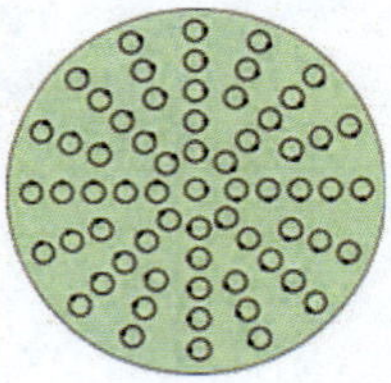

7.1.4 *Conversion of Polarizations by Half Wave Plates*

In vector beams-based applications, it is often necessary to convert their polarization. For instance, the interconversion of radially and azimuthally polarized beam, which can be achieved by means of a half wave plate. The Jones matrix for a half wave plate with the α arranged fast axis is [5].

$$M(\alpha) = \begin{bmatrix} \cos 2\alpha & \sin 2\alpha \\ \sin 2\alpha & -\cos 2\alpha \end{bmatrix} \tag{7.1.37}$$

The conversion of the polarizations of vector beams by the half wave plate can then be expressed as:

$$\vec{E} = M(\alpha)\begin{bmatrix} \cos(p\varphi + \theta_0) \\ \sin(p\varphi + \theta_0) \end{bmatrix} = \begin{bmatrix} \cos(2\alpha - p\varphi - \theta_0) \\ \sin(2\alpha - p\varphi - \theta_0) \end{bmatrix} \tag{7.1.38}$$

Equation (7.1.38) shows that when the vector beam passes through a half wave plate whose fast axis is α arrangement, the output beam is still a vector beam, but its polarization order p becomes opposite. In other words, the positive order becomes the negative and the negative order becomes positive. Additionally, the initial polarization direction becomes $(2\alpha - \theta_0)$, which is a rotation of $2(\alpha - \theta_0)$ compared to the original.

When a vector beam passes through the two half wave plates $M1(\alpha_1)$ and $M_2(\alpha_2)$, the Jones vector of the output beam is

$$\begin{aligned} \vec{E} &= M_2(\alpha_2)M_1(\alpha_1)\begin{bmatrix} \cos(p\varphi + \theta_0) \\ \sin(p\varphi + \theta_0) \end{bmatrix} \\ &= \begin{bmatrix} \cos 2(\alpha_2 - \alpha_1) & -\sin 2(\alpha_2 - \alpha_1) \\ \sin 2(\alpha_2 - \alpha_1) & \cos 2(\alpha_2 - \alpha_1) \end{bmatrix}\begin{bmatrix} \cos(p\varphi + \theta_0) \\ \sin(p\varphi + \theta_0) \end{bmatrix} \\ &= \begin{bmatrix} \cos[p\varphi + \theta_0 + 2(\alpha_2 - \alpha_1)] \\ \sin[p\varphi + \theta_0 + 2(\alpha_2 - \alpha_1)] \end{bmatrix} \end{aligned} \tag{7.1.39}$$

Table 7.2 Jones matrices for several common polarization optics elements in Cartesian and polar coordinates

Polarization optical element	Jones matrix	Polar Jones matrix
Horizontal polarizer	$\begin{bmatrix} 1 & 0 \\ 0 & 0 \end{bmatrix}$	$\begin{bmatrix} \cos^2\varphi & -\sin\varphi\cos\varphi \\ -\sin\varphi\cos\varphi & \sin^2\varphi \end{bmatrix}$
Vertical polarizer	$\begin{bmatrix} 0 & 0 \\ 0 & 1 \end{bmatrix}$	$\begin{bmatrix} \sin^2\varphi & \sin\varphi\cos\varphi \\ \sin\varphi\cos\varphi & \cos^2\varphi \end{bmatrix}$
Horizontally arranged quarter wave plate	$\begin{bmatrix} 1 & 0 \\ 0 & i \end{bmatrix}$	$\begin{bmatrix} \cos 2\varphi & \sin 2\varphi \\ \sin 2\varphi & -\cos 2\varphi \end{bmatrix}$
Horizontally arranged half wave plate	$\begin{bmatrix} 1 & 0 \\ 0 & -1 \end{bmatrix}$	$\begin{bmatrix} \cos^2\varphi + i\sin^2\varphi & -\sin\varphi\cos\varphi + i\sin\varphi\cos\varphi \\ -\sin\varphi\cos\varphi + i\sin\varphi\cos\varphi & \sin^2\varphi + i\cos^2\varphi \end{bmatrix}$
Radial polarizer	$\begin{bmatrix} \cos^2\varphi & \sin\varphi\cos\varphi \\ \sin\varphi\cos\varphi & \sin^2\varphi \end{bmatrix}$	$\begin{bmatrix} 1 & 0 \\ 0 & 0 \end{bmatrix}$
m-th order spatially variable half wave plate	$\begin{bmatrix} \cos m\varphi & \sin m\varphi \\ \sin m\varphi & -\cos m\varphi \end{bmatrix}$	$\begin{bmatrix} \cos(m-2)\varphi & \sin(m-2)\varphi \\ \sin(m-2)\varphi & -\cos(m-2)\varphi \end{bmatrix}$

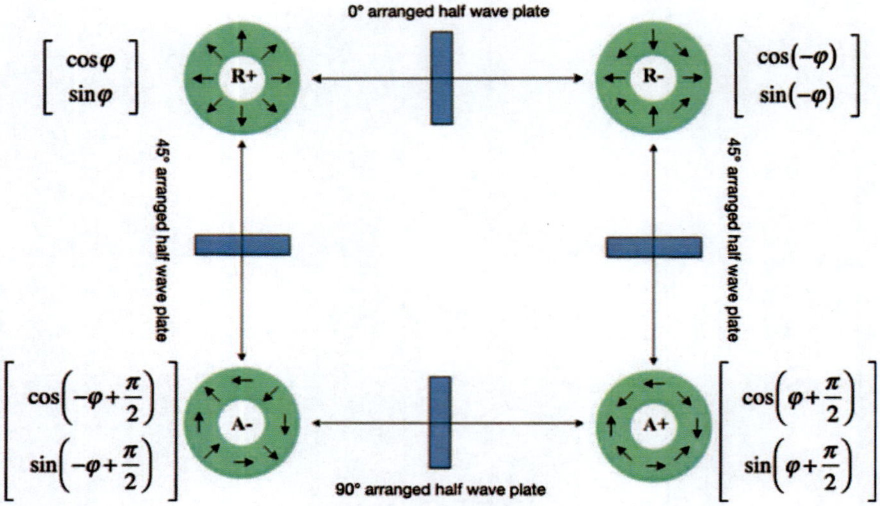

Fig. 7.3 Schematic diagram of polarization state conversion between different vector beams using half wave plates [3]

This indicates that the initial polarization direction of the vector beam is rotated by $2\Delta\alpha$ compared to the original after passing through two half wave plates with an angle $\Delta\alpha = \alpha_2 - \alpha_1$ between two fast axes, but the polarization order remains the same. Thus the polarization order of the vector beam can be changed by one half wave plate. The initial polarization direction of the vector beam can be changed by two half wave plates. In particular, since the initial polarization angle of azimuthally polarized beam is $\pi/2$ more than that of radially polarized beam, radially polarized beam can be converted into azimuthally polarized beam by introducing two half-waveplates with an fast axis angle difference of $\pi/4$.

Taking the conversion between four vector beams of $+1$st order radial polarization (R+), -1st order radial polarization (R$-$), $+1$st order azimuthal polarization (A+) and -1st order azimuthal polarization (A$-$) as an example, Fig. 7.3 gives a schematic representation of their conversion using a half wave plate.

7.2 Generating Vector Beams

7.2.1 Intra-Cavity Generation

Generating vector beam inside a laser cavity is to insert mode selection optical elements in the laser cavity, making vector modes oscillation. These insert optical elements can be axially birefringent, axially dichroic devices etc.

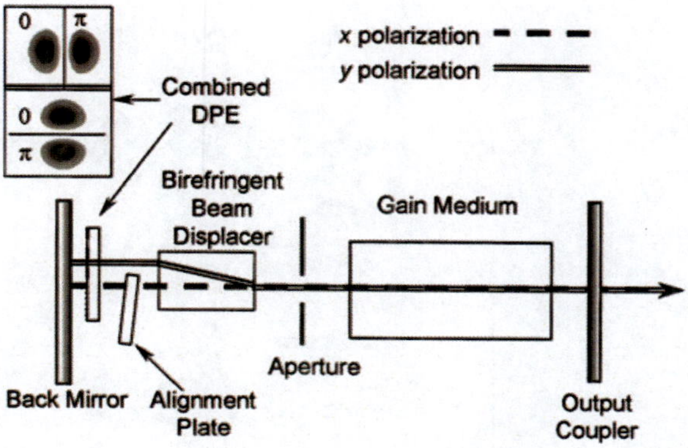

Fig. 7.4 Placing a DPE and a birefringent element inside a laser cavity to generate radially and azimuthally polarized vector beams [6]

Early intracavity schemes can usually produce only two specific vector beams, radial or azimuthal polarizations. By placing a discontinuous phase element (DPE) in front of the end mirror and a birefringent device, two Hermit Gauss mode oscillate stably with orthogonal linear polarizations and orthogonal alignment. These two modes are combined by the birefringent element and then output from the output mirror to obtain a vector beam [6], as shown in Fig. 7.4. However, this scheme is only capable for obtaining two specific vector beams, radial or azimuthal polarizations.

Moreover, the birefringence effect of the c-axis cut Nd:YVO_4 crystal allows the p-polarized and s-polarized beam to undergo different paths in the laser cavity, and the s-polarization loss is increased by adjusting the cavity length so that only the p-polarization is retained in the laser cavity, resulting in a radially polarized vector beam [8]. Using Nd:YAG crystals as the gain and c-cut YVO_4 crystals as the polarization selector, the laser cavity can output radially polarized beam directly [9]. In addition, radially polarized beams can be achieved by using the thermally induced birefringence effect of isotropic laser crystals for polarization state selection under high power pumping [10].

In 2008, a photonic crystal grating mirror is used as an intracavity polarizer. Such polarizer is employed in an end-pumped Nd:YAG microchip laser to generate a radially polarized beam of 610mW with purity 98% [11]. In the same year, a passively Q-switched Nd:YAG ceramic microchip laser is used as the source, a Cr^{4+}:YAG crystal is employed as as the saturable absorber and a sub-wavelength grating to select polarization, and finally output pulsed radially polarized beams [12].

By placing polarization modulation elements such as q-plates and quarter wave plates inside the laser cavity, the oscillated fundamental mode can be converted into a higher-order vector mode, realizing tunable generation of vector beams from a laser cavity [7], whose device and experimental results are shown in Fig. 7.5. In addition, by changing parameters such as the order (q-value) of the q-plate, or by rotating q-plate or quarter wave plate, vector beams with arbitrary polarization distributions can also be generated.

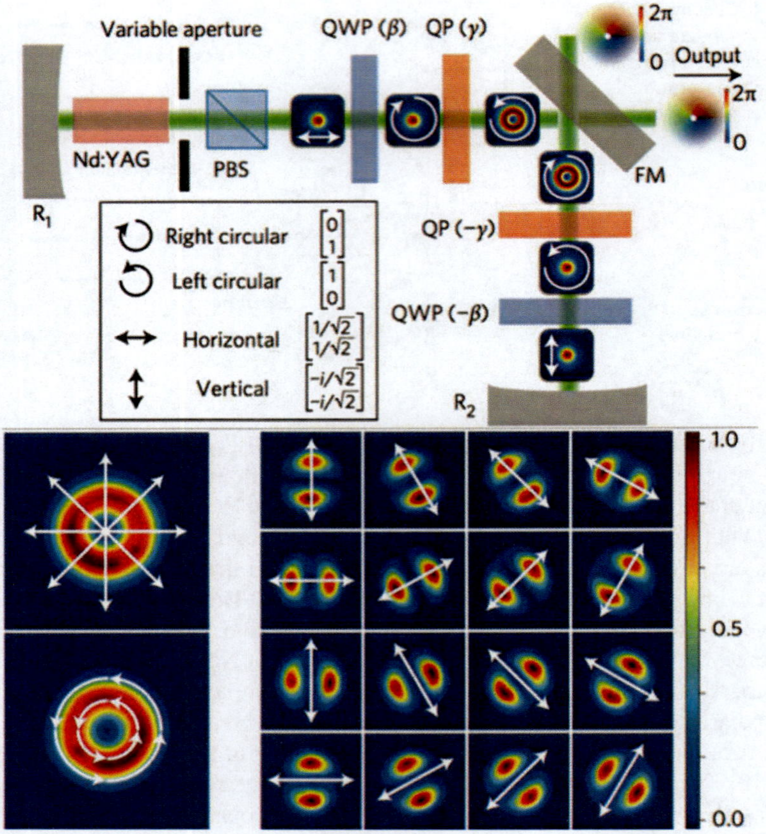

Fig. 7.5 Placing a q-plate inside a laser cavity to generate vector beams. R1 and R2, reflectors; PBS, polarized beam splitter; QWP, quarter wave plate; QP, q-plate; FM, 45° mirror [7]

7.2.2 Sub-wavelength Gratings

Sub-wavelength metal or dielectric gratings are a class of spatially polarized conversion devices that exhibit the birefringent properties of uniaxial crystals when the grating period is less than the wavelength of the incident beam. The incident circularly polarized beam can be converted into a vector beam by designing the period and direction of the inscribed groove. A typical method of generating a vector beam using a subwavelength grating is through a multi-sector structure, where each sector is equivalent to a polarization beam splitter, with most of the p-polarizations perpendicular to the grating bar being transmitted, and most of the s-polarization parallel to the grating bar being reflected. The purity of the resulting radially polarized beam increases as the number of sectors increases [13]. In addition, by setting the spatial distribution of the grating, higher order vector beams can be generated.

7.2.3 Spatial-Varying Polarizers

In addition to detecting vortex beams, spatial-varying polarizers can also be used to generate vector beams. As described in Sect. 5.4.2, a spatial-varying polarizer is a polarization optical device with an angularly anisotropic starting direction, consisting of a combination of M sub-polarizers placed at different angles, as shown in Fig. 7.6. When $M \to \infty$, the Jones matrix of the m-order spatial-varying polarizer can be written as

$$M_{SVP}(\varphi) = \begin{bmatrix} \cos^2(m\varphi + \theta_0) & \sin(m\varphi + \theta_0)\cos(m\varphi + \theta_0) \\ \sin(m\varphi + \theta_0)\cos(m\varphi + \theta_0) & \sin^2(m\varphi + \theta_0) \end{bmatrix} \tag{7.2.1}$$

where φ is the azimuthal angle and θ_0 is the initial direction of the sub-polarizer when $\varphi = 0$. In particular, when $m = 1$ and $\theta_0 = 0$, the spatial-varying polarizer has a radially distributed direction, called a radial polarizer; similarly, when $m = 1$ and $\theta_0 = \pi/2$, the spatial-varying polarizer has an azimuthal distributed direction, called an azimuthal polarizer.

When a right circularly polarized beam passes through a spatial-varying polarizer with $\theta_0 = 0$, the output can be expressed as:

$$\begin{aligned} \vec{E} &= M_{SVP}(\varphi)\frac{1}{\sqrt{2}}\begin{bmatrix} 1 \\ i \end{bmatrix} = \begin{bmatrix} \cos^2(m\varphi) + i\sin(m\varphi)\cos(m\varphi) \\ \sin(m\varphi)\cos(m\varphi) + i\sin^2(m\varphi) \end{bmatrix} \\ &= \begin{bmatrix} \cos m\varphi \\ \sin m\varphi \end{bmatrix} \exp(im\varphi) \end{aligned} \tag{7.2.2}$$

This shows that the electric field distribution of the output beam is axisymmetric and is essentially the same as the Jones vector of the vector beam, except that it has an additional helical phase term $\exp(im\varphi)$. The output beam has both the polarization state of the vortex distribution and the spiral phase, which is the vectorial vortex beams and will be discussed later in this chapter. To generate a vector beam only, a vortex beam with topological charge $-m$ can be used instead of a Gaussian beam to

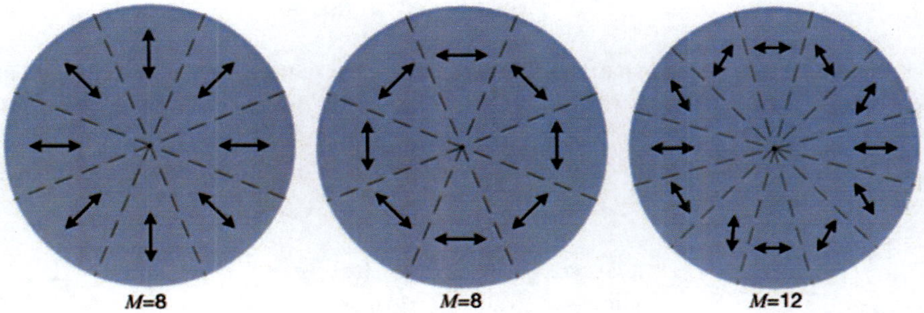

Fig. 7.6 Spatial-varying polarizers

eliminate the helical phase of the output beam:

$$\vec{E} = M_{SVP}(\varphi)\exp(-im\varphi)\frac{1}{\sqrt{2}}\begin{bmatrix} 1 \\ i \end{bmatrix} = \begin{bmatrix} \cos m\varphi \\ \sin m\varphi \end{bmatrix} \tag{7.2.3}$$

Equation (7.2.3) illustrates that a right circularly polarized vortex beam with topological charge $-m$ ($\exp(-im\varphi)[1, i]^T$) is transformed into an m-th order vector beam after passing through an m-order spatial-varying polarizer. Similarly, when a left circularly polarized vortex beam of topological charge m is incident on a m-th order spatial-varying polarizer, it also produces an m-th order vector beam.

$$\vec{E} = M_{SVP}(\varphi)\exp(im\varphi)\frac{1}{\sqrt{2}}\begin{bmatrix} 1 \\ -i \end{bmatrix} = \begin{bmatrix} \cos m\varphi \\ \sin m\varphi \end{bmatrix} \tag{7.2.4}$$

7.2.4 Spatially Variable Half Wave Plates

In the third and fifth chapters of this book, the spatially variable half wave plate has been mentioned for many times, which is a kind of polarization element with azimuthal anisotropy in fast axis direction. It composes M sub-half wave plates placed at different angles, which can be used to generate and detect vortex beams. When $M \to \infty$, the Jone matrix of the m-th order spatially variable half wave plate is:

$$M_{SVHWP}(\varphi) = \begin{bmatrix} \cos m\varphi & \sin m\varphi \\ \sin m\varphi & -\cos m\varphi \end{bmatrix} \tag{7.2.5}$$

When a linearly polarized beam passes through a spatially variable half wave plate, the output can be expressed as:

$$\vec{E} = M_{SVHWP}\begin{bmatrix} \cos\theta_0 \\ \sin\theta_0 \end{bmatrix} = \begin{bmatrix} \cos(m\varphi - \theta_0) \\ \sin(m\varphi - \theta_0) \end{bmatrix} \tag{7.2.6}$$

From Eq. (7.2.6), the spatially variable half wave plate can convert an incident linearly polarized beam into an m-th order vector beam. In other words, the spatially variable half wave plate can be used to generate a vector beam.

When an n-order vector beam is incident on an m-order spatially variable half wave plate, the output is

$$\vec{E} = M_{SVHWP}\begin{bmatrix} \cos(n\varphi + \theta_0) \\ \sin(n\varphi + \theta_0) \end{bmatrix} = \begin{bmatrix} \cos[(m-n)\varphi - \theta_0] \\ \sin[(m-n)\varphi - \theta_0] \end{bmatrix} \tag{7.2.7}$$

Equation (7.2.7) shows that an n-th order vector beam passing through an m-th order spatially variable half wave plate contribute to a (m-n)-th order vector beam. Therefore, spatially variable half wave plates can therefore be used in cascade to generate vector beams of any order. In addition, when the order of the incident vector beam is equal to that of the spatially variable half wave plate ($n = m$), the output beam degenerates to a Gaussian beam. Such property can be used to determine the polarization order of a vector beam.

7.2.5 Q-plates

In Sect. 3.4.2, the concept of a q-plate has been introduced as a polarization modulation element made from nematic liquid crystals whose Jones matrix is:

$$M_q = \begin{bmatrix} \cos 2\alpha & \sin 2\alpha \\ \sin 2\alpha & -\cos 2\alpha \end{bmatrix} \tag{7.2.8}$$

where,

$$\alpha(r, \varphi) = q\varphi + \alpha_0 \tag{7.2.9}$$

with the initial angle θ_0. When linearly polarized beam is incident in, the output reads:

$$E = M_q \begin{bmatrix} \cos\theta_0 \\ \sin\theta_0 \end{bmatrix} = \begin{bmatrix} \cos(2\alpha - \theta_0) \\ \cos(2\alpha - \theta_0) \end{bmatrix} = \begin{bmatrix} \cos[2q\varphi + (\alpha - \theta_0)] \\ \sin[2q\varphi + (\alpha - \theta_0)] \end{bmatrix} \tag{7.2.10}$$

indicating that a vector beam with a polarization order $2q$ and an initial polarization direction (α-θ_0) is generated.

7.3 Generating Vector Beams Through Coherent Combination

Coherent combination scheme, a kind of out-of-cavity method, can generate vector beams by coaxial combining scalar beams. This method is easy to realize in laboratory, and has become one of the mainstream solutions to generate vector beams.

7.3.1 Principles of Coherent Combination

The most commonly used scale beams for coherent combination based vector beams generation are Hermit Gauss beams and vortex beams carrying OAM.

1. **Coherent combination through Hermit-Gauss beams**

The principle of coherent combination of vector beams using Hermit Gauss beams is

$$\vec{E}_r = HG_{10}\vec{e}_x + HG_{01}\vec{e}_y \tag{7.3.1}$$

$$\vec{E}_\varphi = HG_{01}\vec{e}_x + HG_{10}\vec{e}_y \tag{7.3.2}$$

where $\vec{E}_r$ and $\vec{E}_\varphi$ denote radially and azimuthally polarized beams, respectively. Equations (7.3.1) and (7.3.2) show that two Hermit-Gauss modes with both horizontally and vertically linear polarizations and orthogonal spot arrangement orientation can be coherently combined into radially or azimuthally polarized beams. Such scheme is able to generate first-order vector beams. When generating higher-order vector beams, multiple Hermit-Gauss modes need to be superimposed [14].

The generation of arbitrary order vector beams employing different Hermit-Gauss modes draws on the principle that a LG beam can be decomposed into a set of Hermit-Gauss modes [15]

$$LG_{pl}(x, y, z) = \sum_{s=0}^{N} i^s a(m, n, s) \cdot HG_{N-s,s}(x, y, z) \tag{7.3.3}$$

where the weight $a(m, n, s)$ is:

$$a(m, n, s) = \left[\frac{(N-s)!s!}{2^N m!n!}\right]^{0.5} \cdot \frac{1}{s!} \tag{7.3.4}$$

In Eqs. (7.3.3) and (7.3.4), $l = n - m$, $p = m$, and $N = m + n$. The factor i^s in Eq. (7.3.3) indicates the two adjacent beams have a phase difference $\pi/2$. If the i^s term in Eq. (7.3.3) is replaced by the Jones vector $[\cos(\pi s/2), \sin(\pi s/2)]^T$, that is, the phase difference is replaced by the difference in polarization direction, the field after the replacement can be expressed as:

$$U_{pl}(x, y, z) = \begin{bmatrix} E_x \\ E_y \end{bmatrix} = \sum_{s=0}^{N} a(m, n, s) \cdot HG_{N-s,s}(x, y, z)\begin{bmatrix} \cos(\pi s/2) \\ \sin(\pi s/2) \end{bmatrix} \tag{7.3.5}$$

From Euler's formula, we have

$$i^s = \cos\left(\frac{\pi s}{2}\right) + i\sin\left(\frac{\pi s}{2}\right) \tag{7.3.6}$$

It can be found that both i^s and $[\cos(\pi s/2), \sin(\pi s/2)]^T$ are periodic functions with a period of 4. Their function values are 1, i, −1, $-i$, and $[1, 0]^T$, $[0, 1]^T$, $[-1, 0]^T$, $[0, -1]^T$, respectively, when the integer s is 0, 1, 2, and 3. These two functions determine the phase and polarization of the beam, respectively, and also have a one-to-one correspondence. The x-direction component of Eq. (7.3.5) corresponds to the real part of Eq. (7.3.3), and the y-direction component of Eq. (7.3.5) corresponds to the imaginary part of Eq. (7.3.3). Based on the above correspondence between phases and polarizations, and considering Eq. (1.4.1), Eq. (7.3.5) can be written as:

$$U_{pl}(r, \varphi, 0) = |LG_{pl}|\begin{bmatrix}\cos l\varphi \\ \sin l\varphi\end{bmatrix} \tag{7.3.7}$$

where $|LG_{pl}|$ denotes the amplitude distribution of the LG beam. From the above equations, when $l \neq 0$, the combined beam is a vector beam. It has the same intensity distribution as the LG beam, but it does not have a helical phase distribution, carrying no OAM. The beam has an axisymmetric vortex-like polarization distribution in transverse.

Two simulation examples are given below according to Eq. (7.3.5), first is case of $m = 0$ and second ie case of $m \neq 0$. To clearly describe the merging process, six mathematical expressions are given here, of which Eqs. (7.3.8)–(7.3.10) are the first example, and Eqs. (7.3.11)–(7.3.13) are the second example. The Jones vector on the left-hand side of the equation represents the polarization state of the Hermit-Gaussian beam, and the fraction before the Jones vector is the weight of the amplitude. The Jones vector on the right-hand side of the equation describes the polarization distribution of the coherently combined vector beam, and the term before the Jones vector represents the amplitude distribution.

Example 1 $m = 0$, thus $p = 0$

$$\frac{\sqrt{2}}{2}\begin{bmatrix}1 \\ 0\end{bmatrix}HG_{10} + \frac{\sqrt{2}}{2}\begin{bmatrix}0 \\ 1\end{bmatrix}HG_{01} = |LG_{01}|\begin{bmatrix}\cos\varphi \\ \sin\varphi\end{bmatrix} \tag{7.3.8}$$

$$\frac{1}{2}\begin{bmatrix}1 \\ 0\end{bmatrix}HG_{20} + \frac{\sqrt{2}}{2}\begin{bmatrix}1 \\ 0\end{bmatrix}HG_{11} + \frac{1}{2}\begin{bmatrix}-1 \\ 0\end{bmatrix}HG_{02} = |LG_{02}|\begin{bmatrix}\cos 2\varphi \\ \sin 2\varphi\end{bmatrix} \tag{7.3.9}$$

$$\begin{aligned}&\frac{\sqrt{2}}{4}\begin{bmatrix}1 \\ 0\end{bmatrix}HG_{30} - \frac{\sqrt{6}}{4}\begin{bmatrix}1 \\ 0\end{bmatrix}HG_{21} + \frac{\sqrt{6}}{4}\begin{bmatrix}-1 \\ 0\end{bmatrix}HG_{12} \\ &-\frac{\sqrt{2}}{4}\begin{bmatrix}0 \\ -1\end{bmatrix}HG_{03} = |LG_{03}|\begin{bmatrix}\cos 3\varphi \\ \sin 3\varphi\end{bmatrix}\end{aligned} \tag{7.3.10}$$

Several of the above merging processes are displayed in Fig. 7.7, where the polarization and intensity distribution of the Hermit Gauss beam are shown in the left box. And the intensity and polarization distribution of the merged vector beam is shown in the right box.

Example 2 $m \neq 0$, thus $p \neq 0$.

$$\frac{\sqrt{6}}{4}\begin{bmatrix}1\\0\end{bmatrix}HG_{30}-\frac{\sqrt{2}}{4}\begin{bmatrix}1\\0\end{bmatrix}HG_{21}-\frac{\sqrt{2}}{4}\begin{bmatrix}-1\\0\end{bmatrix}HG_{12}+\frac{\sqrt{6}}{4}\begin{bmatrix}0\\-1\end{bmatrix}HG_{03}=|LG_{11}|\begin{bmatrix}\cos\varphi\\\sin\varphi\end{bmatrix} \tag{7.3.11}$$

$$\frac{\sqrt{6}}{4}\begin{bmatrix}1\\0\end{bmatrix}HG_{30}+\frac{\sqrt{2}}{4}\begin{bmatrix}1\\0\end{bmatrix}HG_{21}-\frac{\sqrt{2}}{4}\begin{bmatrix}-1\\0\end{bmatrix}HG_{12}-\frac{\sqrt{6}}{4}\begin{bmatrix}0\\-1\end{bmatrix}HG_{03}=|LG_{11}|\begin{bmatrix}\cos(-\varphi)\\\sin(-\varphi)\end{bmatrix} \tag{7.3.12}$$

$$\frac{\sqrt{5}}{4}\begin{bmatrix}1\\0\end{bmatrix}HG_{50}-\frac{1}{4}\begin{bmatrix}1\\0\end{bmatrix}HG_{41}-\frac{\sqrt{2}}{4}\begin{bmatrix}-1\\0\end{bmatrix}HG_{32}+\frac{\sqrt{2}}{4}\begin{bmatrix}0\\-1\end{bmatrix}HG_{23}+\frac{1}{4}\begin{bmatrix}1\\0\end{bmatrix}HG_{14}-\frac{\sqrt{5}}{4}\begin{bmatrix}0\\1\end{bmatrix}HG_{05}=|LG_{21}|\begin{bmatrix}\cos\varphi\\\sin\varphi\end{bmatrix} \tag{7.3.13}$$

Beam combining processes displayed in Eqs. (7.3.11)–(7.3.13) is given in Fig. 7.8. Figures 7.7 and 7.8 illustrate that the intensity distribution of the vector beam coherently combined from Hermit-Gauss beams are the same as that of a LG beam. The

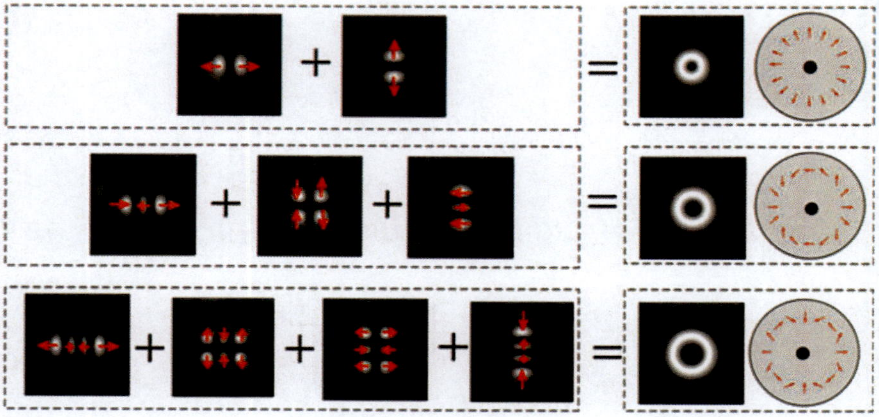

Fig. 7.7 Vector beams combined from Hermit-Gauss beams when $m = 0$ [14]

transverse polarization distribution satisfies Eq. (7.1.1). When $m = 0$, $p = 0$, the vector beam has a single-ring structure. When $m \neq 0, p \neq 0$, the vector beam has a multi-ring structure.

2. **Coherent combination through vortex beams**

The combination scheme introduced previously is currently not widely used since it often requires multiple Hermit-Gauss beams, which is difficult to achieve. Coherent combination of vector beams through vortex beams is more popular.

The principle of coherent combinations through vortex beams is shown in Fig. 7.9, which process reads:

$$\frac{1}{2}\exp(ip\varphi)\begin{bmatrix}1\\-i\end{bmatrix}+\frac{1}{2}\exp(-ip\varphi)\begin{bmatrix}1\\i\end{bmatrix}=\begin{bmatrix}\cos p\varphi\\\sin p\varphi\end{bmatrix} \tag{7.3.14}$$

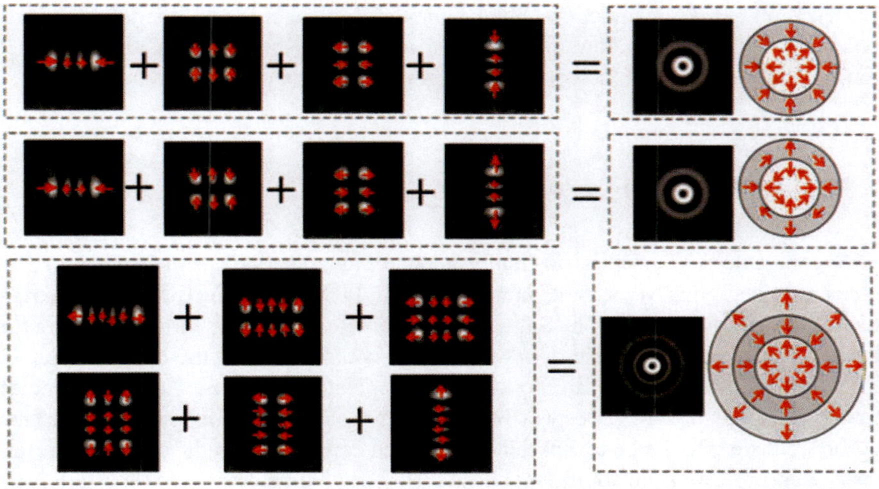

Fig. 7.8 Vector beams combined from Hermit-Gauss beams when $m \neq 0$ [14]

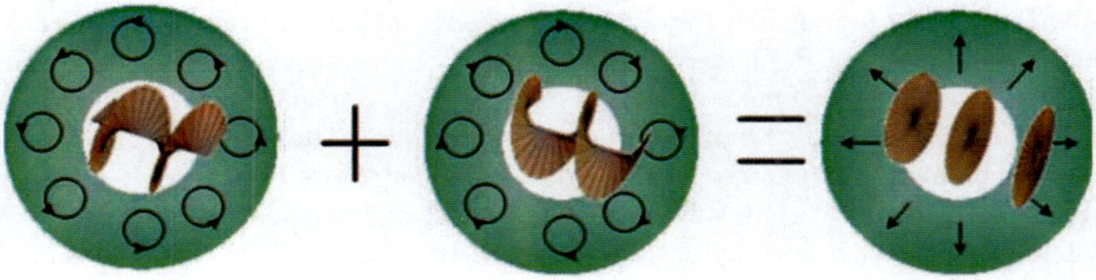

Fig. 7.9 Coherently combining a vector beam through vortex beams

Equation (7.3.14) illustrates that, a left circularly polarized vortex beam with topological charge p and another right circularly polarized vortex beam with topological charge -p can be coherently combined into a vector beam with polarization order p and initial polarization direction $\theta_0 = 0$. If an additional phase difference $\Delta\phi$ is introduced between two orthogonal circularly polarized vortex beams in Eq. (7.3.14), Eq. (7.3.14) can be written in a more general form as:

$$\frac{1}{2}\exp[i(p\varphi + \Delta\phi)]\begin{bmatrix} 1 \\ -i \end{bmatrix} + \frac{1}{2}\exp(-ip\varphi)\begin{bmatrix} 1 \\ i \end{bmatrix} = \exp\left[\frac{1}{2}i\Delta\phi\right]\begin{bmatrix} \cos\left(p\varphi + \frac{1}{2}\Delta\phi\right) \\ \sin\left(p\varphi + \frac{1}{2}\Delta\phi\right) \end{bmatrix} \tag{7.3.15}$$

The initial polarization direction of the resulting vector beam becomes $\theta_0 = \Delta\phi/2$. This means that by controlling the phase difference $\Delta\phi$ between the two circularly polarized vortex beams, the initial polarization direction of the resulting vector beam can be controlled. Unlike combination through Hermit Gauss beams, two vortex beams can be combined to produce a vector beam of arbitrary order. Its polarization order is identical with the same topological charge of the left circularly polarized vortex beam. Note that the OAM carried by the vortex beam annihilates during combination and the generated vector beam has a plane wave front.

7.3.2 *Sagnac Interference*

A Saganc interferometer is originally a quantum optical instrument designed to measure the angular velocity of inertia through the Sagnac effect. Its typical structure is shown in Fig. 7.10. The Sagnac interferometer has a loop structure where the beams from the source S are divided into two components by the beam splitter G. The two components travel in the loop with opposite directions. They are reflected by the mirrors M1and M2 respectively and travel along the loop for one revolution before returning to the beam splitter G, and then combined together, where interference occurs. There is no additional phase difference between them as the beam from the source travels exactly the same distance. If the Sagnac interferometer is suitably upgraded, a stable vector beam generation system can be obtained.

The optical system for the coherent combining vector beams based on the Sagnac interferometer is presented in Fig. 7.11. Compared to the standard Sagnac interferometer presented in Fig. 7.10, there are several major improvements. Firstly, the beam splitter is replaced by a polarized beam splitter (PBS). Secondly, one of the mirrors is replaced by a reflective liquid crystal spatial light modulator (SLM) encoded by a vortex phase grating [exp($ip\varphi$)] to introduce a helical phase. Furthermore, an additional Dove prism and a half wave plate with the fast axis oriented at 45° are placed in the loop. When a horizontally or vertically linearly polarized Gaussian beam is incident from the left side of the system, it is converted into circularly polarized beam by the quarter wave plate (QWP). And then it is divided by a PBS. The reflected beam is

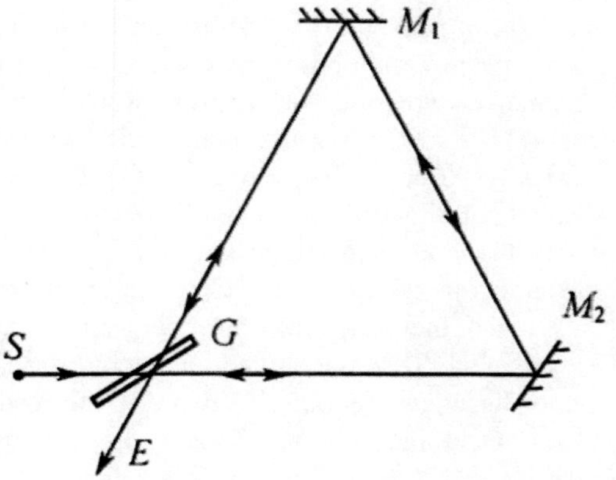

Fig. 7.10 A Sagnac interferometer

vertically linear polarization (s-polarization) and the transmitted beam is horizontally linear polarization (p-polarization). Since the incident beam is circularly polarized, the transmitted and reflected beams are of equal intensity.

The reflected vertically linear polarization is reflected by a mirror and then passed through a half wave plate (HWP) to become a horizontally linear polarization. And then it is converted into a vortex beam with topological charge -*p* by a reflective liquid crystal SLM (the reflection of the beam by the liquid crystal screen of the SLM is equivalent to introducing an additional reflection after the vortex beam is generated, making the topological charge being opposite). The Dove prism introduces an additional reflection. Hence, after the Dove prism, the topological charge of the vortex beam becomes *p* and returns to the PBS for transmission.

The transmitted horizontally linear polarization is reflected by a Dove prism firstly, and is transformed into a vortex beam with topological charge -*p* by the SLM, and then turns to vertically linear polarizaion by the HWP. Then, after the reflection of

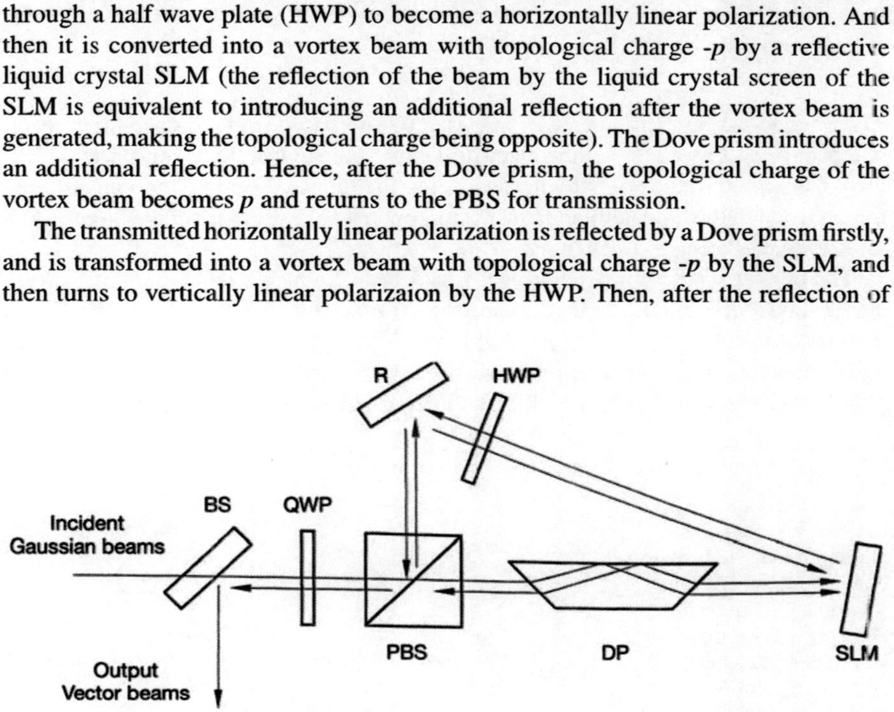

Fig. 7.11 Coherent combining vector beams based on a Sagnac interferometer. BS, beam splitter. QWP, quarter wave plate. PBS, polarized beam splitter. DP, Dove prism. SLM, spatial light modulator. HWP, half wave plate. R, reflector

the total mirror and the PBS twice, the topological charge is unchanged, and the vortex beam with topological charge -*p* is output from the left side of the PBS.

Finally, two vortex beams, one of which is vertically polarized with topological charge -*p*, the other is horizontally polarized with topological charge *p*, are combined coaxially. After passing through the QWP, the combined beam is transformed into the coaxial combination of two vortex beams with orthogonal circular polarizations and opposite topological charges. In addition, because the two beams in Sagnac loop have the same optical path, there is no phase difference between them, thus Eq. (7.3.14) is satisfied, that is, a vector beam is generated.

When the SLM is encoded with a vortex phase grating, a Laguerre Gaussian type vector beam is generated. When a higher order holographic axicon is encoded, a Bessel Gaussian type vector beam is generated. This illustrates that the intensity distribution of the generated vector beam is determined by the hologram encoded on the SLM. Furthermore, since the order of the vector beam is the same as the topological charge of the left circularly polarized vortex beam, which in turn is determined by the order of the vortex phase grating. Therefore, the order of the vector beam also determined by the hologram encoded on the SLM.

7.3.3 Sagnac-Like Interference

The Sagnac-like interferometer is proposed on the basis of the Sagnac interferometer of Fig. 7.12 by changing the place direction of the polarized beam splitter (PBS) [16]. Compared to the Sagnac interferometer, the reflected beam is no longer transmitted through the loop, although the overall optical path structure of the Sagnac-like interferometer remains unchanged.

The optical system for generating a vector beam using a Sagnac-like interferometer is shown in Fig. 7.12. First, the Gaussian beam is transformed into a vortex beam with topological charge *p* after passing through a SLM. Then it is incident in a PBS and divided into two components. Here, the fast axis arranged angle of the half wave plate (HWP) and is adjustable so that the PBS can be split incident beams with equal

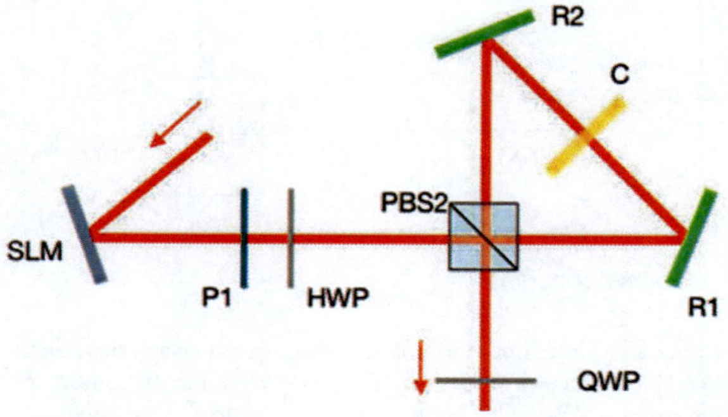

Fig. 7.12 Coherent combining vector beams based on a Sagnac-like interferometer. SLM, spatial light modulator. HWP, half wave plate. QWP, quarter wave plate. P. polarizer. PBS, polarized beam splitter. C, compensator. R1-R2, reflector [16]

intensity. In general, the fast axis of the HWP is usually 45° arranged, as common SLMs can achieve phase-only modulating only for horizontally linear polarizations.

Unlike a Sagnac interferometer, in a Sagnac-like interferometer, only the transmitted component from the PBS transmits along the entire loop, while the reflected component does not enter the loop, as shown in Fig. 7.13. For the reflected beam, it is only reflected once by the PBS, so the topological charge is opposite and becomes -*p*. The transmitted beam is reflected twice by the two mirrors in total. Hence, its topological charge is unchanged as *p*, and returns to the PBS to coaxially combine with the reflected beam. After passing through a 45° fast-axis arranged QWP, Eq. (7.3.15) is satisfied, and a vector beam is generated (Figs. 7.14 and 7.15).

It is worth mentioning that in Sagnac-like interference the two beams split by the PBS undergo different paths. Therefore, there is a phase difference between them. By placing a phase compensator in the transmitted path, the phase difference between

Fig. 7.13 Principle of generating vector beams through a Sagnac-like interferometer [16]

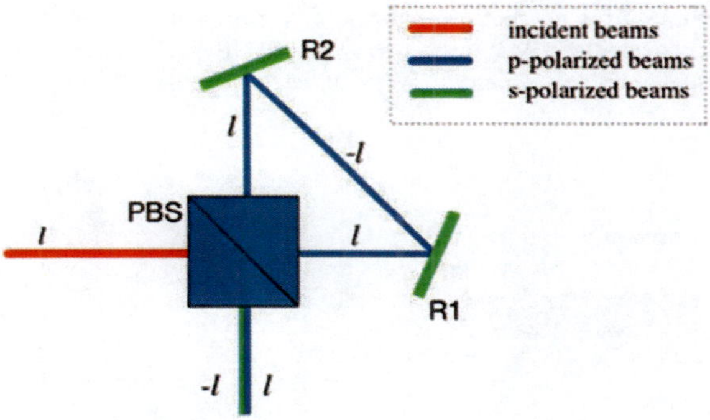

Fig. 7.14 Experimentally generated vector beams through a Sagnac-like interferometer [16]

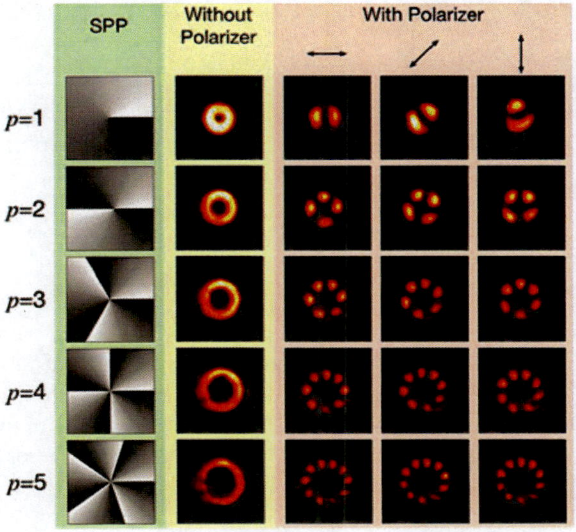

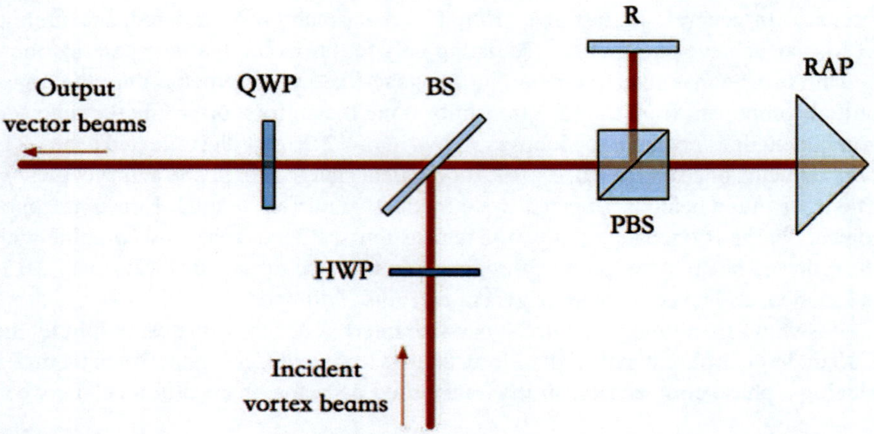

Fig. 7.15 Coherent combination of vector beams based on a Twyman-Green interferometer. BS, beam splitter. QWP, quarter wave plate. PBS, polarized beam splitter. RAP, right angle prism. SLM, spatial light modulator. HWP, half wave plate. R, reflector

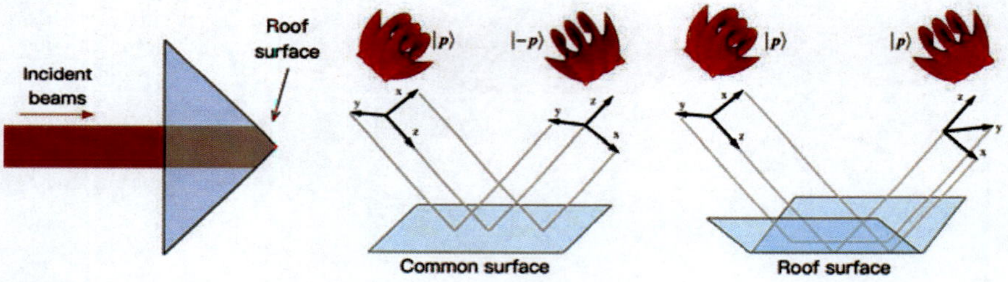

Fig. 7.16 Reflections properties of vortex beams when incident in a common surface and a roof surface

the two split vortex beams can be adjusted, namely, $\Delta\phi$ in Eq. (7.1.4), which can theoretically produce a vector beam with arbitrary polarization distribution, as shown in Fig. 7.16.

7.3.4 Twyman-Green Interference

The Twyman-Green interferometer is a two-arm interferometric device. It uses a beam splitter (BS) to split the incident beam into two beams, which are reflected by two mirrors separately and then returned back to interfere. A modified Twyman-Green interferometer can be used to generate vector beams [17, 18], as shown in Fig. 7.15.

When a vortex beam $|p\rangle$ is incident in the system shown in Fig. 7.15, it is reflected by a beam splitter and then incident on the Twyman-Green system. The role of the half wave plate (HWP) in the path is to adjust polarization of the incident vortex beam to ensure that the PBS can split the beam with equal intensity. For a vertically linearly polarized vortex beam reflected by the PBS, it is reflected by a reflector and then returned to the PBS and reflected again to the left, where it is reflected once by the full reflection mirror and twice by the PBS, resulting in total three reflections. Hence its topological charge is opposite as $|-p\rangle$. The transmitted horizontally linearly polarized vortex beam is incident on the roof surface of a right angle prism and reflected back to merge with the vertical linear polarization. The roof surface is a special reflective surface, which actually consists of two reflective surfaces placed perpendicular to each other, as shown in Fig. 7.16. When a beam hits the roof surface, it is actually reflected twice by the two mutually perpendicular planes. Hence the roof surface does not change the imaging properties, i.e. the right-handed system is still right-handed after being reflected [19]. This indicates that, the topological charge of the vortex beam won't change after being reflected by the roof surface of a right-angled prism. The combined beam therefore contains two linearly polarized vortex beams with orthogonal linear polarizations and opposite topological charges. Finally, it is converted to a vector beam after passing through a 45° fast axis arranged QWP, as shown in Fig. 7.17.

It is easy to understand that the order of the generated vector beam is identical with the topological charge of the incident vortex beam, and the transverse intensity distribution is identical to that of the incident beam. Therefore, by varying the incident vortex beam, vector beams of different orders and field distributions can be generated. In addition, the phase difference between the orthogonal circular polarization components can be adjusted by varying the length of the double arms. Thus Eq. (7.1.1) can be satisfied and arbitrary vector beams can be generated.

7.3.5 Wollaston Prism

A Wollaston prism is made by two right-angled trigonal crystals (e.g., calcite) with orthogonal optical axes, as shown in Fig. 7.18. It works by using the difference in the orientation of the optical axes of the crystals on either side of the interface, to make beams of a certain polarization direction undergoes changes from ordinary to extraordinary or extraordinary to ordinary as it passes through the glued surface, thus allowing beams of different polarizations to be refracted differently.

Since a Wollaston prism can split the incident beam into two components with orthogonal linearly polarization, it can also be used to generate a vector beam by combining orthogonal linearly polarized vortex beams, due to the reversibility of the optical path. The key to generating a vector beam using a Wollaston prism is to design a diffraction grating so that it can generate vortex beams with opposite topological charge at ± 1 diffraction orders, and then turns the polarization of one of the diffraction orders to form orthogonal polarizations, and then converge to match

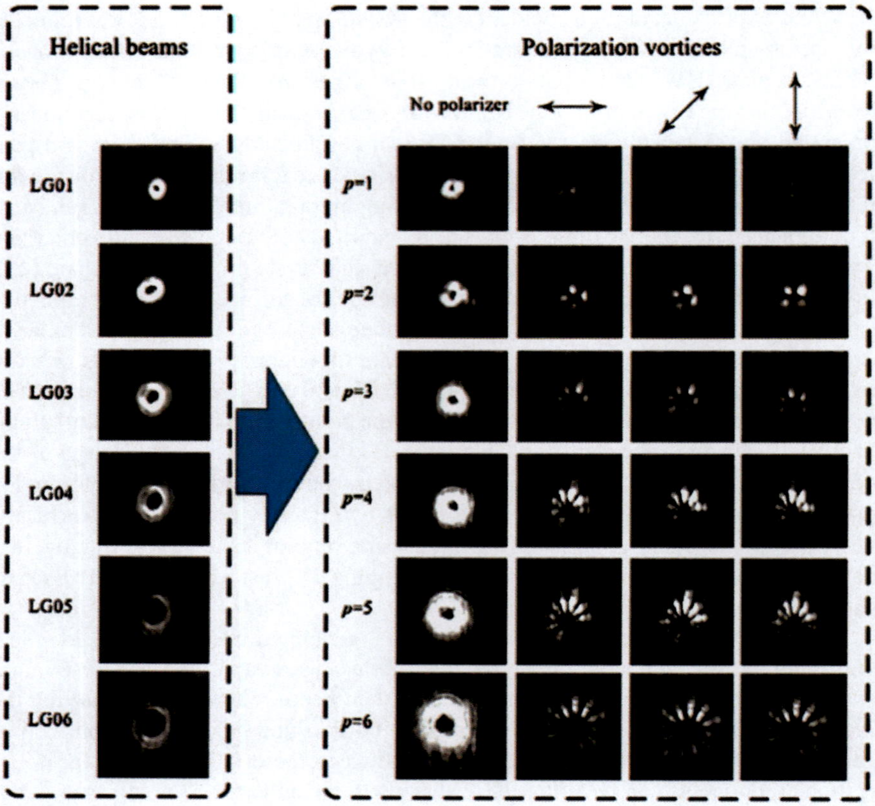

Fig. 7.17 Experimentally generated vector beams through a Twyman-Green interferometer [17]

Fig. 7.18 Wollaston prism

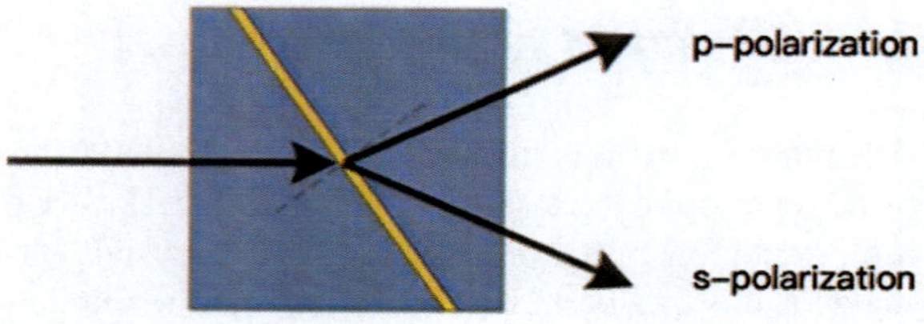

the splitting angle of the Wollaston prism to achieve coaxial combination, as shown in Fig. 7.19. The nature of the diffraction grating here can regarded as producing a 1×2 vortex beams array, as the Dammann vortex grating introduced in Chap. 4.

First, a horizontally linearly polarized Gaussian beam from the laser and a polarizer is expanded by a beam expander (BE) and is then incident in the liquid crystal spatial light modulator (SLM) encoded by a 1×2 Dammann vortex grating, to produce two vortex beams with equal intensity and opposite topological charges.

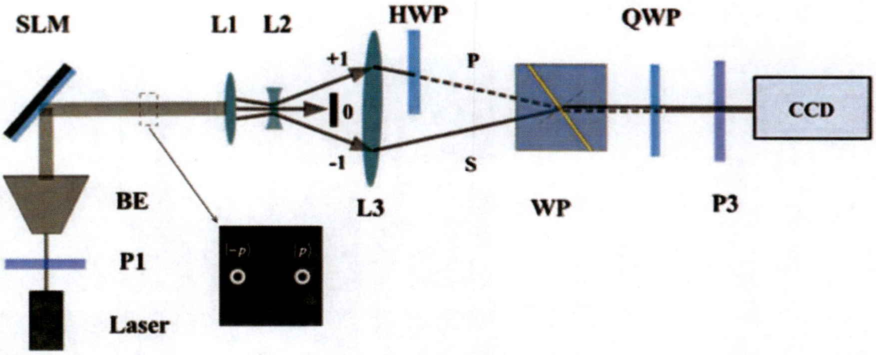

Fig. 7.19 Coherent combining vector beams based on Wollaston prisms. P1 and P2, polarizer; BE, beam expander; SLM, liquid crystal spatial light modulator; L1–L3, lenses; HWP, half wave plate; WP, Wollaston prism; QWP, quarter wave plate. [20]

Three lenses L1–L3 are employed to adjust the convergence angle of the two vortex beams to meet with the divergence angle of the Wollaton prism (WP). A half wave plate (HWP) is placed to convert the polarization of one of diffraction orders into vertically linear polarization. Similar with previous introduced schemes, two vortex beams with opposite topological charges and orthogonal polarizations are combined coaxially, and then is a vector beam is generated after passing through a 45° fast axis arranged quarter wave plate (QWP), as shown in Fig. 7.20.

Fig. 7.20 Experimentally generated vector beams using a Wollaston prism [20]

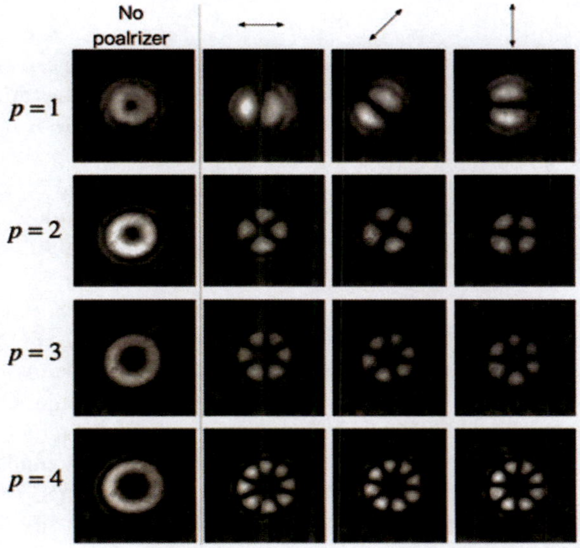

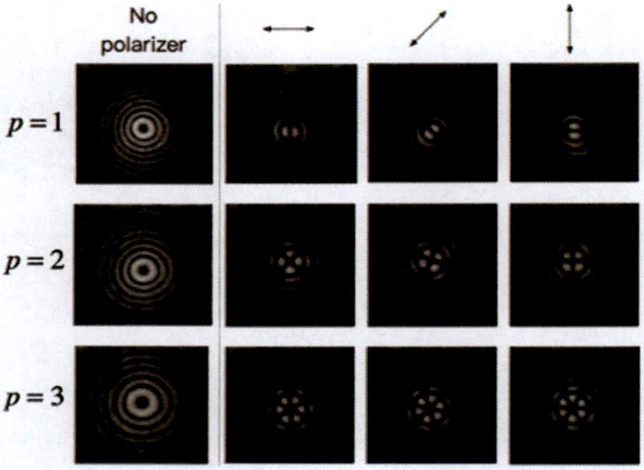

Fig. 7.21 Experimentally generated Bessel Gauss type vector beam using a Wollaston prism [20]

A Bessel Gauss type vector beam can also be generated. By superposing a 1 × 2 Dammann vortex grating with a holographic axicon introduced in Sect. 3.7.1 together, one can obtain a 1 × 2 Bessel Gauss type vortex beam array. Figure 7.21 gives the experimental results. This shows that the intensity and polarization distribution of vector beams generated using a Wollaston prism are determined by the diffraction grating encoded on the SLM.

7.3.6 Cascading Liquid Crystal Spatial Light Modulators

The liquid crystal spatial light modulator (SLM) can only achieve phase only modulation for a linearly polarized beam whose polarization direction is parallel to the optical axis of the liquid crystal molecule. There is no modulations on the linearly polarized beam whose polarization direction is perpendicular to the optical axis of the liquid crystal molecule. Therefore, the liquid crystal SLM can be regarded as a grating with polarization sensitivity, and the modulation is various for beams with different polarizations. Based on such property of liquid crystal SLMs, vector beam can be well generated.

Figure 7.22 shows a scheme for generating vector beams through a liquid crystal SLMs, where a transmitted liquid crystal SLM is employed and its liquid crystal screen is divided into two parts encoded with two vortex phases of opposite order. The principle of this scheme is that, a fundamental Gaussian beam is incident and become 45° linear polarization after passing through a linear polarizer (LP). Such beam can be decomposed orthogonally into horizontally linear polarization and vertically linear polarization with equal intensity. After passing through the bottom half of the liquid crystal screen of the SLM, horizontally linear polarization component is transformed into a vortex beam $|p\rangle$, while the vertically polarization component is unaffected and remains a fundamental mode. After the first modulation, the beam passes through a quarter wave plate (QWP) 1, a focusing lens, a mirror, and then back to the upper half of the liquid crystal screen. The role of the lens is to work with the mirror to form a reflective 4-f system, to image the bottom half of the beam on SLM to the upper half with equal magnification. QWP1 is 45° arranged and works in conjunction with the mirror to exchange the horizontal and vertical polarization components, the principle of which is the same as that of the common optical isolator. Thus, when the beam is returned to the SLM, the originally modulated horizontally polarized component $|p\rangle$ is converted to vertical polarization and is unchanged, while the previously unmodulated vertically linear polarization component is converted to horizontal polarization and is turned to $|-p\rangle$. This means that the twice modulated beam is actually the combination of two linearly polarized vortex beams with opposite topological charges and orthogonal polarizations. Such combined beam is then converted into a vector beam after passing through QWP2 with 45° arranged fast axis.

The nature of this optical setup is to modulate the two orthogonal linear polarizations of one beam separately by two polarization sensitive gratings. Here, the liquid crystal screen is employed cleverly in combination with a 4-f imaging system, which cleverly combines two liquid crystal spatial light modulators into one, reducing the cost of the system and miniaturising the whole system [22]. Its disadvantage is that the position of the center of the split liquid crystal screen is uneasy to control.

Another common solution is cascading two reflective liquid crystal SLMs, as shown in Fig. 7.23. Firstly, the horizontally linear polarized Gaussian beam is modulated by the liquid crystal SLM 1. SLM1 is encoded by a $-p$-order vortex phase

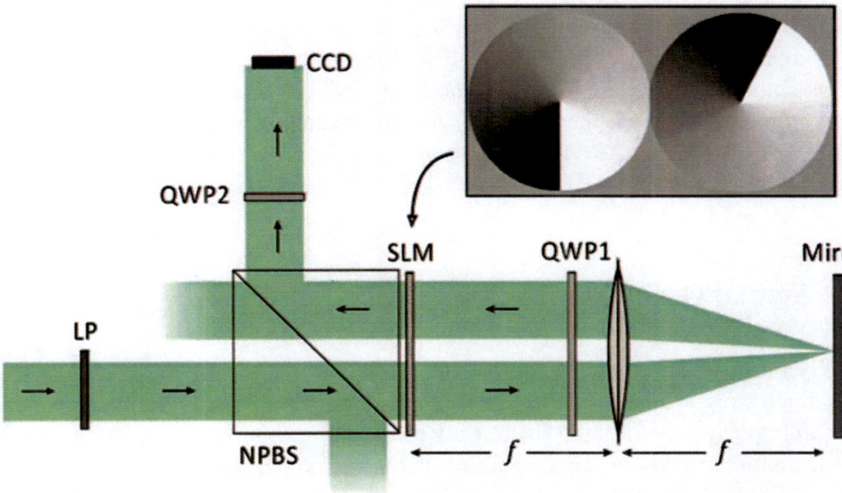

Fig. 7.22 Coherent combining vector beams through a transmissive liquid crystal SLM. LP, linear polarizer; NPBS, non-polarized beam splitter; SLM, spatial light modulator; QWP1–QWP2, quarter wave plates [21]

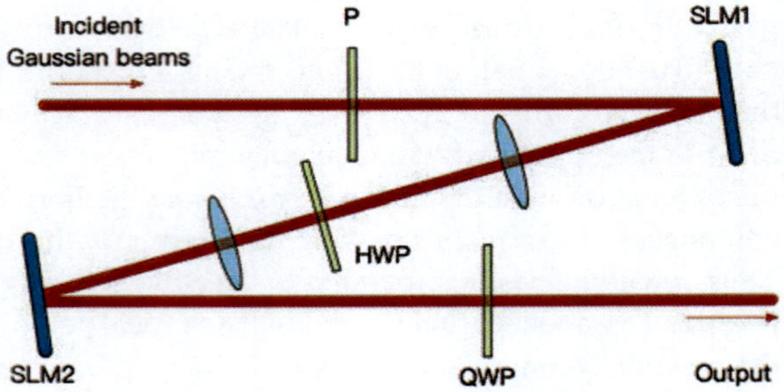

Fig. 7.23 Coherent combining vector beams based on cascading reflective liquid crystal SLMs. P, polarizer; SLM1 and SLM2, spatial light modulators; HWP, half wave plate; QWP, quarter wave plate

grating, and the incident Gaussian beam is transformed into a p-order vortex beam $|p\rangle$. The fast axis of the HWP is arranged at 22.5°, turning the polarization direction of the horizontally polarized vortex beam to 45°. By now, the vortex beam can be decomposed into two parts with equal intensity but orthogonal linear polarizations (horizontal and vertical polarization). The liquid crystal SLM2 is encoded by a $-2p$-order vortex phase grating. When passing through SLM2, the vertical polarization component is unmodulated, but reflected once. Hence the topological charge of vertical component is ropposite and becomes $|-p\rangle$. The horizontal polarization component is modulated and its topological charge becomes $-(p-2p)=p$, forming the vortex beam $|p\rangle$. Therefore, the beam after SLM2 is now the combination of two linearly polarized vortex beams with orthogonal polarizations and opposite topological charges p and $-p$, which are then converted into a vector beam by a QWP with 45° arranged fast axis.

Cascading reflective liquid crystal SLM to generate vector beams is relatively simple. Figure 7.24 shows various vortex phase gratings that encoded on the two SLMs and the experimentally generated vector beams. The polarization orders of the generated vector beamsare determined only by the vortex gratings encoded on the two SLM. Also, since there are only two SLMs and a few polarization optics in the system, the efficiency of the generation is mainly determined by the diffraction efficiency of the employed SLMs.

7.4 Spatial Oscillating Polarization

Usually, the generated vector beams only have transverse vortex polarization distributions. Such structured vortex polarization won't change as its propagation. A spatial oscillating polarization vector beam is a Bessel-type vector beam whose transverse polarization distribution changes periodically when propagation [24, 25], as shown in Fig. 7.25. Such beam can present various vortex polarization distributions.

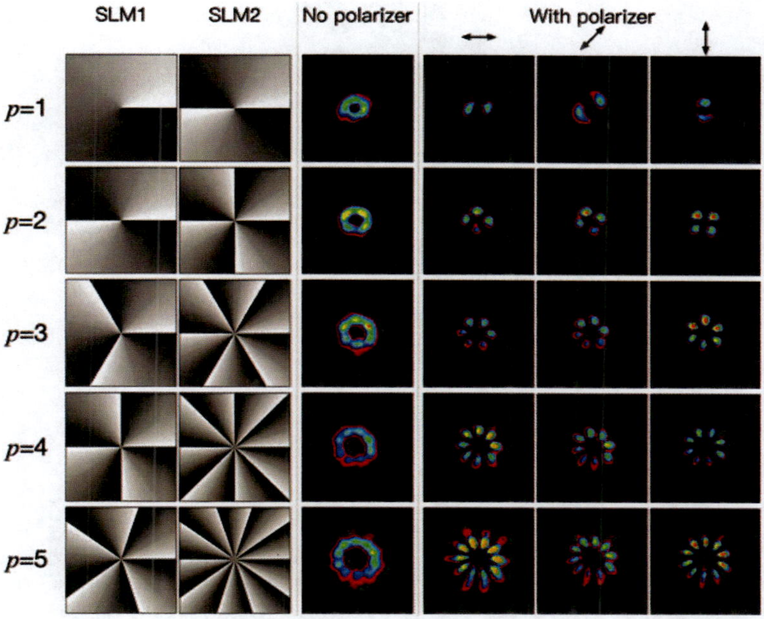

Fig. 7.24 Experimentally generated vector beams through cascading reflective liquid crystal SLMs [23]

7.4.1 Principles of Spatial Oscillating Polarization

For the spatial oscillating polarization vector beam given in Fig. 7.25, its polarization distribution rotates as the transmission distance z varies. Combined with Eq. (7.1.1), it is easy to understand that the initial polarization direction θ_0 of such a vector beam is z-dependent. Equation (7.3.15) shows that when a vector beam is generated through vortex beam combining, the initial polarization direction of the resulting vector beam is determined by the phase difference $\Delta\phi$ between the two scalar vortex beams. Therefore, by expressing $\Delta\phi$ as a function of position z, different additional phases can be introduced at different transmission distances, and the axial modulation of polarization can be achieved.

A BG beam can be viewed as a conical wave. It is formed by the superposition of multiple planar subwaves of equal amplitude whose wave vector point to optical axis (z-axis) in equal angles. The axicon introduced in Sect. 3.7.1 can be employed to generate BG beams. In the overlapping region, the field at a point with coordinate z on the optical axis is generated by superposing rays on a circle of an axicon with radius r. Thus from a simple geometric relationship (Fig. 3.7.2), the relationship between z and r satisfies:

$$z = \frac{rd}{\lambda} \tag{7.4.1}$$

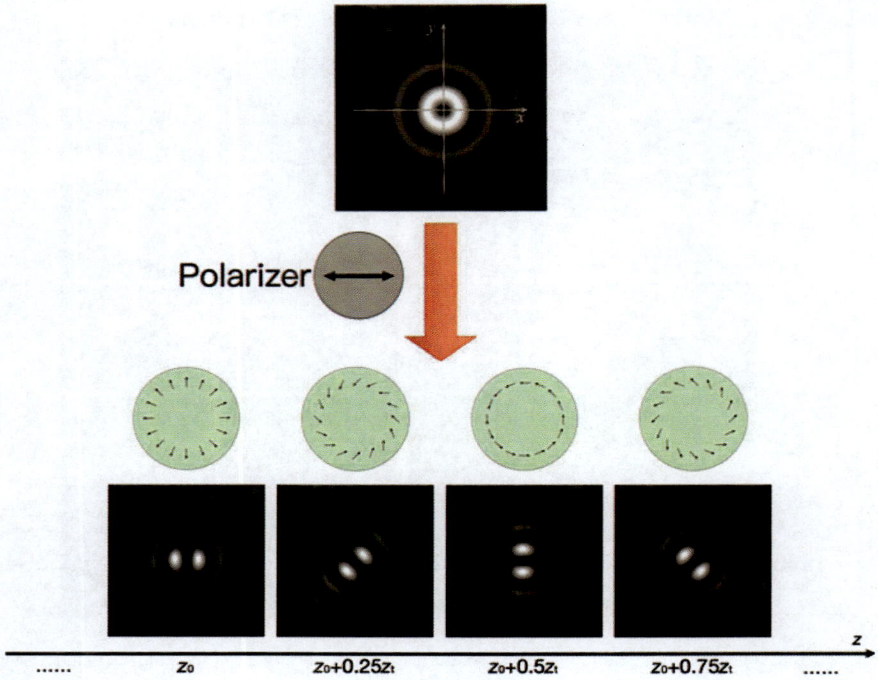

Fig. 7.25 Spatial oscillating polarization vector beams

where d is the axicon period and λ is the wavelength of light.

In the overlapping region behind the axicon, a phase-modulated device whose phase distribution varying with the radial coordinate r is placed, as shown in Fig. 7.26. Let the radial period of the device be D and its phase distribution function be

$$\phi(r) = \frac{2\pi r}{D} \tag{7.4.2}$$

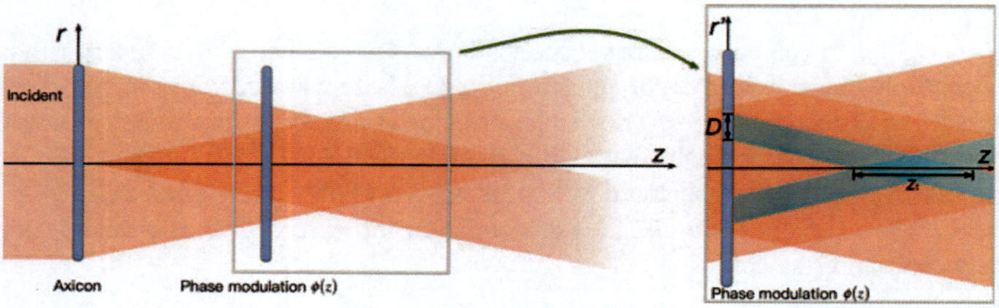

Fig. 7.26 Introducing additional phases along optical axis [24]

Then substituting Eq. (7.4.1) into Eq. (7.4.2) gives

$$\phi(z) = \frac{2\pi \lambda z}{Dd} \tag{7.4.3}$$

The phase in Eq. (7.4.3) is a function of coordinate z, indicating that the phase-modulated device introduces different phases at different positions slong the optical axis. This phase-modulated device is actually also an axiconatic grating. When Eq. (7.4.3) is used as the phase difference between the two BG beams, a vector beam with spatial oscillating polarization is generated as:

$$\frac{1}{2}\exp\left[i(p\varphi + \phi(z))\right]\begin{bmatrix} 1 \\ -i \end{bmatrix} + \frac{1}{2}\exp(-ip\varphi)\begin{bmatrix} 1 \\ i \end{bmatrix} = \exp\left[\frac{1}{2}i\phi(z)\right]\begin{bmatrix} \cos\left(p\varphi + \frac{1}{2}\phi(z)\right) \\ \sin\left(p\varphi + \frac{1}{2}\phi(z)\right) \end{bmatrix} \tag{7.4.4}$$

An important parameter of a spatial oscillating polarization beam is the spatial period of polarization. For vector beams, the initial polarization direction θ_0 and (θ_0 + π) are the same. Therefore, when $\phi(z)$ is changed by 2π, the spatial polarization oscillating vector beam experiences a spatial period z_t as:

$$z_t = \frac{dD}{\lambda} \tag{7.4.5}$$

Equation (7.4.5) shows that the spatial period of the spatial oscillating polarization is only determined by the wavelength λ, the axicon constant d, and the radial period D of the additional axicon phase. Since the BG beam exists only within a certain range ($z < z_{\max}$) behind the axicon, to observe a complete spatial period, it must satisfy $z_t < z_{\max}$. Combined with Eq. (3.7.3), it follows that $D < \omega$, with ω the radius of the Gaussian beam incident in the axicon.

7.4.2 Generating Spatial Oscillating Polarization Vector Beams

The optical system for generating a spatial oscillating polarization vector beam is the same as the system given in Fig. 7.23. For the system shown in Fig. 7.23, to generate a vector beam, it is necessary to satisfy the requirement that the vortex phase grating order encoded on SLMs is two times that that on SLM1. When generating an spatial oscillating polarization vector beam, in addition to encoding these two gratings, two separate axicon phases need to be superposed. Note that for a reflective liquid crystal spatial light modulator, anti-axicon phases should be superimposed, as shown in Fig. 7.27. The first axicon phase is used to generate a BG beam and the second axicon phase is used to introduce an axial additional phase. As the horizontally linearly polarized Gaussian beam passes through SLM1, it is converted

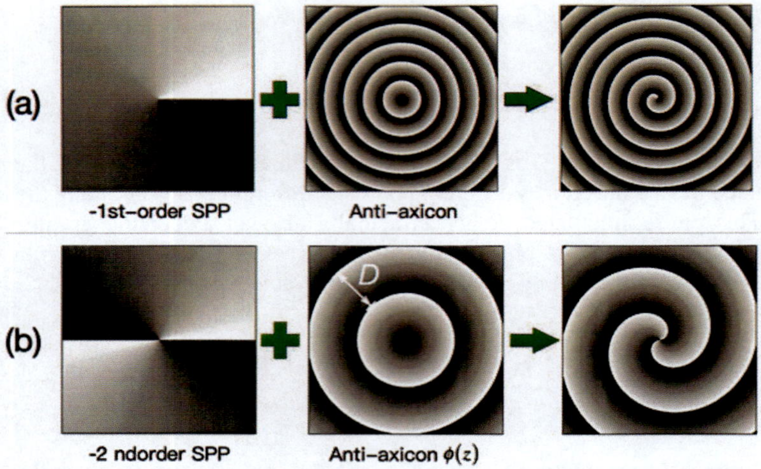

Fig. 7.27 Examples of generating a first-order spatial oscillating polarization vector beam, **a** the phase grating encoded on the first SLM and **b** the phase grating encoded on the second SLM. The phase of the anti-axicon in (**a**) is used to generate a Bessel Gauss beam, and the phase of the anti-axicon in (**b**) is used to introduce a *z* related additional phase difference [24]

into a *p*-order BG beam, whose polarization direction is then converted to 45° to the horizontal by the half wave plate. After passing through SLM2, the vertically polarized component is only reflected once without being modulated, and its topological charge becomes *-p*. The horizontally polarized component is both reflected and modulated, and its topological charge is still $p[-(p-2p)=p]$. However, since the second modulator is superimposed on the axicon phase $\phi(z)$, it introduces an additional phase difference to the vertically polarized component compared to the horizontally polarized component, and after passing through a 45° fast axis arranged quarter wave plate, Eq. (7.4.4) is satisfied, that is, a spatial oscillating polarization vector beam is generated. The generating results are given in Fig. 7.28.

7.5 Polarization Poincare Sphere

The polarization Poincare sphere, first proposed by the mathematician Poincare from France, is a theoretical model for describing the polarization states of beam, initially limited to a few common isotropic polarization states including linear, elliptical and circular polarization [26]. Later, the definition of the Poincare sphere is extended. The high-order Poincare sphere was proposed to describe a vector beam with an anisotropic vortex polarization distribution in transverse [27]. Based on the high-order Poincare sphere, the concept of hybrid-order Poincare sphere was introduced to describe vectorial vortex beams with both vortex polarizations and helical phases [28]. Thus, any vector beam or vectorial vortex beams can find its corresponding

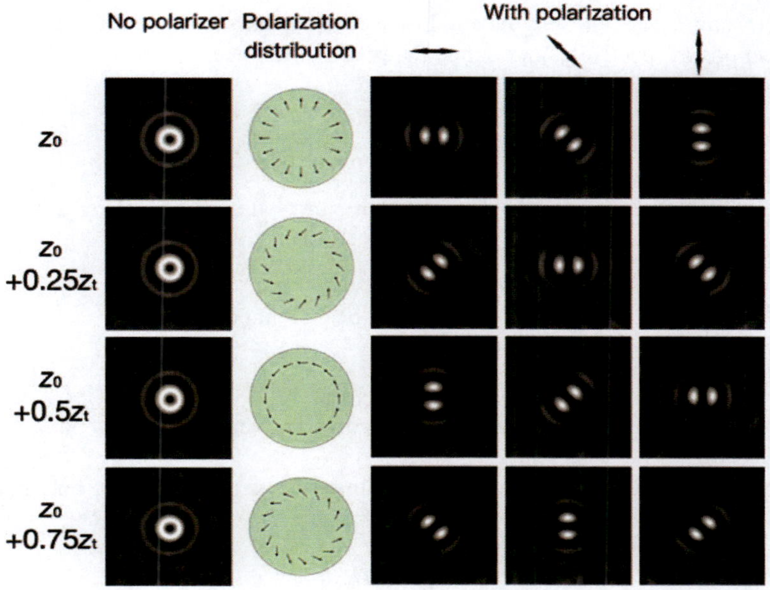

Fig. 7.28 Experimentally generated first-order spatial oscillating polarization vector beams [24]

point on the surface of a Poincare sphere, and the polarization state of the beam can be expressed in terms of the longitude and latitude coordinates (θ, σ), which greatly simplifies the representation of complex vectorial vortex beams. Polarization Poincare sphere is an important basis for the study of complex vectorial vortex beams. Recently, researchers have proposed the generalized Poincare sphere, which extends the study of polarization states from surface to inner and is used to represent more complex polarized fields [29]. This book will focus more on the high-order Poincare sphere and the hybrid-order Poincare sphere.

7.5.1 Stokes Vector and Fundamental Poincare Sphere

First, Let's review the concept of the Stokes vector. In 1852, Stokes proposed a way to describe the intensity and polarization of light waves using four parameters, which are the time average of the intensity and form a mathematical vector:

$$S = \begin{bmatrix} I \\ Q \\ U \\ V \end{bmatrix} = \begin{bmatrix} \langle E_x^2(t) \rangle + \langle E_y^2(t) \rangle \\ \langle E_x^2(t) \rangle - \langle E_y^2(t) \rangle \\ 2\langle E_x(t)E_y(t)\cos\phi \rangle \\ 2\langle E_x(t)E_y(t)\sin\phi \rangle \end{bmatrix} \tag{7.5.1}$$

$E_x(t)$ and $E_y(t)$ represent the amplitude components of the electric field in x and y directions, respectively, and define the phase difference between the components of the electric field in both directions:

$$\phi = \phi_y(t) - \phi_x(t) \tag{7.5.2}$$

In Eq. (7.5.1), I, Q, U and V have intensity dimensions. Their physical meanings are:

I Total intensity of the beam;
Q Horizontally (x-axis) linearly polarized beam component;
U 45° linearly polarized component;
V Right circularly polarized component.

Q, U and V are negative for a vertically linear polarizations, −45° linear polarizations, and left circular polarizations.

The Stokes vector can represent any of the various homogeneous polarizations. Under quasi-monochromatic non-coherent conditions, usually there are:

$$I^2 \geq Q^2 + U^2 + V^2 \tag{7.5.3}$$

For partially polarized beam,

$$I^2 > Q^2 + U^2 + V^2 \tag{7.5.4}$$

For fully polarized beam such as linearly polarized beam, elliptically polarized beam and circularly polarized beam, there are:

$$I^2 = Q^2 + U^2 + V^2 \tag{7.5.5}$$

It is shown that the Stokes vector contains four components that are not independent of each other. When the total light intensity I is constant, the polarization state of light can be fully represented by Q, U and V.

From the Stokes vector, the degree of polarization P can also be defined as

$$P = \frac{\sqrt{Q^2 + U^2 + V^2}}{I} \tag{7.5.6}$$

P = 0 indicates natural light. P ∈ (0, 1) indicates partially polarized beam. P = 1 indicates fully polarized beam. Table 7.3 gives the Stokes vectors and polarizations of common homogeneous polarization states.

Let $S_1 = U$, $S_2 = V$, $S_3 = Q$. In Cartesian coordinate with $(S_1,\ S_2,\ S_3)$ being the coordinate axes, making a sphere whose center is the origin and taking I as the radius, as shown in Fig. 7.29. The relationship from Eqs. (7.5.3)–(7.5.6) shows that on the surface, the degree of polarization $P = 1$, so any point on the surface indicates fully polarized beam. At the center of the sphere, $P = 0$, indicating natural light.

Table 7.3 Some of the common polarization state and its Stokes parameters

Polarization state	Stokes vector $[I, Q, U, V]^T$	Polarization degree P
Natural light	$[1, 0, 0, 0]^T$	0
Horizontally linear polarization	$[1, 1, 0, 0]^T$	1
Vertically linear polarization	$[1, -1, 0, 0]^T$	1
45° linearly polarization	$[1, 0, 1, 0]^T$	1
−45° linearly polarization	$[1, 0, -1, 0]^T$	1
Right circular polarization	$[1, 0, 0, 1]^T$	1
Left circular polarization	$[1, 0, 0, -1]^T$	1

Inside the sphere, $0 < P < 1$, indicating partially polarized beam. Such mathematical model based on the Stokes vector representing the polarization state of the beam is the polarization Poincare sphere.

Since any point on the surface of the Poincare sphere can represent fully polarized beam, the polarization state can be described by the longitude and latitude coordinates (θ, σ) on the surface. From the geometric relationship, it follows that the longitude θ

$$\theta = \arctan \frac{S_2}{S_1} \tag{7.5.7}$$

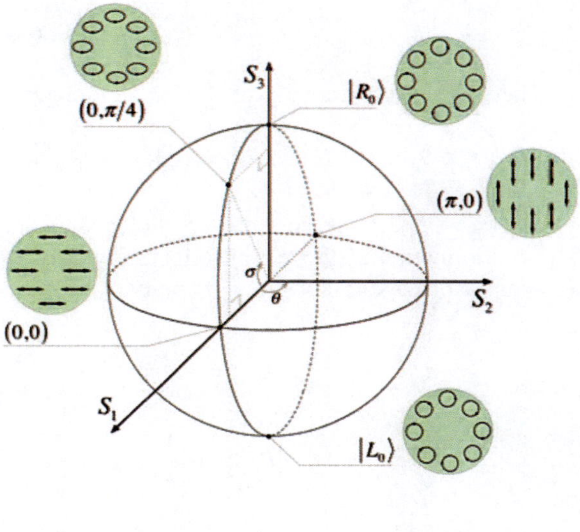

Fig. 7.29 Fundamental Poincare sphere

and latitude σ:

$$\sigma = \arcsin \frac{S_3}{S_0} \tag{7.5.8}$$

The point at the equator of the Poincare sphere, satisfying $\sigma = 0$, $S_3 = 0$, indicates that there is no circularly polarized component in the beam, which represents linearly polarized beam. At the north and South Poles, $\sigma = \mp\pi/2$, $S_3 = \mp 1$, indicates that the beam has only circularly polarized components, so the points on the north and south poles represent the left and right circularly polarized beams, respectively. The points on the north and south hemispheres have both linear and circular polarization components, so they represent the left and right elliptically polarized beams, respectively. Since the Poincare sphere given in Fig. 7.29 can only represent the common homogeneous polarizations, but cannot represent the complex inhomogeneous polarizations such as vector beams, thus called the fundamental Poincare sphere. In order to represent vector beams, the Stokes vector and the fundamental Poincare sphere must be extended, namely, the higher-order Stokes vector and the higher-order Poincare sphere.

7.5.2 Higher-Order Stokes Vector and Higher-Order Poincare Sphere

From Eq. (7.3.15), the vector beam can be regarded as the combination of two left- and right circularly polarized beams with opposite topological charges. For the convenience of representation, let:

$$|\psi_p\rangle \equiv \vec{E}(r, \varphi) = A(r)\begin{bmatrix} \cos(p\varphi + \theta_0) \\ \sin(p\varphi + \theta_0) \end{bmatrix} \tag{7.5.9}$$

Then the process can be expressed in a more general form:

$$|\psi_p\rangle = \psi_L^p |L_p\rangle + \psi_R^{-p} |R_{-p}\rangle \tag{7.5.10}$$

where ψ_L^p and ψ_R^{-p} are complex coefficients containing the amplitude and initial phase information of the left and right circularly polarizations, respectively. $\{L_p, R_p\}$ form a orthogonal substrate of circularly polarization.

$$|L_p\rangle = \frac{1}{\sqrt{2}} \exp(ip\varphi)\begin{bmatrix} 1 \\ -i \end{bmatrix} \tag{7.5.11}$$

$$|R_{-p}\rangle = \frac{1}{\sqrt{2}} \exp(-ip\varphi)\begin{bmatrix} 1 \\ i \end{bmatrix} \tag{7.5.12}$$

and contain helical phase terms with opposite topological charges, respectively. By now, the higher-order Stokes vector can be defined [27] as

$$\begin{bmatrix} S_0^p \\ S_1^p \\ S_2^p \\ S_3^p \end{bmatrix} = \begin{bmatrix} |\langle R_{-p}|\psi_p\rangle|^2 + |\langle L_p|\psi_p\rangle|^2 \\ 2\mathrm{Re}(\langle R_{-p}|\psi_p\rangle^* \langle L_p|\psi_p\rangle) \\ 2\mathrm{Im}(\langle R_{-p}|\psi_p\rangle^* \langle L_p|\psi_p\rangle) \\ |\langle R_{-p}|\psi_p\rangle|^2 - |\langle L_p|\psi_p\rangle|^2 \end{bmatrix} = \begin{bmatrix} \left|\psi_R^{-p}\right|^2 + \left|\psi_L^p\right|^2 \\ 2\left|\psi_R^{-p}\right|\left|\psi_L^p\right| \cos\phi \\ 2\left|\psi_R^{-p}\right|\left|\psi_L^p\right| \sin\phi \\ \left|\psi_R^{-p}\right|^2 - \left|\psi_L^p\right|^2 \end{bmatrix} \tag{7.5.13}$$

where p is the order of the higher-order Stokes vector and ϕ is the initial phase difference between the right and left circularly polarized vortex components:

$$\phi = \arg\left(\psi_R^{-p}\right) - \arg\left(\psi_L^p\right) \tag{7.5.14}$$

The higher-order Stokes vector is consistent with the Stokes vector given in Eq. (7.5.1): the first term S_0^p is the sum of the intensities of the right and left circular polarization components, which represents the total intensity of the vector beam. The second term S_1^p is the inner product of the complex amplitude terms of the right and left circular polarization components, which represents the vector beam at the initial polarization direction $\theta_0 = 0$. The third term S_2^p is the module of the outer product of right and left circular polarization components, which represents the vector beam at the initial polarization direction $\theta_0 = \pi/4$. The fourth term S_3^p is the intensity difference between the right and left circular polarization components, which represents the circularly polarized beam. However, the higher-order Stokes vector given in Eq. (7.5.13) has a form that is not exactly the same as the Stokes vector given in Eq. (7.5.1). The reason is that the Stokes vector in Eq. (7.5.1) is obtained by decomposing the beam into horizontal and vertical linear polarization components with orthogonal directions, while the higher-order Stokes vector in Eq. (7.5.13) is obtained by decomposing the beam into left and right circular polarizations. When p equals to 0, the higher-order Stokes vector will degenerate to the fundamental Stokes vector in Eq. (7.5.1), so Eq. (7.5.1) is the 0th-order Stokes vector.

In Cartesian coordinate with $(S_1,\ S_2,\ S_3)$ being the coordinate axes, making a sphere whose center is the origin and taking S_0 as the radius, we get a higher-order Poincare sphere, as shown in Fig. 7.30. The points on the spherical surface of the higher-order Poincare sphere can represent any vector beam. The longitudinal and latitudinal coordinates (θ, σ) of the points on the surface can be used to represent the polarization state of the beam under a certain order p. From the geometric relationship:

$$\theta = \arctan\frac{S_2}{S_1} = \phi \tag{7.5.15}$$

$$\sigma = \arcsin\frac{S_3}{S_0} = \arcsin\frac{\left|\psi_R^{-p}\right|^2 - \left|\psi_L^{p}\right|^2}{\left|\psi_R^{-p}\right|^2 + \left|\psi_L^{p}\right|^2} \tag{7.5.16}$$

Equations (7.5.15) and (7.5.16) show that the longitude coordinate θ on the high-order Poincare sphere is identical to the initial phase difference between the right and left circular polarization vortex components, while the latitude coordinate σ is determined by the intensity of the right and left circular polarization vortex components. At the equator, $\sigma = 0$, then $\left|\psi_R^{-p}\right|^2 = \left|\psi_L^{p}\right|^2$, at which point Eq. (7.5.10) is then identical to Eq. (7.3.15), so that the point on the equator of the pth-order Poincare sphere represents the pth-order vector beam with linear polarization anisotropy in transverse. In particular, on the first-order Poincare sphere, $(\theta, \sigma) = (0, 0)$ means radially polarized beam and $(\theta, \sigma) = (\pi, 0)$ means azimuthally polarized beam. At the north and south poles, $\sigma = \mp\pi/2, S_3 = \mp 1$, indicating that $\left|\psi_R^{-p}\right|^2$ or $\left|\psi_L^{p}\right|^2$ must be 0. In other words, there is only right or left circular polarization components, so the north and south poles indicate the left and right circularly polarized vortex beams with topological charges p and $-p$, respectively. The point located on the northern and southern hemispheres has a non-zero latitude coordinate σ, $\left|\psi_R^{-p}\right|^2 \neq \left|\psi_L^{p}\right|^2$, so it can be regarded as the result of the joint action of the linear polarization vector beam at the equator and the circularly polarized vortex beam at the north and south poles, which represents the beam with left and right elliptical polarization anisotropy in transverse.

The higher-order Poincare sphere is the extension of the fundamental Poincare sphere, which introduces the helical phase term with opposite topological charges to the left and right circular polarizations. When the $p = 0$, the helical phase term

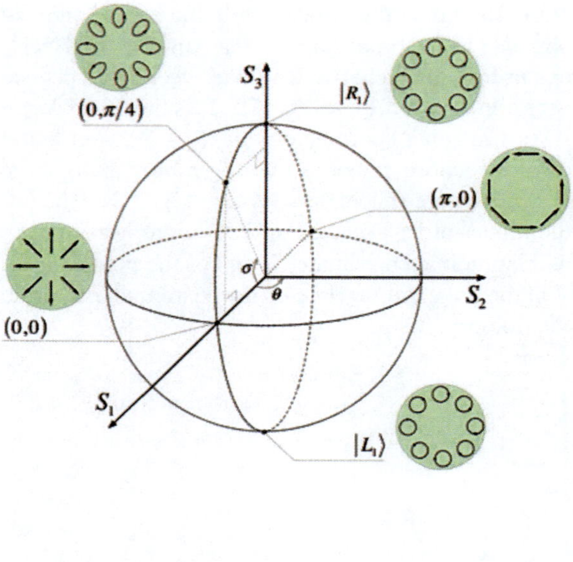

Fig. 7.30 The higher-order Poincare sphere (order $p = 1$)

disappears, and the higher-order Poincare sphere will degenerate to the fundamental Poincare sphere. Therefore, the Poincare sphere representing each homogeneous polarization state given in Fig. 7.30 is actually a 0th-order Poincare sphere.

7.5.3 Hybrid-Order Poincare Sphere

Although the higher-order Stokes vector and higher-order Poincare sphere can represent the vector beam very well, they cannot represent the vectorial vortex beams with both transverse vortex polarizations and helical phases. Therefore, further expansion of the higher-order Stokes vector and higher-order Poincare sphere is essencial. In Eq. (7.5.10), the topological charges of the two circularly polarized vortex components are opposite to each other. If such astriction of opposite topological charges is removed, Eq. (7.5.10) can be extended to a more general form:

$$|\psi\rangle = \psi_L^n|L_n\rangle + \psi_R^m|R_m\rangle \tag{7.5.17}$$

where ψ_L^n and ψ_R^m are complex coefficients and $\{L_n, R_m\}$ form orthogonal substrates of vortex circular polarization with topological charges n and m:

$$|L_n\rangle = \frac{1}{\sqrt{2}}\exp(in\varphi)\begin{bmatrix}1\\-i\end{bmatrix} \tag{7.5.18}$$

$$|R_m\rangle = \frac{1}{\sqrt{2}}\exp(im\varphi)\begin{bmatrix}1\\i\end{bmatrix} \tag{7.5.19}$$

A beam that satisfies Eq. (7.5.17) is the vectorial vortex beams. Then according to Eqs. (7.5.17)–(7.5.19), define the hybrid-order Stokes vector [28]:

$$\begin{bmatrix}S_0\\S_1\\S_2\\S_3\end{bmatrix} = \begin{bmatrix}|\langle R_m|\psi\rangle|^2 + |\langle L_n|\psi\rangle|^2\\ 2\mathrm{Re}\big(\langle R_m|\psi\rangle^*\langle L_n|\psi\rangle\big)\\ 2\mathrm{Im}\big(\langle R_m|\psi\rangle^*\langle L_n|\psi\rangle\big)\\ |\langle R_m|\psi\rangle|^2 - |\langle L_n|\psi\rangle|^2\end{bmatrix} = \begin{bmatrix}|\psi_R^m|^2 + |\psi_L^n|^2\\ 2|\psi_R^m||\psi_L^n|\cos\phi\\ 2|\psi_R^m||\psi_L^n|\sin\phi\\ |\psi_R^m|^2 - |\psi_L^n|^2\end{bmatrix} \tag{7.5.20}$$

where,

$$\phi = \arg\big(\psi_R^m\big) - \arg\big(\psi_L^n\big) \tag{7.5.21}$$

Similar to the high-order Poincare sphere, in Cartesian coordinate with (S_1, S_2, S_3) being the coordinate axes, making a sphere whose center is the origin and taking S_0 as the radius, the hybrid-order Poincare sphere is established, as shown in Fig. 7.31. The nature of the hybrid-order Poincare sphere and the formula for calculating the latitude and longitude coordinates are exactly the same as that of the high-order

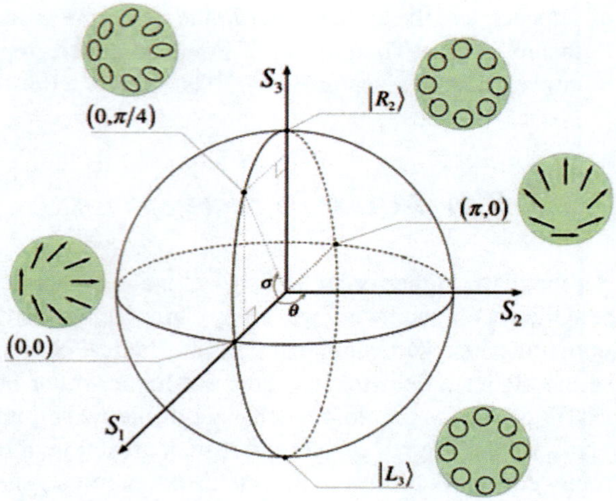

Fig. 7.31 Hybrid order Poincare sphere ($m = 2$, $n = 2$)

Poincare sphere. The difference between the hybrid-order Poincare sphere and the higher order Poincare sphere is that the topological charges of left and right circularly polarized beam at the north and south poles are independent of each other and are not constrained by the opposite topological charges.

The fundamental Poincare sphere, the higher-order Poincare sphere and the hybrid-order Poincare sphere are collectively called the polarization Poincare sphere. Any homogeneous polarizations like linear polarization and circular polarization, or complex polarization states like vector beams or vectorial vortex beams, all can be represented by points on the surface of the polarization Poincare sphere. Therefore, fundamental polarized beam, vector beam, and vectorial vortex beams, are collectively referred to as Poincare beams [30].

7.6 Vectorial Vortex Beams

7.6.1 Overview of Vectorial Vortex Beams

A vectorial vortex beam is a new type of structured beams, which is a universal form containing the properties of both vector and vortex beams with transverse anisotropic vortex polarizations and helical phase, thus carrying OAM. Equation (7.5.17) shows that a vectorial vortex beam can be decomposed into two orthogonal circularly polarized vortex beams with different absolute values of topological charges. Vectorial vortex beams can be represented by the points (θ, σ) on the surface of the hybrid-order Poincare sphere.

A vectorial vortex beam located on the equator of the hybrid-order Poincare sphere reads

$$|\psi\rangle = A(r)\exp(il_p\varphi)\begin{bmatrix}\cos(p\varphi+\theta_0)\\ \sin(p\varphi+\theta_0)\end{bmatrix} \tag{7.6.1}$$

where $A(r)$ is the amplitude, p is the polarization order indicating the number of azimuthal periods of transverse polarization change in one circle, and l_P is called the Pancharatnam topological charge. Pancharatnam topological charges originate from Pancharatnam-Berry phases. Different spatial positions in a vectorial vortex beam possess different polarization states, and the phase delay between the different spatial positions cannot be understood in terms of ordinary scalar phases (such as the spiral phase of a vortex beam, etc.). Therefore, Pancharatnam introduced a new geometric phase to represent the additional phase delay introduced by the polarization state of a scalar (transverse polarization state of each homogeneous) beam undergoing a cyclic transformation during propagation, namely, the Pancharatnam phase [31]. The concept of Pancharatnam phase was later extended to vector fields to represent the phase difference ϕ_P between different polarized beam fields $|\vec{A}\rangle$, $|\vec{B}\rangle$ at various spatial points in a complex vectorial vortex beams:

$$\phi_P = \arg\langle\vec{A}|\vec{B}\rangle \tag{7.6.2}$$

ϕ_P is called the Pancharatnam-Berry phase [32]. The Pancharatnam topological charge l_P is defined as the number of azimuthal phase changes period around the phase singularity in the equiphase surface of the Pancharatnam-Berry phase:

$$l_P = \frac{1}{2\pi}\oint_C d\phi_P \tag{7.6.3}$$

where C is a closed integral loop enclosing the phase singularity in transverse. From Eq. (7.6.3), the helical phase term $\exp(il_P\varphi)$ in Eq. (7.6.1) is the Pancharatnam-Berry phase.

Similar to the topological charge l (topological charge, order) of scalar vortex beams, the Pancharatnam topological charge l_P in vectorial vortex beams is also associated with the angular momentum of the beam. The difference is that in a vortex beam, l determines the magnitude of OAM, while in the vectorial vortex beams, l_P determines the magnitude of the total angular momentum (SAM+OAM).

When $l_P = 0$, Eq. (7.6.1) is exactly the same as Eq. (7.1.1), indicating that the vector beam is actually a vectorial vortex beam with 0 Pancharatnam topological charge, ans is a special form of vectorial vortex beams.

In addition, from Eq. (7.5.17), the vectorial vortex beams can be decomposed into two circularly polarized vortex beams with opposite spins, whose topological charges are m and n, respectively, satisfying $|m| \neq |n|$. Since the beam sizes of vortex beams with different topological charges are various, $|\psi_L^n| \neq |\psi_R^m|$, the radial intensity distributions of the two vortex components to combine vectorial vortex beams are also different, indicating that the polarization state of the vectorial vortex beams is a function of both the azimuth and radial. This is different from the vector beam, for which the intensity distributions of the two orthogonal circularly polarized vortex components are identical at any point in transverse because their topological charges are opposite.

The relationship between the topological charges m and n of the two orthogonal circularly polarized vortex components and the polarization order p, and also Pancharatnam topological charge l_P is discussed below. The complex coefficients ψ_L^n and ψ_R^m only determine the amplitude and the initial phase distribution, and are independent of helical phases of the each vortex component, hence they are fixed to 1 here for convenience:

$$\psi_L^n|L_n\rangle + \psi_R^m|R_m\rangle = \exp\left(i\varphi\cdot\frac{m+n}{2}\right)\left(\left|L_{\frac{n-m}{2}}\right\rangle + \left|R_{\frac{m-n}{2}}\right\rangle\right) \tag{7.6.4}$$

Considering Eqs. (7.5.10)–(7.5.12), Eq. (7.6.4) can be expresses as:

$$\psi_L^n|L_n\rangle + \psi_R^m|R_m\rangle = \exp\left(i\varphi\cdot\frac{m+n}{2}\right)\left|\psi_{\frac{n-m}{2}}\right\rangle = \exp\left(i\varphi\cdot\frac{m+n}{2}\right)\begin{bmatrix}\cos\frac{n-m}{2}\varphi\\ \sin\frac{n-m}{2}\varphi\end{bmatrix} \tag{7.6.5}$$

Comparing with Eq. (7.6.1), we have

$$l_P = \frac{n+m}{2} \tag{7.6.6}$$

$$p = \frac{n-m}{2} \tag{7.6.7}$$

It is shown that the Pancharatnam topological charge of the vectorial vortex beams is the arithmetic average of the topological charges of the two orthogonal circularly polarized vortex components. And the polarization order is half of the difference between two topological charges. In particular, when $m = -n$, $l_P = 0$ and $p = n$, this is the vector beam located on the sphere of the high-order Poincare sphere. These phenomena also reconfirm that vector beam is a special form of vectorial vortex beams. Vectorial vortex beams are more universal extension of vector beams.

For vectorial vortex beams, although the point (θ, σ) on the surface of the hybrid-order Poincare sphere can represent the phase difference and intensity specificity between the two orthogonal circularly polarized vortex beam components, the Pancharatnam topological charge and polarization order of the vectorial vortex beams cannot be accurately obtained from this coordinate alone, and the mode distribution of the vectorial vortex beams cannot be completely characterized. A mode vector $[m, n, \theta, \sigma]$ is given here, whose first two parameters m, n characterize the Pancharatnam topological charge and the polarization order, while the last two parameters θ, σ characterize the position on the surface of the hybrid-order Poincare sphere. Using the mode vectors $[m, n, \theta, \sigma]$, the full properties of the vectorial vortex beams can be obtained.

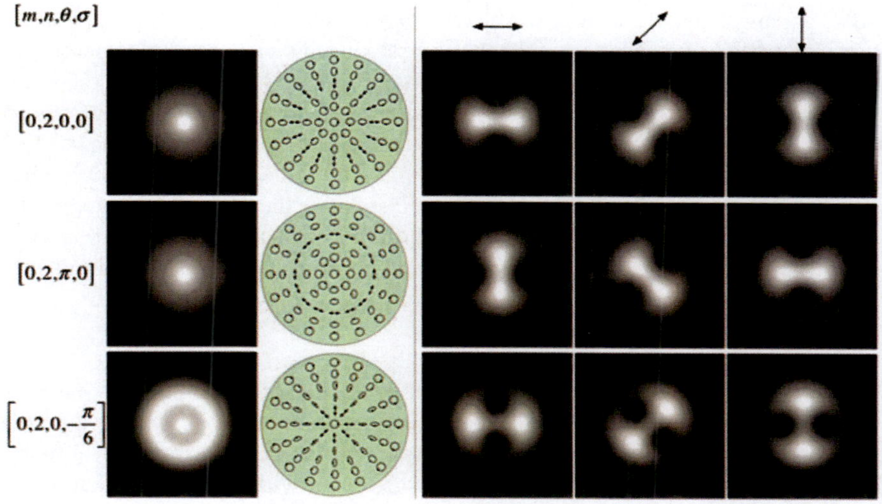

Fig. 7.32 Vectorial vortex beams

Figure 7.32 shows several vectorial vortex beams of different modes, together with their polarization distributions and the intensity distributions after passing through a polarizer.

7.6.2 Generating Vectorial Vortex Beams

Since a vectorial vortex beam has both transverse anisotropic vortex polarization and helical phase, it can be generated as producing a vector beam firstly and then introducing a Pancharatnam-Berry phase [33], as illustrated in Fig. 7.33.

Figure 7.34 gives a typical set-up for generating a vectorial vortex beam by introducing a spiral phase to the vector beam. A fundamental Gaussian mode is incident in, and is converted into linaer polarization after passing through a polarizer (P).

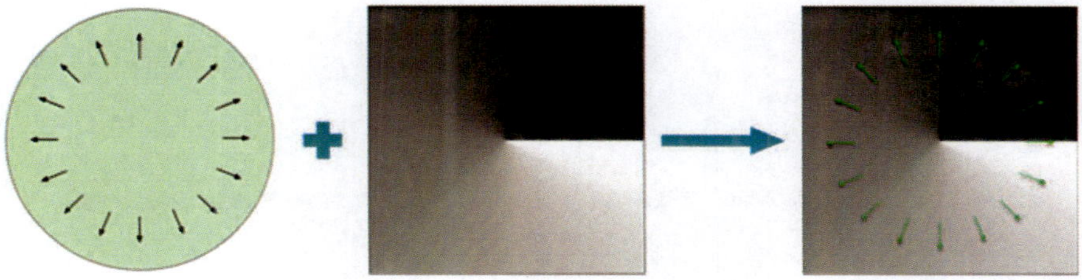

Fig. 7.33 Introducing a helical phase for a vector beam to generate a vectorial vortex beam

After passing through a quarter wave plate (QWP) and a q-plate (QP) in sequence, it is transformed to a vector beam. When the fast axis of QWP is parallel or perpendicular to the main axis of polarizer, the output is still linearly polarized, and the vector beam generated after the q-plate is located at the equator of the high-order Poincare sphere, with transverse anisotropic linear polarization. When the fast axis of QWP is not parallel or perpendicular to the main axis of polarizer, , the output is elliptical polarization, and the vector beam generated after the q-plate is located on the northern and southern hemispheres of the high-order Poincare sphere with transverse anisotropic elliptically polarization. The spiral phase plate (SPP) then introduces a helical phase to the generated vector beam, transforming the vector beam into a vectorial vortex beam.

The principle of the above method can also be understood by Eq. (7.5.10) and Eq. (7.5.17). From the discussion in Sect. 7.2.5, it is clear that the polarization order of a vector beam generated by q-plates is $2q$. According to Eq. (7.5.10), it can be decomposed into a left circularly polarized vortex component with topological charge $2q$ and a right circularly polarized vortex component with topological charge $-2q$. Let the order of the spiral phase plate be l_s. When the vector beam passes through the spiral phase plate, the topological charge of the left circularly polarized component of the beam becomes $(2q + l_s)$, while the topological charge of the right circularly polarized component of the beam becomes $(-2q + l_s)$. Both q and l_s are non-zero, hence $|2q + l_s| \neq |-2q + l_s|$, satisfying Eq. (7.5.17), and a vectorial vortex beam is generated.

Similar to vector beams, vectorial vortex beams can be generated by cascading liquid crystal spatial light modulators. The optical system for this approach is identical to that of Fig. 7.23. However, in order to precisely generate vectorial vortex beams of different modes $[m, n, \theta, \sigma]$, the vortex phase encoded on the modulator and the arranged angle of the spiral phase plate should be reconsidered. As the principles and methods are identical to those presented in Sect. 7.3.6, they will not be repeated here.

In order to achieve selective generation of vectorial vortex beams, researchers develop a scheme for generating vectorial vortex beams by hologram coding [34]. Its setup is similar but not identical to that of Fig. 7.19, as shown in Fig. 7.35. First, a laser combined with a polarizer generates a horizontally linearly polarized Gaussian beam, which is expanded by a beam expander and is then incident into a liquid crystal spatial light modulator (SLM). The SLM is encoded with a special-designed

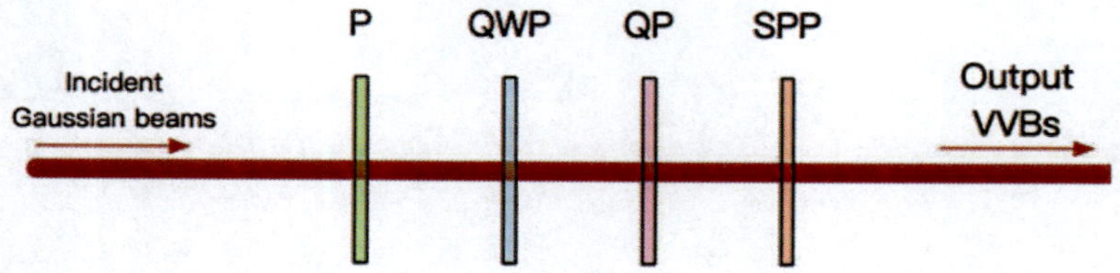

Fig. 7.34 Setup for generating a vectorial vortex beam by introducing a helical phase to the vector beam. P, polarizer; QWP, quarter wave plate; QP, q-plate; SPP, spiral phase plate

hologram that split incident beam into two beams with different topological charges, intensities and initial phases. The two beams are converged by a convex lens L1, and the convergence angle is set at same as the splitting angle of the Wollaston prism. One of the beams is converted to vertical linear polarization by placing a half wave plate with 45° arranged fast axis. Here the two beams can be combined by the Wollaston prism and then transformed by a 45° fast axis arranged quarter wave plate (QWP) into a beam combined from two orthogonal circular vortex beams with different topological charges, namely, a vectorial vortex beam. The convex lenses L2 and L3 are used to form a 4-f imaging system, which images the field at the back focal plane of L1 to the back focal plane of L3 in equal proportion to observe the generated vectorial vortex beam in far field.

There are two keys in this technique. Firstly, a reasonable diffraction grating should be designed to produce a 1 × 2 vortex beam array containing two beams with different topological charges, intensities and initial phases. Secondly, a reasonable grating constant should be chosen to make the convergence angle of the convex lens L1 identical with the splitting angle of the Wollaston prism.

The method introduced in Sect. 4.2.1 can be used to design the grating. In order to obtain a better diffraction effect, only ±1 diffraction orders are present and all the other diffraction orders are missed. According to Eqs. (4.2.4) and (4.2.6), the transmittance function of the diffraction grating can be obtained as

$$\begin{aligned} \exp[iP(x)] = |a_{-1}|\exp\left[i\left(\phi_{-1} - m\varphi - 2\pi xT^{-1}\right)\right] \\ +|a_{+1}|\exp\left[i\left(\phi_{+1} - n\varphi + 2\pi xT^{-1}\right)\right] \end{aligned} \tag{7.6.8}$$

where $|a_{+1}|$ and $|a_{-1}|$, ϕ_{+1} and ϕ_{-1} are the amplitudes and initial phases of the ± 1 diffraction order, respectively. $-m$ and $-n$ are the topological charges of the vortex beams located at the ±1 diffraction order, respectively. T is the grating constant. According to Eq. (7.6.8), a diffraction grating can be designed by defining the amplitudes, initial phases and topological charges of beams located at ±1 diffraction orders. However, as discussed in Chap. 4, for the diffraction grating designed by this method, the desired array cannot be generated by phase-only modulation and there must be extraneous diffraction orders. Therefore, the grating needs to be optimized using

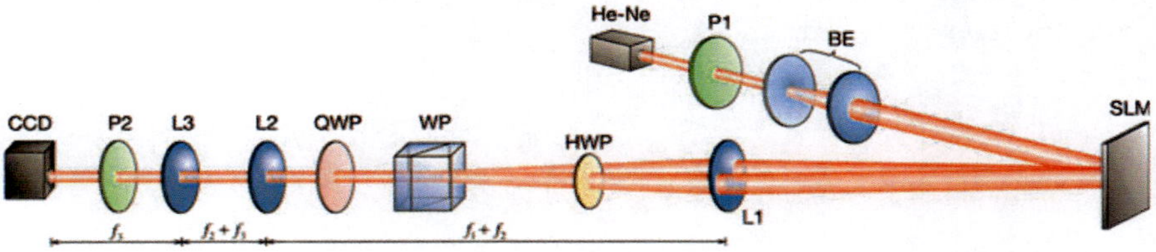

Fig. 7.35 System for tunable generation of vectorial vortex beams with different modes, L1–L3: convex lenses with focal length distributions of f_1, f_2 and f_3; P1–P3, polarizers; SLM, liquid-crystal spatial light modulator; HWP, half wave plate; WP, Wollaston prism; QWP, quarter wave plate [34]

algorithms such as GS algorithm to achieve the desired amplitudes, initial phases and topological charges of ±1 diffraction orders.

The choice of each parameter in Eq. (7.6.8) determines the generated vectorial vortex beams. Firstly, since a reflective liquid crystal SLM is employed, the topological charges of the two orthogonal circularly polarized vortex components of the vectorial vortex beams are m and n respectively; Eq. (7.5.15) shows that the longitude coordinate θ can be expressed as the initial phase difference between the two vortex components, and therefore as:

$$\theta = \phi_{-1} - \phi_{+1} \tag{7.6.9}$$

From Eq. (7.5.16), the spherical latitudinal coordinate σ can be deduced as:

$$\sigma = \arcsin \frac{|a_{-1}|^2 - |a_{+1}|^2}{|a_{-1}|^2 + |a_{+1}|^2} \tag{7.6.10}$$

Equations (7.6.9) and (7.6.10) show that the meridional coordinates (θ, σ) are determined by the intensity and initial phase of the two diffraction orders only. When designing a grating, the relative values of $|a_{+1}|$, $|a_{-1}|$, ϕ_{+1} and ϕ_{-1} can be derived from the latitudinal and longitudinal coordinates (θ, σ) of the desired vectorial vortex mode.

Clearly, the vector $[m, n, \theta, \sigma]$ can be determined by just one phase grating, and the vector $[m, n, \theta, \sigma]$ represents the pattern of the generated vectorial vortex beams. This means that in practice, this technique can be used to generate a vectorial vortex beam at any point on the surface of an arbitrary hybrid-order Poincare sphere without changing any hardware. By encoding the different holographic grating on the liquid crystal spatial light modulator, and vectorial vortex beams can be generated.

In Fig. 7.19, beam combining is achieved by shifting the position of the lens so that the convergence angle of the beam is exactly equal to the divergence angle of the Wollaston prism. Here, the angle matching is achieved through reasonable design of the grating constant. As shown in Fig. 7.36, u is the distance between the convex lens and the liquid crystal SLM. v is the distance between the Wollaston prism and the convex lens. f_1 is the focal length of the lens. α is the angle between ±1 diffraction orders. β is the splitting angle of the Wollaston prism. Then, from principles of geometrical optics, we can obtain:

$$u^{-1} + v^{-1} = f_1^{-1} \tag{7.6.11}$$

Under paraxial approximation:

$$\alpha u = \beta v \tag{7.6.12}$$

If b is the diffraction order, according to the grating equation at normal incidence:

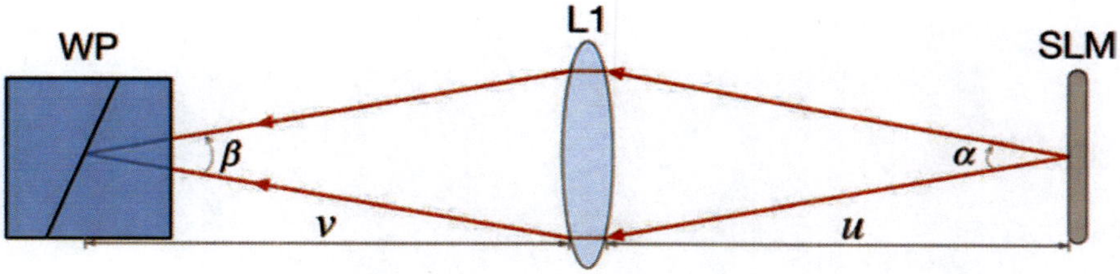

Fig. 7.36 Convergence and divergence angle matching of a Wollaston prism

$$T \sin \alpha_b = b\lambda \tag{7.6.13}$$

Hence the angle α between ± 1 diffraction orders can be obtained as:

$$\alpha = \alpha_{+1} - \alpha_{-1} = \frac{2\lambda}{T} \tag{7.6.14}$$

Substituting Eq. (7.6.14) into Eq. (7.6.13) gives

$$T = \frac{2\lambda u}{\beta v} \tag{7.6.15}$$

The above derivation shows the distances u and v must satisfy Eq. (7.6.11). The grating constant T is calculated from Eq. (7.6.15) and substituted into Eq. (7.6.8) to design the grating.

Figure 7.37 presents the simulation results of generating a vectorial vortex beam using the system shown in Fig. 7.35. From left to right are, the encoded holograms, the intensity distributions of the generated vectorial vortex beams without and with a polarizer, respectively. The simulations show that vectorial vortex beams can be generated through this method. As mentioned earlier, the greatest advantage of this vectorial vortex beams source is that, vector vortex modes at any point on the surface of arbitrary hybrid-order Poincare sphere can be generated without changing any hardware. What you should do is just change the encoded holographic grating, enabling the continuously selective generation of arbitrary vectorial vortex beams.

7.7 Vectorial Vortex Beams Arrays

Vectorial vortex beams arrays, a complex form of vector field in which multiple various vectorial vortex beams are arranged in space with a certain location [35–37]. Two common types of arrays are rectirectilinear arrays [35, 36] and rectangular arrays [37]. In fact, vectorial vortex beams arrays can be viewed as generating multiple vectorial vortex beams with selective modes at different diffraction orders simultaneously. Its diffraction orders corresponding to the spatial positions of beams

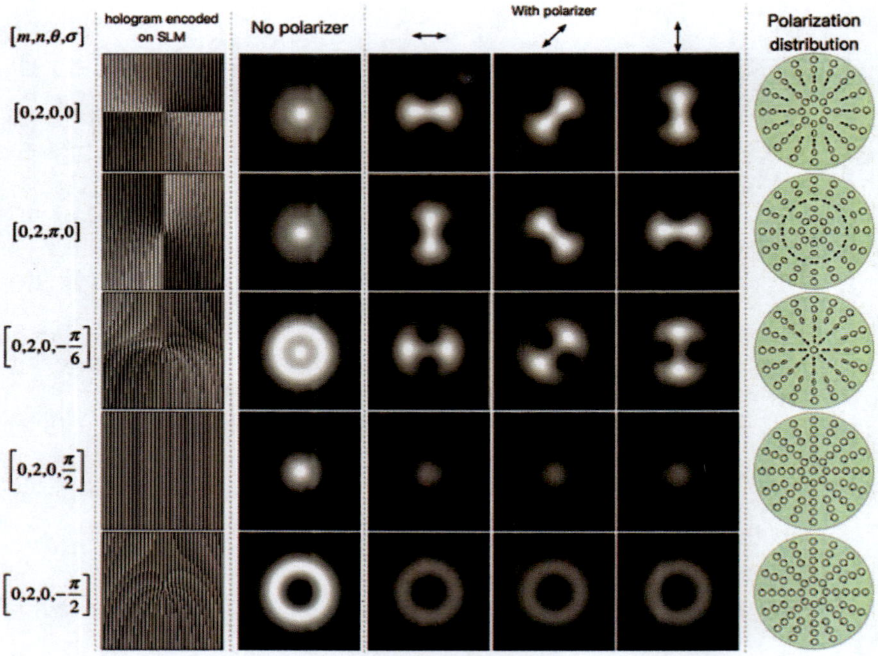

Fig. 7.37 Simulation results for generating a vectorial vortex beam using the system shown in Fig. 7.35

in the array. Vectorial vortex beams arrays are of great importance scenarios where need multiple different vectorial vortex beams, for instance, vector-based optical communications and multipoint laser processing. Since vector beams are a special case of vectorial vortex beams with non-zero Pancharatnam topological charges, this section will only discuss vectorial vortex beams arrays.

7.7.1 Principles of Generating Vectorial Vortex Beams Arrays

Chapter 4 has described generating (scalar) vortex beam arrays through designing special diffraction gratings to convert Gaussian beams. The intensity, initial phase and topological charge of each vortex beams in the array is selective. For (scalar) vortex beam arrays, only the phase modulation needs to be considered when designing gratings, and ignore other degrees of freedom as polarization. However, since vectorial vortex beams have a transverse anisotropic polarization distribution, in addition to a spiral phase. Therefore, when generating a vectorial vortex beam array, both phase and polarization modulation must be accomplished simultaneously, which is much more complex than when generating a (scalar) vortex beam array.

Equation (7.5.17) has shown that a vectorial vortex beam can be decomposed into two orthogonal circularly polarized vortex beams. Representing the $M \times N$ vectorial vortex beams array as an $M \times N$ matrix, where each element of the matrix is a vectorial vortex beam and its position in the matrix characterizes the relative position in the array, then according to Eq. (7.5.17), we have

$$\begin{bmatrix} |\psi_{(1,1)}\rangle & \cdots & |\psi_{(1,N)}\rangle \\ \vdots & & \vdots \\ |\psi_{(M,1)}\rangle & \cdots & |\psi_{(M,N)}\rangle \end{bmatrix} = \begin{bmatrix} (\psi_L^n|L_n\rangle)_{(1,1)} & \cdots & (\psi_L^n|L_n\rangle)_{(1,N)} \\ \vdots & & \vdots \\ (\psi_L^n|L_n\rangle)_{(M,1)} & \cdots & (\psi_L^n|L_n\rangle)_{(M,N)} \end{bmatrix} + \begin{bmatrix} (\psi_R^m|L_m\rangle)_{(1,1)} & \cdots & (\psi_R^m|L_m\rangle)_{(1,N)} \\ \vdots & & \vdots \\ (\psi_R^m|L_m\rangle)_{(M,1)} & \cdots & (\psi_R^m|L_m\rangle)_{(M,N)} \end{bmatrix} \quad (7.7.1)$$

Equation (7.7.1) indicates that two arrays of orthogonal circularly polarized (scalar) vortex beams can be generated firstly using the principles in Chap. 4. Then combined the two arrays coaxially with corresponding diffraction orders to form a vectorial vortex beam array, as shown in Fig. 7.38.

Figure 7.39 presents a typical vectorial vortex beams array generation systems based on the above principles. First, the laser is combined with a 45° arranged polarizer produce a 45° linear polarized Gaussian beam, which can be orthogonally decomposed into horizontal and vertical linear polarization components of equal intensity. The first liquid crystal spatial light modulator (SLM1) is encoded by a diffraction grating which converts the horizontal linear polarization component into

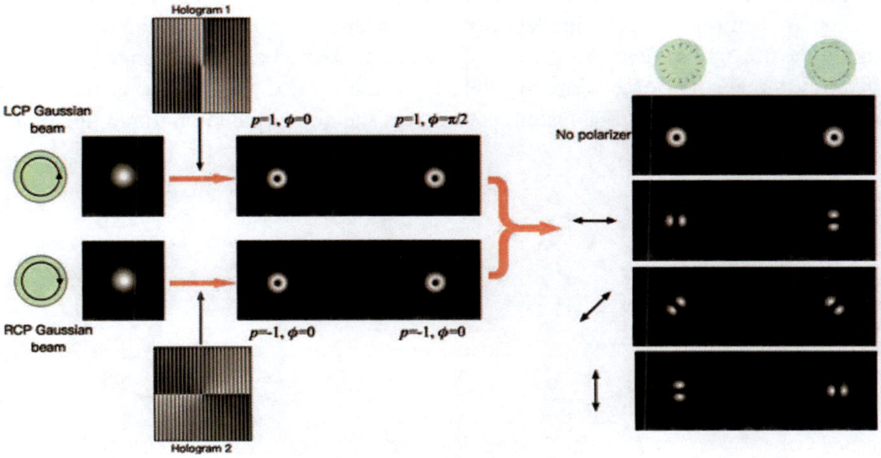

Fig. 7.38 Principles of vectorial vortex beams array generation

a (scalar) vortex beam array, while the vertical linear polarization component is unaffected. The fast axis of the half wave plate (HWP) is 45° arranged and is used to exchange the horizontal and vertical linear polarization components. The vertical component unmodulated by SLM1 is then modulated by SLM2 as it is converted to the horizontal polarization, while the horizontal component modulated by SLM1 is unaffected when passing through SLM2 as it is converted to the vertical polarization. SLM2 is encoded by another diffraction grating which converts the unmodulated vertical linearly polarizations into another (scalar) vortex beam array. A 45° arranged quarter wave plate (QWP) is placed to transform the horizontal and vertical linear polarizations into two orthogonal circular polarizations, respectively. Finally, a vectorial vortex beam array is generated.

The process described above is almost identical to that of generating vectorial vortex beams through cascading SLMs described previously. Nevertheless, adjustment must be introduced. After passing through SLM1, the horizontal component has been transformed into the form of a beam array with non-zero diffraction orders that propagate outwards. Thus, when the beam array of horizontally polarized components arrives at SLM2, the multiple diffraction orders except 0-th order are off the optical axis, indicating that diffraction orders of horizontally and vertically polarized components cannot be one-to-one combined. The solution for this is, two SLMs must be placed in the same plane, which is impossible to achieve in practice.

To address such issue, a 4-f imaging system is introduced between the two SLMs. 4-f imaging systems are coherent optical information processing systems consisting of two convex lenses with focal lengths f. The two lenses are placed on the optical axis at a distance of $2f$ from each other, with the first lens at a distance f from the object and the second lens at a distance f from the image plane. According to Fourier optics, the field at the back focal plane of a convex lens is the Fourier transformation of that at the front focal plane. Thus, a modulation device such as a diaphragm is placed at position $2f$ (midpoint between the two lenses) of the 4-f imaging system to modulate the optical field spectrum, and the modulated spectrum is then transformed back by the second lens, allowing filtering of the optical field. Another important property of the 4-f imaging system is to image the object plane on the image plane

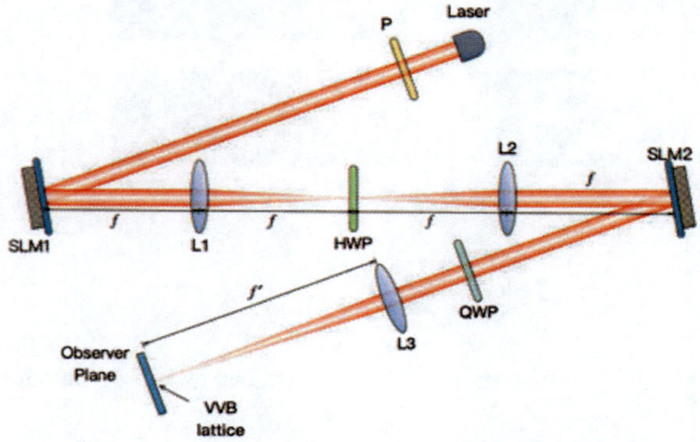

Fig. 7.39 Generation system of vectorial vortex beams array. P, polarizer; SLM1–SLM2, liquid-crystal spatial light modulators; L1–L3, convex lenses; HWP, half wave plate; QWP, quarter wave plate [35–37]

with equal proportion. Thus a 4-f system (L1 and L2) images the grating of SLM1 on SLM2 to achieve modulating the two linear polarization components at the same plane.

As with the (scalar) vortex beam array, the vectorial vortex beams array is located in the far field, so a convex lens L3 is placed behind the quarter wave plate (QWP). Then at the back focal plane of L3, we can observe the far field diffraction field, which is the generated vectorial vortex beams array.

7.7.2 Mode Control of Vectorial Vortex Beams Arrays

In a vectorial vortex beam array, the vectorial vortex beams at each diffraction order are superposed by two (scalar) vortex beams with orthogonal circular polarizations at the corresponding diffraction orders. The mode distribution of the vectorial vortex beams array is therefore determined by the two (scalar) vortex beam arrays, which are derived from the diffraction of phase gratings encoded in two SLMs. Therefore, the mode distribution of the vectorial vortex beams array is determined by the parameters of the two diffraction gratings.

From Eq. (4.3.3), the transmittance functions for the two diffraction gratings generating the right and left circular vortex beam arrays are

$$\exp\left[iP_R(x,y)\right] = \sum_{a,b=-\infty}^{+\infty} \left|c_{a,b}^R\right| \exp\left[i\left(\phi_{a,b}^R + m_{a,b}\varphi + a\gamma_x x + b\gamma_y y\right)\right] \tag{7.7.2}$$

$$\exp\left[iP_L(x,y)\right] = \sum_{a,b=-\infty}^{+\infty} \left|c_{a,b}^L\right| \exp\left[i\left(\phi_{a,b}^L + n_{a,b}\varphi + a\gamma_x x + b\gamma_y y\right)\right] \tag{7.7.3}$$

where (a, b) is the diffraction order, $\left|c_{a,b}^R\right|$ and $\left|c_{a,b}^L\right|$, $\phi_{a,b}^R$ and $\phi_{a,b}^L$, $m_{a,b}$ and $n_{a,b}$ are the amplitudes, initial phases and topological charges of the right and left circular vortex components located at diffraction order (a, b), respectively. γ_x and γ_y are the spatial angular frequencies of the grating in tx and y directions, respectively, defined by Eq. (4.3.4).

Since the right circularly polarized component of the vectorial vortex beams is converted from the vertical linearly polarized component after a QWP, and the vertical linearly polarized component just after SLM2 is modulated by SLM1. The right circularly polarized component, after being modulated into a vortex beam by SLM1, is reflected twice by SLM1 and SLM2. Hence its topological charge distributions are unchanged as $m_{a,b}$. The left circularly polarized vortex component, modulated by SLM2, is only reflected once by SLM2, with opposite topological charges, becoming $-n_{a,b}$. Considering the one-to-one correspondence between the diffraction orders when beam combining, from Eqs. (7.5.15) and (7.5.16) one can obtain longitude and latitude coordinates respectively:

$$\theta_{a,b} = \phi^R_{a,b} - \phi^L_{a,b} \tag{7.7.4}$$

$$\sigma_{a,b} = \arcsin\left(\frac{\left|c^R_{a,b}\right|^2 - \left|c^L_{a,b}\right|^2}{\left|c^R_{a,b}\right|^2 + \left|c^L_{a,b}\right|^2}\right) \tag{7.7.5}$$

Here, the vectorial vortex beams at the diffraction order (a, b) can be represented by the mode vector as

$$[m, n, \theta, \sigma] = \left[m_{a,b}, -n_{a,b}, \left(\phi^R_{a,b} - \phi^L_{a,b}\right), \arcsin\left(\frac{\left|c^R_{a,b}\right|^2 - \left|c^L_{a,b}\right|^2}{\left|c^R_{a,b}\right|^2 + \left|c^L_{a,b}\right|^2}\right)\right] \tag{7.7.6}$$

References

1. Zhan Q. Cylindrical vector beams: from mathematical concepts to applications. Adv Opt Photon. 2009;1:1–57.
2. Hall DG. Vector-beam solutions of Maxwell's wave Eq. Opt Lett 1996;21(1):9–11.
3. Xin J, Study on the generation of vector beams and their applications. Doctoral dissertation of Beijing Institute of Technology, 2013 (in Chinese).
4. Tovar AA. Production and propagation of cylindrically polarized Laguerre–Gaussian laser beams. J Opt Soc Am A 1998;15(10):2705–2711.
5. Xie J, Zhao D, Yan J. Physical optics. Beijing: Beijing Institute of Technology Press; 2012. (in Chinese).
6. Oron R, Blit S, Davidson N, et al. The formation of laser beams with pure azimuthal or radial polarization. Appl Phys Lett. 2000;77(21):3322–4.
7. Naidoo D, Roux FS, Dudley A, et al. Controlled generation of higher-order Poincaré sphere beams from a laser. Nat Photonics. 2016;10:327–32.
8. Yonezawa K, Kozawa Y, Sato S. Generation of a radially polarized laser beam by use of the birefringence of a c-cut Nd:YVO4 crystal. Opt Lett. 2006;31(14):2151–3.
9. Kozawa Y, Yonezawa K, Sato S. Radially polarized laser beam from a Nd:YAG laser cavity with a c-cut YVO4 crystal. Appl Phys B. 2007;88(1):43–6.
10. Machavariani G, Lumer Y, Moshe I, et al. Birefringence-induced bifocusing for selection of radially or azimuthally polarized laser modes. Appl Opt. 2007;46(16):3304–10.
11. Li JL, Ueda K, Zhong LX, et al. Efficient excitations of radially and azimuthally polarized Nd^{3+}:YAG ceramic microchip laser by use of subwavelength multilayer concentric gratings composed of Nb_2O_5/SiO_2. Opt Express. 2008;16(14):10841–8.
12. Li JL, Ueda KI, Musha M, et al. Radially polarized and pulsed output from passively Q-switched Nd:YAG ceramic microchip laser. Opt Lett 2008;33(22):2686–2688.
13. Zhou Z, Tan Q, Li Q, et al. Achromatic generation of radially polarized beams in visible range using segmented subwavelength metal wire gratings. Opt Lett. 2009;34(21):3361–3.
14. Xin J, Gao C, Li C. Combination of Hermit-Gaussian beams to arbitery order vector beams. Sci Sin Phys Mech Astron. 2012;10:1017–21 (in Chinese).
15. Beijersbergen MW, Allen L, van der Veen HELO, et al. Astigmatic laser mode converters and transfer of orbital angular momentum. Opt Commun 1993;96(1–3):123–132.
16. Wang T, Fu S, Zhang S, et al. A Sagnac-like interferometer for the generation of vector beams. Appl Phys B. 2016;122(9):231.
17. Fu S, Gao C, Yang S, et al. Generating polarization vortices by using helical beams and a Twyman Green interferometer. Opt Lett. 2015;40(8):1775–8.

18. Gao C, Fu S, Dai K. The method and setup for generating vector beams based on Twyman-Green interferometers. Chinese patent: ZL. 2017; 201510069408.9.
19. Li L, Huang YF, Wang YT. Applied optics. Beijing: Beijing Institute of Technology Press, 2005.
20. Xin J, Gao C, Li C, et al. Generation of polarization vortices with a Wollaston prism and an interferometric arrangement. Appl Opt. 2012;51(29):7094–7.
21. Moreno I, Davis JA, Cottrell DM, et al. Encoding high-order cylindrically polarized light beams. Appl Opt. 2014;53(24):5493–501.
22. Zheng X, Lizana A, Peinado A, et al. Compact LCOS–SLM based polarization pattern beam generator. J Lightwave Technol. 2015;33(10):2047–55.
23. Fu S, Wang T, Gao C. Generating perfect polarization vortices through encoding liquid-crystal display devices. Appl Opt 2016;55(23):6501–6505.
24. Fu S, Zhang S, Gao C. Bessel beams with spatial oscillating polarization. Sci Rep. 2016;6:30765.
25. Gao C, Fu S, Zhang S. A setup for the generation of three-dimensional vector beams. Chinese Patent: ZL201610007355.2, 2016-01-06 (in Chinese) .
26. H. Poincare, Theorie Mathematique de la Lumiere. Paris: Gauthiers-Villars; 1892, Vol. 2.
27. Milione G, Sztul HI, Nolan DA, et al. Higher-order Poincaré sphere, stokes parameters, and the angular momentum of light. Phys Rev Lett. 2011;107(5): 053601.
28. Yi X, Liu Y, Ling X, et al. Hybrid-order Poincare sphere. Phys Rev A. 2015;91: 023801.
29. Ren ZC, Kong LJ, Li SM, et al. Generalized Poincare sphere. Opt Express. 2015;23(20):26585–95.
30. Galvez EJ. Light beams with spatially variable polarization[M]// Photonics: scientific foundations, technology and applications, vol. 1. John Wiley & Sons: Inc.; 2015. p. 61–76.
31. Pancharatnam S. Generalised theory of interference and its applications. Proc Ind Acad Sci Sect A. 1957;45(6):402–11.
32. Biener G, Niv A, Kleiner V, et al. Formation of helical beams by use of Pancharatnam-Berry phase optical elements. Opt Lett. 2002;27(21):1875–7.
33. Liu Z, Liu Y, Ke Y, et al. Generation of arbitrary vector vortex beams on hybrid-order Poincaré sphere. Photon Res. 2017;5(1):15–21.
34. Fu S, Zhai Y, Wang T, et al. Tailoring arbitrary hybrid Poincaré beams through single hologram. Appl Phys Lett. 2017;111: 211101.
35. Fu S, Zhang S, Wang T, et al. Rectilinear lattices of polarization vortices with various spatial polarization distributions. Opt Express. 2016;24(16):18486–91.
36. Fu S, Gao C, Wang T, et al. Simultaneous generation of multiple perfect polarization vortices with selective spatial states in various diffraction orders. Opt Lett. 2016;41(23):5454–7.
37. Fu S, Wang T, Zhang Z, et al. Selective acquisition of multiple states on hybrid Poincare sphere. Appl Phys Lett. 2017;110: 191102.

Chapter 8
Perfect Optical Vortices

Once the fundamental mode waist is fixed, the beam size (transverse spot diameter) of a vortex beam is determined by its topological charge. The larger the absolute value of the topological charge, the larger the spot diameter and the larger the hollow region. This property makes it very limited in some applications. For example, in optical tweezers, a larger topological charge but a smaller spot diameter are usually required to achieve better capture performance, which can only be achieved by minimising the radius of the fundamental mode waist for conventional vortex beams. Besides, in optical communication, it is difficult to couple vortex beams of different topological charges into fibers after multiplexing because of the different spot radius. In order to solve these problems, the concept of perfect optical vortices (POV) is proposed, whose transverse spot diameter is independent of the topological charge and can be controlled by several other parameters. For different orders of perfect vortices converted from the same Gaussian beam by the same system, they all have identical transverse diameters.

8.1 Overview of Perfect Optical Vortices

8.1.1 Theoretical Model of Perfect Optical Vortices

The ideal perfect optical vortex is a vortex beam whose transverse beam diameter is independent of the topological charge, with ring width tending to zero and power density over the ring tending to infinity [1]. A perfect optical vortex can be expressed as

$$POV_1(r, \varphi) \equiv \delta(r - r_0)\exp(il\varphi) \tag{8.1.1}$$

where r_0 is the spot radius of the perfect optical vortices, l is the topological charge and $\delta(\zeta)$ is the Dirac function.

S. Fu and C. Gao, *Optical Vortex Beams*, Advances in Optics and Optoelectronics,
https://doi.org/10.1007/978-981-99-1810-2_8

Any function $g(r)$ can be Bessel expanded as [2]:

$$g(r) = \sum_{n=1}^{\infty} c_{l,n} J_l\left(\alpha_{1,n}\frac{r}{a}\right), \quad 0 \le r \le a, \quad l \ge -1 \tag{8.1.2}$$

where

$$c_{l,n} = \frac{2}{a^2\left[J_{l+1}\left(\alpha_{l,n}\right)\right]^2} \int_0^a g(r) J_l\left(\alpha_{l,n} \cdot \frac{r}{a}\right) r dr \tag{8.1.3}$$

In Eqs. (8.1.2) and (8.1.3), $J_l(\zeta)$ is an l-th order Bessel function of first kind, $\alpha_{l,n}$ is the n-th null point of $J_l(\zeta)$, and a is an upper bound on the radial coordinate r. Assuming $a > r_0$, substituting POV_l in Eq. (8.1.1) as $g(r)$ in Eq. (8.1.3) and taking into account the screening properties of the Dirac function, the perfect optical vortices can be expressed as [1]

$$POV_l(r, \varphi) \propto circ\left(\frac{r}{a}\right) \exp(il\varphi) \sum_{n=1}^{N} \frac{J_l\left(\alpha_{l,n} \cdot \frac{r_0}{a}\right)}{\left[J_{l+1}\left(\alpha_{l,n}\right)\right]^2} J_l\left(\alpha_{l,n} \cdot \frac{r}{a}\right) \tag{8.1.4}$$

where circ(ζ) is a binary function of an inseparable variable, and when ζ denotes radial coordinates, the function is defined as:

$$circ(\varsigma) = \begin{cases} 1, & \varsigma \le 1 \\ 0, & others \end{cases} \tag{8.1.5}$$

Figure 8.1 gives two perfect optical vortices of different orders obtained from the simulation at $N = 40$ according to Eq. (8.1.4). Both of their spot radii are $r_0 = 0.5a$ and their topological charges are $l = 1$ and $l = 10$ separately. Obviously, two beams have exactly the same transverse beam diameters, which are independent of their topological charge.

Equation (8.1.4) is much more complicated than Eq. (8.1.1), but the perfect optical vortices given in Fig. 8.1 is calculated by Eq. (8.1.4), not Eq. (8.1.1). The reason is that Eq. (8.1.1) describes the ideal theoretical model of a perfect optical vortex, whereas in practice a ring with zero width and infinite power density cannot exist and therefore can only be approximated. The value of N in Fig. 8.1 is 40. In fact, as the further increase of N, the perfect optical vortices getting closer to the ideal (Eq. 8.1.1). When N approaches infinity, Eq. (8.1.4) turns to Eq. (8.1.1) exactly.

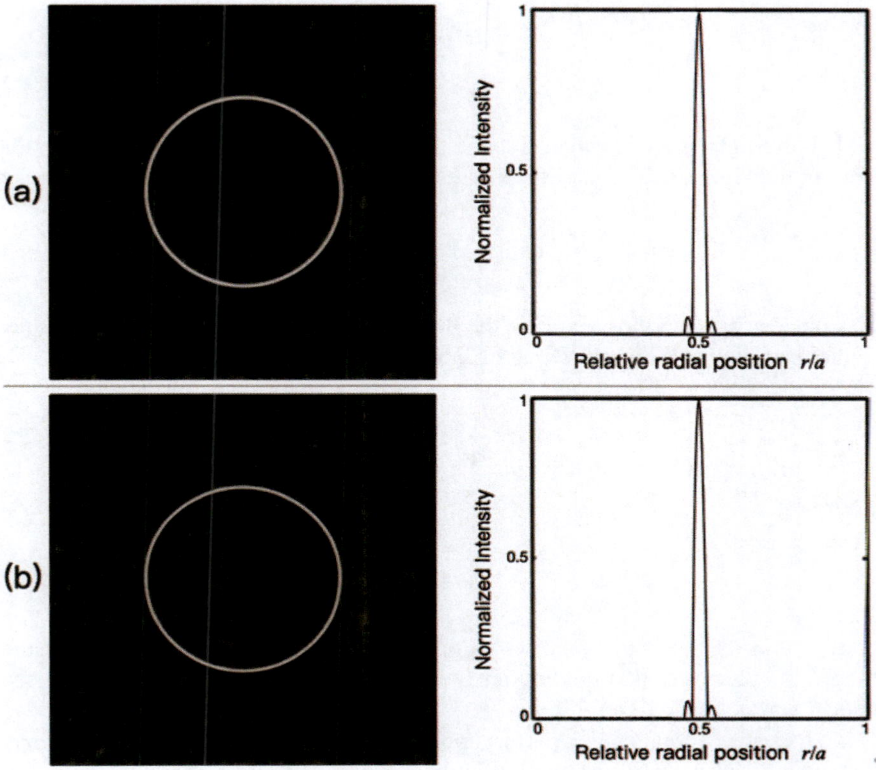

Fig. 8.1 Perfect optical vortices of different orders. **a** $l = 1$; **b** $l = 10$

8.1.2 Relationships Between Perfect Optical Vortices and Bessel Gauss Beams

Without considering the constant and propagation terms, a Bessel beam in cylindrical coordinates given in Eq. (2.4.13) can be written as

$$E(r, \varphi) = J_l(k_r r)\exp(il\varphi) \tag{8.1.6}$$

The individual parameters are defined by Eqs. (2.4.14) and (2.4.15). According to the Fourier optics, the Fourier transformationation of an optical field can be achieved by a convex lens with focal length f. The optical field at the back focal plane of the lens is the Fourier transformationation of the optical field at the front focal plane. In cylindrical coordinates, the relationship between an arbitrary field $E(r', \varphi')$ and its Fourier transformationation $E(r, \varphi)$ is

$$E(r,\varphi)=\frac{k}{2\pi if}\int_0^{\infty}\int_0^{2\pi}E\big(r',\varphi'\big)r'dr'd\varphi'\times\exp\left[-\frac{ik}{f}r'r\cos\big(\varphi-\varphi'\big)\right] \tag{8.1.7}$$

with k the light wave number. Substituting Eq. (8.1.6) into Eq. (8.1.7) gives the Fourier transformationation of the Bessel beam as [3]:

$$E(r,\varphi)=\frac{k}{f}i^{l-1}\exp(il\varphi)\int_0^{\infty}J_l\big(k_r r'\big)J_l\left(\frac{k_r rr'}{f}\right)r'dr' \tag{8.1.8}$$

Considering the orthogonality of the Bessel function, Eq. (8.1.8) can be simplified to the form expressed by the Dirac function as:

$$E(r,\varphi)=\frac{i^{l-1}}{k_r}\delta(r-r_0)\exp(il\varphi) \tag{8.1.9}$$

where,

$$r_0=\frac{k_r f}{k} \tag{8.1.10}$$

Equation (8.1.9) is identical with Eq. (8.1.1) except for an additional constant term. This illustrates that the Fourier transformation of a Bessel beam is a perfect optical vortex.

As mentioned previously, a perfect optical vortex satisfying Eq. (8.1.1) is nonexistent, which is also confirmed by the relationship between a perfect optical vortex and a Bessel beam. A Bessel beam, with its infinitely extended transverse distribution, is only an ideal mathematical model and nonexist in practice. Therefore the ideal perfect optical vortex derived from its Fourier transformation does not exist either.

In practical scenarios, Bessel-Gauss beams are usually employed as an approximation of Bessel beams. According to Eq. (2.4.17), a Bessel Gauss beam can be expressed as

$$E(r,\varphi)=J_l(k_r r)\exp(il\varphi)\exp\left(-\frac{r^2}{\omega_0^2}\right) \tag{8.1.11}$$

where ω_0 is the waist radius of Gaussian beams that confines Bessel beams. Substituting Eq. (8.1.11) into Eq. (8.1.7) yields the Fourier transformation of a Bessel Gauss beam as: [3]

$$E(r,\varphi)=\frac{k}{f}i^{l-1}\exp(il\varphi)\int_0^{\infty}J_l\big(k_r r'\big)J_l\left(\frac{k_r rr'}{f}\right)\exp\left(-\frac{r'^2}{\omega_0^2}\right)r'dr' \tag{8.1.12}$$

Using the standard integral given in literature [4], Eq. (8.1.12) can be reduced to

$$E(r,\varphi) = i^{l-1}\frac{\omega_0}{\omega_a}\exp(il\varphi)\exp\left(-\frac{r^2+r_0^2}{\omega_a^2}\right)I_l\left(\frac{2rr_0}{\omega_a^2}\right) \quad (8.1.13)$$

In Eq. (8.1.13), r_0 is the spot radius, which can be expressed as Eq. (8.1.10). ω_a is the half-ring width (half of the ring width).

$$\omega_a = \frac{2f}{k\omega_0} \quad (8.1.14)$$

$I_l(\zeta)$ is a l-th order modified Bessel function of the first kind defined by the integral of a closed curve containing the origin in the counterclockwise direction:

$$I_l(\zeta) = \frac{1}{2\pi i}\oint \exp\left[\frac{\zeta}{2}\left(t+\frac{1}{t}\right)\right]t^{-l-1}dt \quad (8.1.15)$$

$I_l(\zeta)$ and the l-th order Bessel function of the first kind $J_l(\zeta)$ satisfy:

$$I_l(\zeta) = i^{-l}J_l(i\zeta) = \exp\left(-\frac{il\pi}{2}\right)J_l\left[\zeta\exp\left(\frac{i\pi}{2}\right)\right] \quad (8.1.16)$$

In general, when $r_0 \gg \omega_a$, I_l can be approximated as:

$$I_l\left(\frac{2rr_0}{\omega_a^2}\right) \sim \exp\left(\frac{2rr_0}{\omega_a^2}\right) \quad (8.1.17)$$

Substitute Eq. (8.1.17) into Eq. (8.1.13), we have

$$E(r,\varphi) = i^{l-1}\frac{\omega_0}{\omega_a}exp(il\varphi)\exp\left(-\frac{(r-r_0)^2}{\omega_a^2}\right) \quad (8.1.18)$$

Equation (8.1.18) is the field obtained from the Fourier transformation of a Bessel Gauss beam through a convex lens, with a transverse intensity distribution:

$$I(r,\varphi) = |E(r,\varphi)|^2 = \frac{\omega_0^2}{\omega_a^2}exp\left[-\frac{2(r-r_0)^2}{\omega_a^2}\right] \quad (8.1.19)$$

The above Eq. shows that the transverse beam size is independent of the topological charge l. According to Eq. (8.1.10), its spot radius r_0 is only related to the radial wave number k_r, the optical wave number k and the focal length f of the convex lens. Furthermore, the Bessel Gauss beam becomes a pure Bessel beam when the fundamental mode beam waist ω_0 is infinite. And thus $\omega_a \to 0$ from Eq. (8.1.14). Then Eq. (8.1.18) is the same as Eq. (8.1.1), which shows that the beam shown in Eq. (8.1.18) is a practical approximation of the beam shown in Eq. (8.1.1). Equation (8.1.18) is the perfect optical vortices obtained in practice from the Fourier transformation of the Bessel Gauss beam.

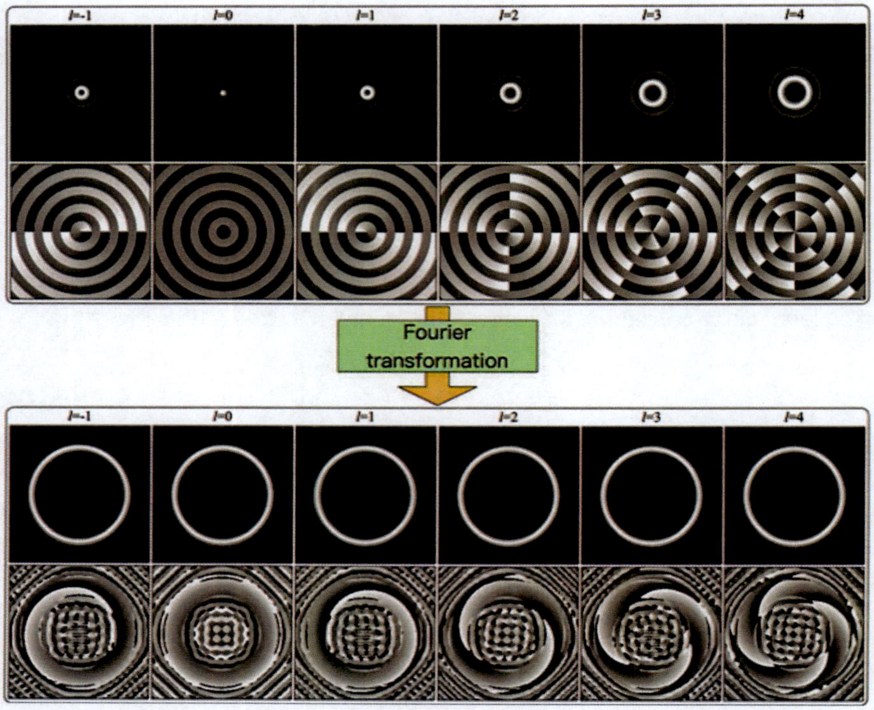

Fig. 8.2 Intensity and phase distributions of Bessel Gauss beams of different orders and the corresponding perfect optical vortices after Fourier transformation

Figure 8.2 gives the intensity and phase distributions of the Bessel Gauss beams of different orders, as well as the intensity and phase distributions of the perfect optical vortices produced by Fourier transformation. The transverse beam diameter of the Bessel Gauss beam becomes larger as the topological charge l increases, while the transverse beam diameter of the perfect optical vortex obtained after the Fourier transformation for different orders l remains consistent and independent of the topological charge l.

8.1.3 Free Space Transport Properties of Perfect Optical Vortices

Since Eq. (8.1.18) is obtained under the condition $r_0 \to \omega_a$, we still use the complex amplitude of the perfect optical vortices given in Eq. (8.1.13) here for analytical accuracy. if

$$A = i^{l-1}\frac{\omega_0}{\omega_a} \tag{8.1.20}$$

then Eq. (8.1.13) reads

$$E(r, \varphi) = A\exp(il\varphi)\exp\left(-\frac{r^2 + r_0^2}{\omega_a^2}\right)I_l\left(\frac{2rr_0}{\omega_a^2}\right) \tag{8.1.21}$$

Substituting into Fresnel diffraction integral, the perfect optical vortex at position z is acquired as:

$$E(r, \varphi, z) \propto \exp\left[-F_z\left(r^2 + r_0^2\right)\right]I_l(2rr_0F_z)\exp(il\varphi) \tag{8.1.22}$$

where,

$$F_z = \frac{1}{\omega_z} - \frac{ik}{2R(z)} \tag{8.1.23}$$

$$R(z) = z + \frac{z_R^2}{z} \tag{8.1.24}$$

$$\omega_z = \omega_0\sqrt{1 + \left(\frac{z}{z_R}\right)^2} \tag{8.1.25}$$

and z_R is the Rayleigh length of the fundamental mode.

$$z_R = \frac{\pi\omega_0^2}{\lambda} \tag{8.1.26}$$

When $z \to z_R$, $F_z \approx \omega_0^{-2}$, Eq. (8.1.22) agrees exactly with Eq. (8.1.21), indicating that the perfect optical vortex has a very stable transverse intensity distribution over a limited transmission distance. As the propagation distance z continues to grow, the first kind modified Bessel function transforms, causing a quite obvious diffraction effect on the perfect optical vortex. When $z \to z_R$, $\omega_z \to \infty$, , $F_z = -ikl(2z)$, from Eq. (8.1.16), we get:

$$I(2rr_0F_z) = I_l\left(\frac{-ikrr_0}{z}\right) = i^l J_l\left(\frac{krr_0}{z}\right) \tag{8.1.27}$$

Equation (8.1.27) indicates the appearance of l-order Bessel functions of first kind in the complex amplitude. The perfect optical vortex evolves into a vortex beam with a regular distribution of Bessel functions in amplitude, as shown in Fig. 8.3.

The above analysis shows that in free space perfect optical vortices exist only at finite distances. When a perfect optical vortex propagates for longer distances, it will return back to a Bessel-type vortex beam, but the topological charge still unchange.

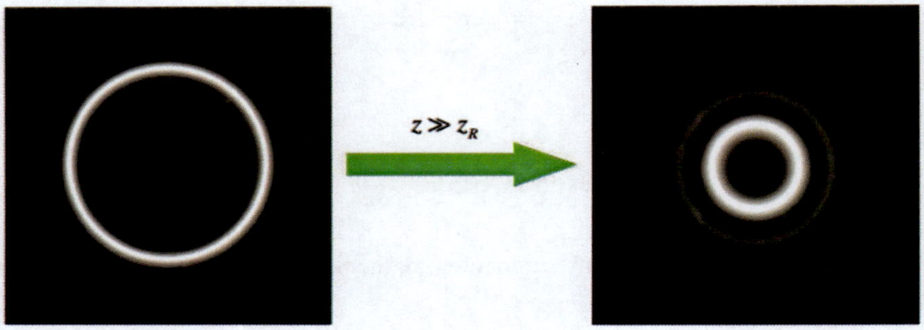

Fig. 8.3 Variation of the intensity distribution of a perfect optical vortex under transmission distance $z \to z_R$

8.1.4 Conversion Between Different Vortex Beams

Single-ring Laguerre Gauss beams are the most common form of vortex beams and the most widely used. Bessel Gauss beams are also of great interest in the fields of free-space optical communication, spin detection and quantum entanglement due to their diffraction-free properties. Perfect optical vortices are a new type of vortex beams that has emerged in recent years. Their beam sizes are independent of their topological charges, and they have great potential for applications in the field of fiber vortex optics. The difference between the three vortex beams carrying identical OAM, is mainly in the intensity distribution of the light field, which can be interconverted by means of diffraction optics.

The axicon introduced in Sect. 3.7.1 introduces a tapered wave front for the incident beam. It is therefore capable for converting a single-ring Laguerre Gaussian beam into a Bessel Gauss beam. Conversely, to convert a Bessel Gauss beam into a single-ring Laguerre Gauss beam, an anti-axicon can be used to remove the tapered wavefront. The relationship between an anti-axicon prism and an axicon prism can be likened to that between a concave lens and a convex lens. The transmittance function of an anti-axicon prism is obtained by inverting the phase of the transmittance function of an axicon:

$$T_{aa}(r, \varphi) = \exp\left(\frac{2\pi r}{d}\right) \tag{8.1.28}$$

where d is the radial phase period of the split-axis prism.

It has already been mentioned that a perfect optical vortex can be obtained from the Fourier transformation of a Bessel Gauss beam, which can be achieved by a convex lens. Equation (8.1.22) already shows that a Bessel Gauss beam is the far-field diffraction of a perfect optical vortex. The relationship between the far field and the near field is a Fourier transformation. Hence, the Fourier transformation of a perfect optical vortex accomplished by a convex lens gives a Bessel Gauss beam.

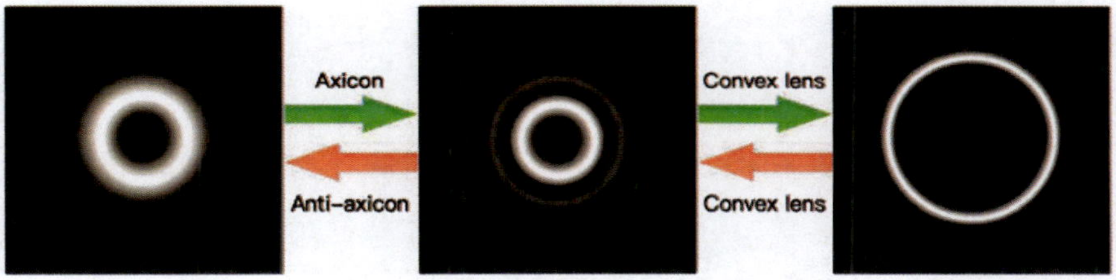

Fig. 8.4 Conversion between a single-ring Laguerre Gaussian beam, a Bessel Gauss beam, and a perfect optical vortex

So far, we have obtained the transformation relations between the single-ring Laguerre Gauss beam, the Bessel Gauss beam and the perfect optical vortex, as shown in Fig. 8.4. In a vortex beam-based optical system, the transformation relations given in Fig. 8.4 can be used to achieve their flexible transformation, by using axicons or convex lenses.

8.2 Generating Perfect Optical Vortices

8.2.1 Axicon

The perfect optical vortices can be obtained by Fourier transforming the Bessel Gauss beam. Therefore, the perfect optical vortices can be produced simply by introducing the Fourier transformation for the generated Bessel Gauss beam through a convex lens.

Figure 8.5 shows a typical system for generating a perfect optical vortex using an axicon. *l*-order Laguerre Gaussian beams are incident into the system from left, transformed into *l*-order Bessel Gauss beams by an axicon, and then Fourier transformed by a convex lens to perfect optical vortices at the back focal plane. Note that, according to the relevant principles of Fourier optics, the beam at the back focal plane of a convex lens is a Fourier transformation of tthat at the front focal plane, and the Bessel Gauss beam exists only in a finite range behind the axicon ($z < z_{\max}$). Hence the front focal plane of the convex lens must be made to lie in this limited range. The distance between the axicon and the convex lens d_z and lens focal length f must meet with:

$$d_z \leq f + z_{\max} \tag{8.2.1}$$

to generate perfect optical vortices.

The transverse beam radius of a perfect optical vortex is independent of its topological charge, but is known from Eq. (8.1.10) to be proportional to the focal length of the lens f, the radial light wave number k_r, and also inversely proportional to the

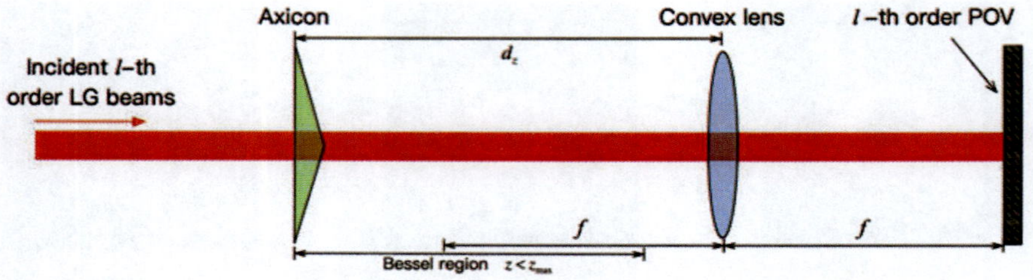

Fig. 8.5 System for generating a perfect optical vortex using an axicon

light wave number k. As shown in Fig. 8.6, it is not difficult to understand from the triangular similarity relationship:

$$\frac{k_r}{k} = \sin\beta \tag{8.2.2}$$

Substituting Eq. (3.7.2) into Eq. (8.2.2) gives

$$\frac{k_r}{k} = \frac{\lambda}{d} \tag{8.2.3}$$

where d is the radial phase period of the axicon. Substituting into Eq. (8.1.10) we have

$$r_0 = \frac{\lambda f}{d} \tag{8.2.4}$$

Equation (8.2.4) shows that the transverse beam radius of a perfect optical vortex generated using an axicon combined with a convex lens is determined by the focal length of the lens f, the wavelength λ and the radial phase period d of the axicon,

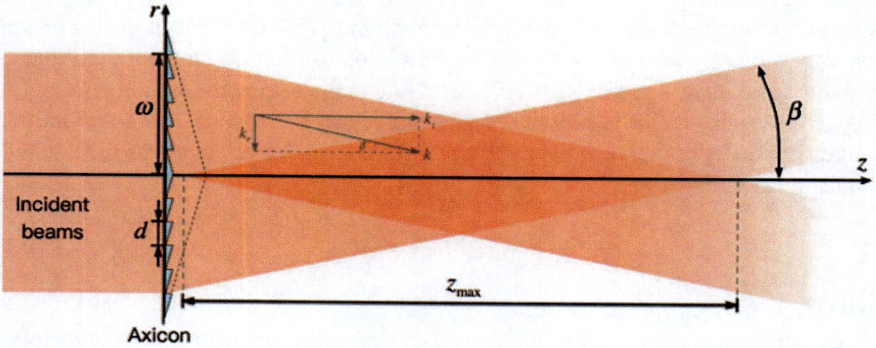

Fig. 8.6 Relationship between the radial wave number k_r, wave number in the direction of propagation k_z, and wave number k for a Bessel Gauss beam

and is independent of the radius ω_0 of the fundamental mode waist of the incident beam. Figure 8.7 shows the intensity distribution of perfect optical vortices of different orders generated from axicons for the same f and λ and different d. The transverse beam diameter of the perfect optical vortices decreases significantly with the increasing radial phase period d. For a fixed optical system, f and λ are constant and r_0 is inversely proportional to d. Therefore, the transverse beam diameter of the generated perfect optical vortices can be manuscripted by adjusting the parameter d of the axicon.

Alternatively, the ring width can be obtained from Eq. (8.1.14) as

$$2\omega_a = \frac{2\lambda f}{\pi \omega_0} \tag{8.2.5}$$

It is shown that for a fixed optical system, the ring width can be controlled by the radius ω_0 of the base mode beam waist, as shown in Fig. 8.8. If a finer ring is desired, a Laguerre Gaussian beam with as large ω_0 as possible should be used for incidence.

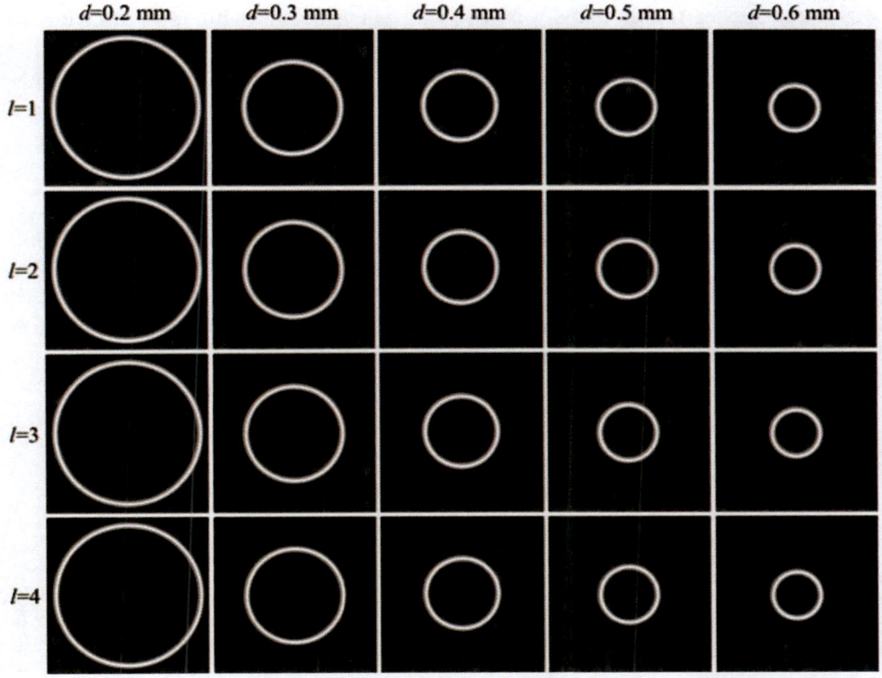

Fig. 8.7 Intensity distribution of perfect optical vortices of different orders generated from axicons for the same f and λ but different d. The individual parameters for the simulation are set as: $\lambda = 1.55\ \mu\text{m}, f = 300\ \text{mm}, \omega_0 = 1.5\ \text{mm}$

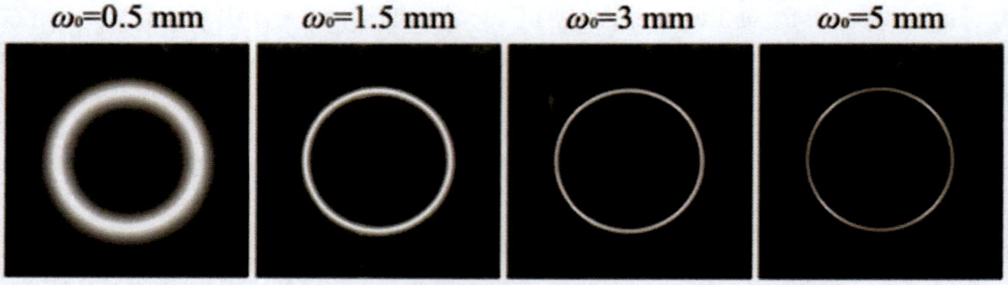

Fig. 8.8 Effect of fundamental mode waist radius ω_0 on the ring width of the generated perfect optical vortices. Parameters of the simulation are: $l = 1$, $\lambda = 1.55\ \mu\text{m}$, $f = 300$ mm, $\text{d} = 0.3$ mm

8.2.2 Bessel Beam Kinoform

A Bessel Beam Kinoform (BBK) is a phase-only diffraction optical element that converts fundamental mode into a perfect optical vortex [5]. Compared to the axicon method, the BBK method produces a perfect optical vortex with a narrower ring width and higher power density on the ring when beams with same fundamental radius are incident.

Supposing that the complex amplitude of an l-order perfect optical vortex with a transverse beam radius of r_0 is

$$E(r, \varphi) = F(r)\exp(il\varphi) \tag{8.2.6}$$

where $F(r)$ is the amplitude distribution, characterizing the presence of an amplitude spike at $r = r_0$. The Gaussian beam has an approximately planar phase, with a planar phase at the beam waist. Therefore, for simplicity of calculation, just the amplitude distribution of the Gaussian beam at the beam waist is considered, and the fundamental mode Gaussian beam incident here can be abbreviated as

$$G(r) = \exp\left(-\frac{r^2}{\omega_0^2}\right) \tag{8.2.7}$$

with ω_0 the waist radius.

The transmittance function of BBK is

$$T_{BBK}(r, \varphi) = \exp[i\beta(r)]\exp(il\varphi) \tag{8.2.8}$$

where l is the order of BBK. The transmittance function given by Eq. (8.2.8) has both radial and azimuthal components. The azimuthal component, known as the vortex phase term, determines the OAM carried by the beam after diffraction. The radial component $\beta(r)$ is the term we currently need to determine.

The fundamental Gaussian mode is diffracted by a BBK to obtain an l-th order perfect optical vortex in the far field. According to the scalar diffraction theory, the far-field diffractions in the Fraunhofer diffraction region can be simply expressed as a Fourier transformation of the initial field. Therefore, the perfect optical vortices given by Eq. (8.2.6) can be expressed as

$$E(r,\varphi) = \mathcal{F}[G(r) \cdot T_{BBK}(r,\varphi)] \tag{8.2.9}$$

where F denotes the Fourier transformation. Substituting Eqs. (8.2.7) and (8.2.8) into Eq. (8.2.9), after eliminating the constant phase term [5]:

$$F(r) = 2\pi \int_0^\infty \rho \exp[i\beta(\rho)] \exp\left(-\frac{\rho^1}{\omega_0^2}\right) J_l(2\pi\rho r) d\rho \tag{8.2.10}$$

Note that the integral given in Eq. (8.2.10) can be regarded as the l-th order Hankel transformation of the radial part of $G(r) \cdot T_{BBK}(r,\varphi)$. The $F(r)$ given in Eq. (8.2.10) determines the amplitude distribution in the far field of the Gaussian beam after diffraction by BBK. For a perfect optical vortex, there should be an amplitude spike at $r = r_0$ and zero elsewhere. so that $F(r)$ appears as a maximum at $r = r_0$, then Eq. (8.2.10) leads to

$$|F(r_0)| = 2\pi \left| \int_0^\infty f_{pos}(\rho) \exp[i\beta(\rho)] sgn[J_l(2\pi r_0 \rho)] d\rho \right| \tag{8.2.11}$$

where

$$f_{pos}(\rho) = \rho \exp\left(-\frac{\rho^2}{\omega^2}\right) |J_l(2\pi r_0 \rho)| \tag{8.2.12}$$

is a function that is constantly greater than zero in the region of integration. sgn(ζ) is a sign function, defined as

$$\text{sgn}(\zeta) = \begin{cases} 1 & \zeta > 0 \\ 0 & \zeta = 0 \\ -1 & \zeta < 0 \end{cases} \tag{8.2.13}$$

From the triangle inequality, it follows that

$$|F(r_0)| \le 2\pi \left| \int_0^\infty f_{pos}(\rho) d\rho \right| \tag{8.2.14}$$

In Eq. (8.2.14), the right-hand is the upper limit (maximum) of $|F(r_0)|$. It is easy to understand that at this point, for Eq. (8.2.11), only if

$$\exp[i\beta(\rho)] = \text{sgn}[J_l(2\pi r_0 \rho)] \tag{8.2.15}$$

Fig. 8.9 BBK phase-only diffraction gratings of different orders l

$|F(r_0)|$ can take the upper limit. So far, substituting Eq. (8.2.15) into Eq. (8.2.8), the transmittance function of BBK can be obtained as

$$T_{\mathrm{BBK}}(r,\varphi) = sgn[J_l(2\pi r_0 r)]\exp(il\varphi) \tag{8.2.16}$$

Several BBK phase-only diffraction gratings obtained according to Eq. (8.2.16) are shown in Fig. 8.9, which is very similar to the phase distribution of the Bessel beam at the beam waist.

8.3 Mode Recognition of Perfect Optical Vortices

Like mode recognitions of other common vortex beams, the mode recognition of perfect optical vortices refers to the measurement of the OAM component they carry. However, since perfect optical vortices exist only at finite distances and the beam size is independent of the topological charge, the mode recognition methods presented in Chap. 5 are not fully applicable.

8.3.1 Interference Method

A perfect optical vortex has spiral phase, and its main difference from the common Laguerre-Gaussian vortex beam is the amplitude distribution. As discussed in Chap. 5, when the Laguerre Gaussian beam coaxially interferes with the fundamental Gaussian mode, the spiral phase of the Laguerre Gaussian beam determines the distribution of the petal shaped field of the interference field, and its amplitude term only defines the outline of the interference region which has no decisive effect on the special intensity distribution of the interference field. Therefore, when a perfect optical vortex in the non-waist position interferes coaxially with the reference fundamental mode, a vortex petal-type interference spot pattern similar to Fig. 5.2.5 in Sect. 5.2.1 is produced, as shown in Fig. 8.10. The absolute value of the topological

Fig. 8.10 Intensity distributions of the interferometric field of the coaxial interference between a perfect optical vortex and a Gaussian beam at the non-waist position

charge l can be determined from the number of petals of the interference spot. And the sign of the topological charge l can be obtained from the direction of the vortex.

When using interferometry to detect a perfect optical vortex, we must ensure that the transverse beam size of the reference base mode Gaussian beam is larger than that of the perfect optical vortices. Otherwise, there will be a "large ring over a small dot" and no interference is present.

8.3.2 Diffraction Method

The diffraction mode recognition method for perfect optical vortices can be understand as diffracting a perfect optical vortex through a diffraction optical element. Then its topological charge can be acquired by observing the particular morphological distribution of the diffraction pattern.

The methods described in Sect. 5.3 for detecting vortex beams using diffraction gratings all require observation of the far-field diffracted light field. For perfect optical vortices, as they travel to infinity, they degenerate into Bessel Gauss beams, i.e. the amplitude distribution of the light field changes, so these methods are not fully applicable. The three main methods currently available are composite fork-shaped grating, standard Dammann vortex grating and integrated Dammann vortex grating. As their mode recognition principles are exactly the same, only the continuous detectable range of topological charges is different, so here only 5×5 standard Daman vortex grating as an example, to introduce the diffraction mode recognition of perfect optical vortices.

The far-field diffraction field of a 0-th order perfect optical vortex through a 5×5 standard Dammam vortex grating is a toroidal beam with no helical phase and topological charge $l = 0$. When considering only the topological charge of the diffraction field, the properties of the 5×5 standard Dammam vortex grating indicate that the topological charges of the 25 main diffraction orders are -12 ~ + 12 from the lower left to the upper right, respectively. When considering only the amplitude distribution of the diffraction field, it is clear from the nature of the perfect optical vortices that each diffraction beam in the far-field turns to a Bessel Gauss beam. In summary, it can be deduced that the 0th order perfect optical vortex diffracted by a Dammann vortex grating forms a 5 × 5 Bessel Gauss beam array, and their topological charges are $-12 \sim +12$ from the lower left to the upper right respectively. The simulation results of the far-field diffraction of a 0-th order perfect optical vortex passing through a 5 × 5 standard Dammann vortex grating are given in Fig. 8.11, which is fully consistent with the previous discussion.

When a l_0-th order perfect optical vortex passing through a 5 × 5 standard Dammann vortex grating, the topological charge of the far-field diffracted Bessel Gauss beam located at the (b_x, b_y) diffraction order can be calculated by Eq. (5.3.24) as

$$l_{b_x,b_y} = l_0 + b_x + 5b_y \tag{8.3.1}$$

When satisfy:

$$l_0 = -b_x - 5b_y \tag{8.3.2}$$

a 0-th order Bessel Gauss beam with a solid central bright spot will present at diffraction order (b_x, b_y). Hence, from the position where the 0-th order Bessel Gauss beam

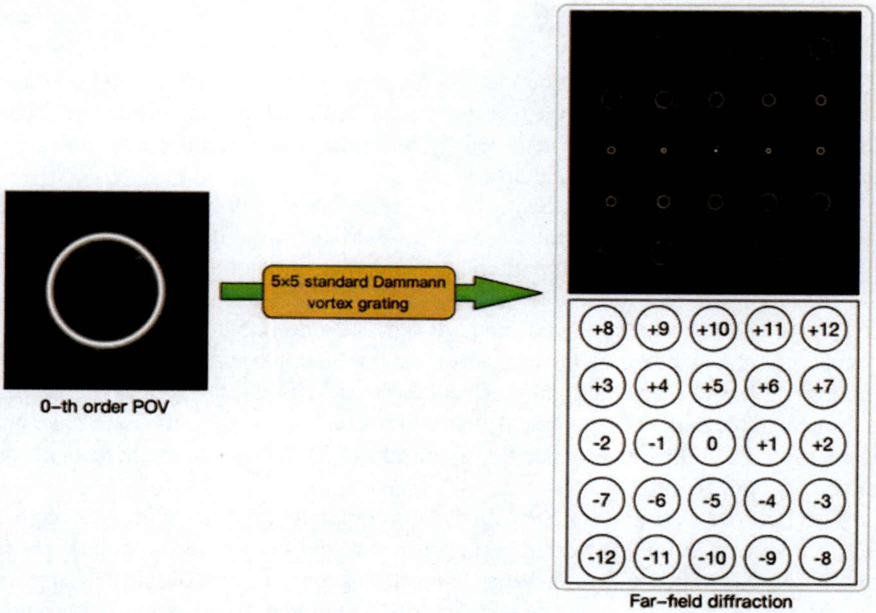

Fig. 8.11 Far-field diffraction patterns and topological charge distribution when a 0-th order perfect optical vortex passing through a 5 × 5 standard Dammann vortex grating

appears, in association with Eq. (8.3.2) [6], the order or topological charge of the incident perfect optical vortices can be deduced.

Based on this approach, a perfect optical vortices mode recognition system shown in Fig. 8.12 is constructed, where the diffraction grating can be a composite forked grating, standard Dammann vortex grating, integrated Dammann vortex grating, etc. The array detector (e.g. CCD camera) must be placed at the back focal plane of the convex lens to observe the far-field diffraction. This method of detecting perfect optical vortices is valid for single-mode or multi-mode mixed perfect optical vortices, providing that the topological charge of the beam to be measured does not exceed the continuous detectable range of the grating. Examples of mode recognition are shown in Figs. 8.13 and 8.14 respectively.

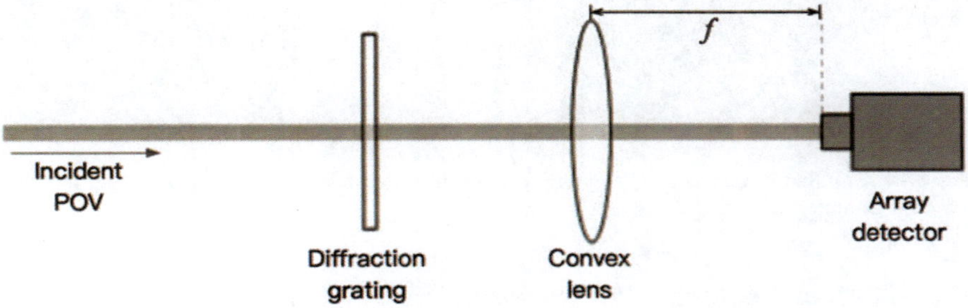

Fig. 8.12 Mode recognition system for a perfect optical vortex

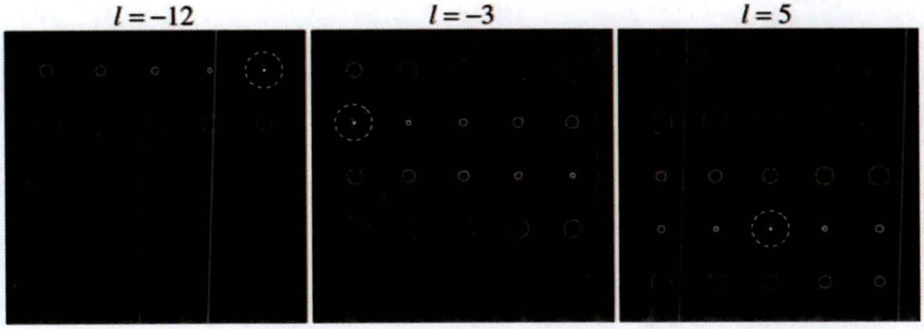

Fig. 8.13 Simulation results for detecting a single-mode perfect optical vortices using a 5 × 5 standard Dammann vortex grating

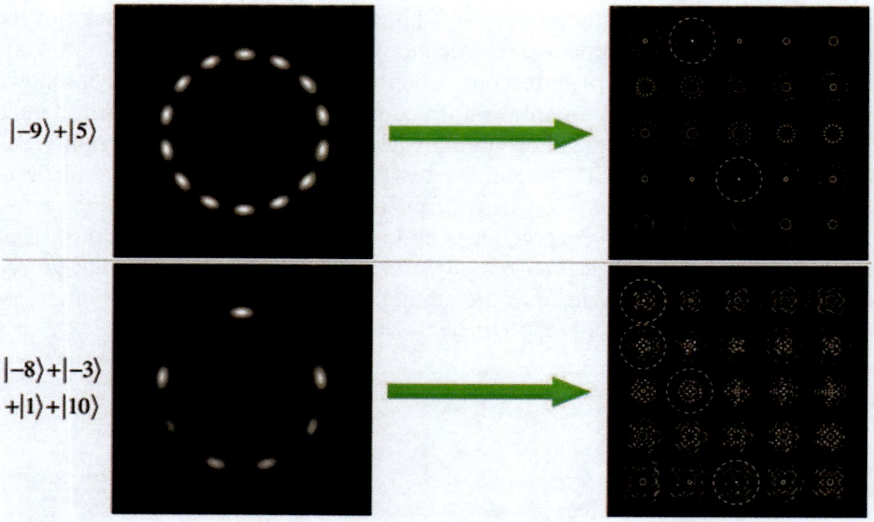

Fig. 8.14 Simulation results for detecting a multiplexed perfect optical vortices using a 5 × 5 standard Dammann vortex grating

8.4 Perfect Vectorial Vortex Beam

8.4.1 *Features of the Perfect Vectorial Vortex Beams*

Perfect vectorial vortex beams are a special type of vectorial vortex beam whose transverse beam size in independent of their polarization order or Pancharatnam topological charge. Their polarization distribution only varies along azimuth but not along radius, as shown in Fig. 8.15.

8.4.2 *Principles of Generating Perfect Vectorial Vortex Beams*

As a particular vectorial vortex beam, the perfect vectorial vortex beam has exactly the same properties as a conventional vectorial vortex beam except for a transverse intensity distribution independent of the polarization order and the Pancharatnam topological charge. A perfect vectorial vortex beam can be expressed as

$$|\psi_{\text{Perfect}}\rangle = A(r)\exp\left(il_p\varphi\right)\begin{bmatrix}\cos(p\varphi+\theta_0)\\ sin(p\varphi+\theta_0)\end{bmatrix} \tag{8.4.1}$$

where $A(r)$ is the term representing the amplitude distribution, p is the order of polarization and l_p is the Pancharatnam topological charge. A perfect vectorial vortex

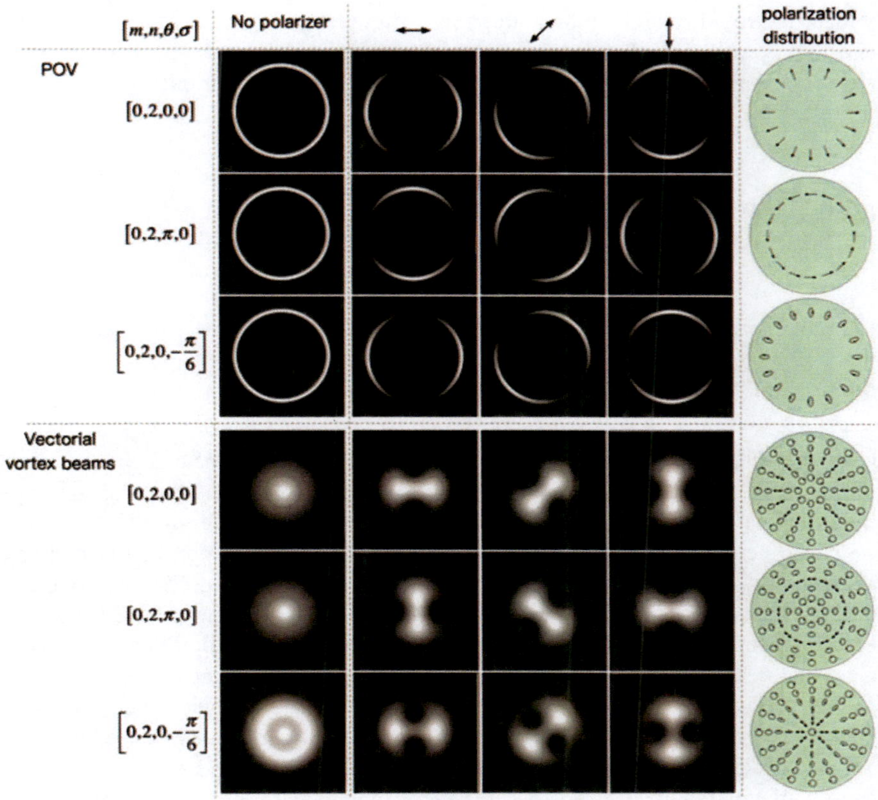

Fig. 8.15 Comparison of perfect vectorial vortex beams with vectorial vortex beams

beam can also be decomposed into a linear combination of two perfect optical vortices of different orders.

$$|\psi_{\text{Perfect}}\rangle = \psi_L^n |L_n\rangle + \psi_R^m |R_n\rangle \tag{8.4.2}$$

where ψ_L^n and ψ_R^m are complex coefficients whose modulus and argument characterize the amplitude and initial phase of the perfect vortex components of the left and right circular polarizations. In this case, together with Eq. (8.1.18), the amplitude term $A(r)$ in Eq. (8.4.1) can be expressed as

$$A(r) \propto \frac{\omega_0}{\omega_a} \exp\left[-\frac{(r - r_0)^2}{\omega_a^2}\right] \tag{8.4.3}$$

The above suggests that a perfect vectorial vortex beam can be generated by coaxially coherent combining of two circularly polarized perfect optical vortices with opposite spin angular momentums and different topological charges. Comparing Eq. (8.4.2) with Eq. (8.4.1) gives the polarization order p and the Pancharatnam topological charge l_p of the resulting perfect vectorial vortex beam:

$$p = \frac{n - m}{2} \tag{8.4.4}$$

$$l_p = \frac{n + m}{2} \tag{8.4.5}$$

8.4.3 Generating Perfect Vectorial Vortex Beams

Current techniques for generating perfect vectorial vortex beams are all based on Eq. (8.4.2), and common systems include cascaded liquid crystal spatial light modulators [7], Sagnac interference [8], and Sagnac-like interference [9].

The system of the cascaded liquid crystal spatial light modulator (SLM) method is shown in Fig. 8.16. The principle is that, SLM1 encoding an m-order anti-axiconatic grating (here an anti-axiconatic grating is used instead of an axiconatic grating because a reflective LC-SLM is used). The incident horizontally linearly polarized Gaussian beam is converted into an -m-order Bessel Gauss beam after modulator 1. The fast axis of the half wave plate (HWP) is 22.5° arranged, which turns the polarization direction of the horizontally linearly polarized Bessel Gauss beam into 45° and decomposed orthogonally into two parts, horizontal and vertical line polarization, of equal intensity. The SLM2 is encoded with an (m-n)-th order vortex phase grating, and when passing through LC-SLM2, the vertical polarization component is not modulated, but the order is transformed to m because od reflection. The horizontal polarization component is modulated and the order becomes $-(-m + m - n) = n$, namely, an n-order perfect optical vortices. The beam after SLM2 is now a combination of two linearly polarized Bessel Gauss beams with horizontal and vertical polarization and topological charges n and m, respectively. The combined beam is transformed into a Bessel Gauss vectorial vortex beam by a quarter wave plate (QWP) whose fast axis is 45° arranged at. Then it passes through a convex lens L3 for Fourier transformation to produce a perfect vectorial vortex beam on the back focal plane.

Here SLM2 must be encoded with a vortex phase grating, not an axicon grating. If SLM2 were also encoded with an axicon grating, it would inevitably result in two orthogonal linearly polarized Bessel Gauss beam components with different radial wave numbers $k_r(k_r = 2\pi/d$, d being the radial period of the axicon), which would ultimately result in two perfect optical vortices components with different spot radius r_0 after the Fourier transformation, forming a double concentric ring structure and preventing coherent beam combining. It must also be ensured that the front focal plane of the convex lens L3 lies within the finite region where the Bessel Gauss

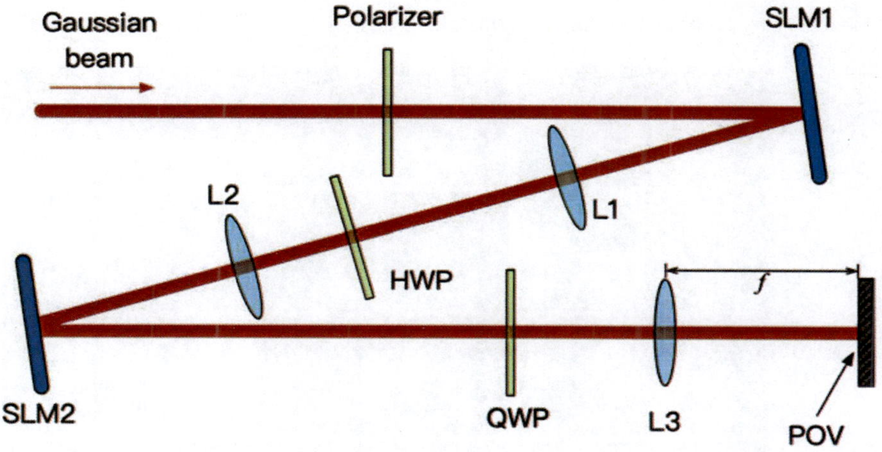

Fig. 8.16 Setup for generating a perfect vectorial vortex beam through cascading liquid crystal spatial light modulator (SLM). L1–L3, lenses. HWP, half wave plate. QWP, quarter wave plate

beam exists. In other words, the distance between modulator 1 and L3 should be less than $(f + z_{\max})$.

Figure 8.17 shows the phase gratings loaded on the two modulators for the generation of a perfect vectorial vortex beam using cascaded liquid crystal SLM, and also the simulation results of the generated perfect vectorial vortex beam. It is worth noting that the system given in Fig. 8.16 can only be used to generate perfect vectorial vortex beams located at the equator of the hybrid order Poincare sphere. To generate modes on the northern and southern hemispheres, the angle between the fast axis of the half-wave plate and the horizontal plane must be adjusted so that the intensity ratios of the two orthogonal polarization components are different.

Perfect optical vortices can also be generated using conventional interferometric systems. Common interferometers include the Sagnac interferometer [8] and the Sagnac-like interferometer [9]. Figure 8.18 shows a system diagram for generating a perfect optical vortex using a Sagnac interferometer. Firstly, the beam from the laser is incident into the polarized beam splitter (PBS) after beam expansion and half wave plate (HWP) 1, and is split into horizontally and vertically linear polarizations with equal intensities. The vertically and horizontally linearly polarization are reflected by reflector 1 (R1) and reflector 2 (R2) respectively and then directed to a phase-only reflective liquid crystal SLM. By reasonably adjusting HWP2 before SLM, some of the horizontal and vertical polarization components can be modulated, but the polarization after modulation being unchanged. The two modulated beams are then reflected back to the PBS where they are combined coaxially, and converted into a vectorial vortex beam after passing through the quarter wave plate (QWP). In addition, the convex lenses L1 and L2 form a 4-f system, and the SLM and CCD are placed the two ends. A pinhole aperture is placed on the spectral plane of the 4-f system to filter stray light.

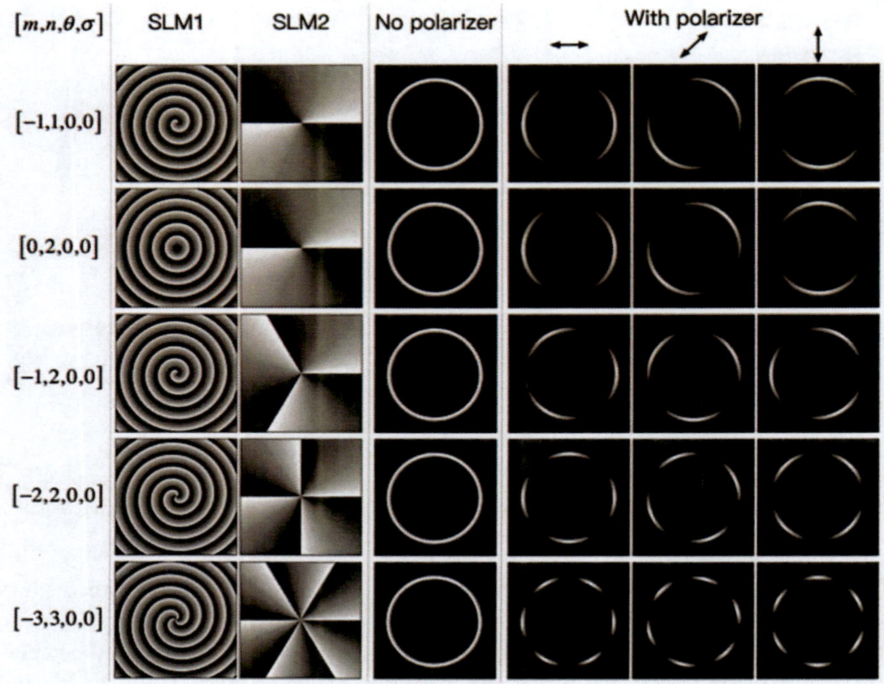

Fig. 8.17 Perfect vectorial vortex beams generated through cascading liquid crystal spatial light modulator

It is crucial to design a diffraction grating properly to generate a vectorial vortex beam using the Sagnac interferometer. As only part of the two beams injected into the SLM are phase modulated, the phase modulated part must be separated from the unmodulated part. The grating should also be capable of generating a Bessel Gauss beam, in order to synthesise a perfect vectorial vortex beam. The ring-slit grating introduced in Sect. 3.7.2 is used here. Designing an annular slit which satisfies:

$$T(r,\varphi) = \begin{cases} 1, |r - R| \le \dfrac{\Delta}{2} \\ 0, |r - R| > \dfrac{\Delta}{2} \end{cases} \tag{8.4.6}$$

A grating that can generate a 2 × 2 vortex beam array is superimposed in the slit, as shown in Fig. 8.18a. The separation of the modulated and unmodulated components is achieved since the four vortex beams of the 2 × 2 vortex beam array are located at the four diffraction orders (−1, −1), (−1, 1), (1, −1) and (1, 1), respectively. While the unmodulated components are all at the (0, 0) diffraction order. In addition, the effect of the ring slit contributes to four Bessel Gauss beams in the far field, namely, the spectral plane of the 4-f system. The mirroring effect of the optical path

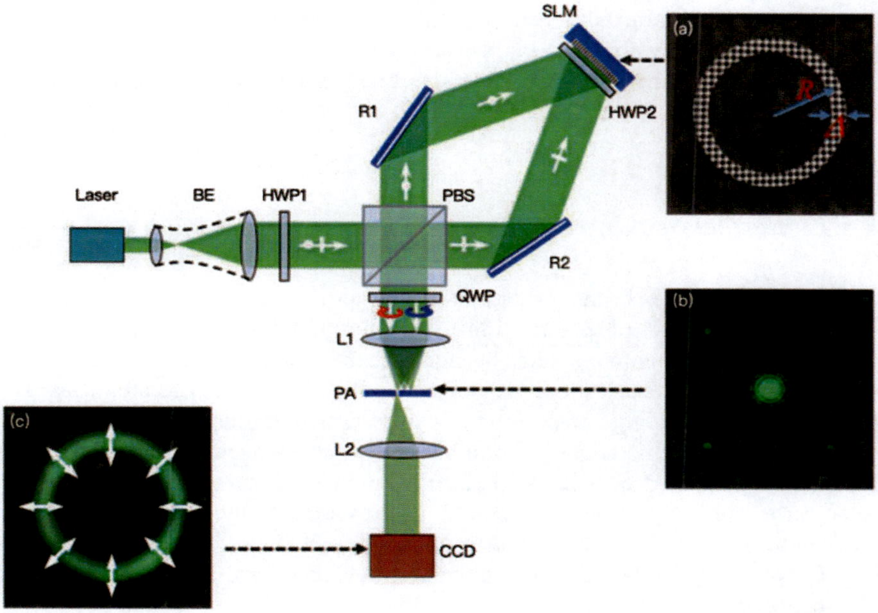

Fig. 8.18 System to generate a perfect vectorial vortex beam through a Sagnac interferometer. BE, beam expander. HWP1-2, half wave plate. R1-2, reflector. QWP, quarter wave plate. L1-2, lenses. PBS, polarized beam splitter. PA, pinhole aperture. CCD, CCD camera. **a** Phase grating encoded on a liquid crystal SLM; **b** diffraction in the spectral plane of the 4-f system; **c** perfect vectorial vortex beam captured by the surface array detector [8]

causes the two beams with two orthogonally polarized components in a position mirror relationship to each other to coaxially merge (Fig. 8.18b). In other words, with the proper design of the diffraction grating, it is possible to achieve left and right circularly polarized Bessel Gauss beams of any orders, and then generate the desired vectorial vortex beams (Fig. 8.18c).

The system of generating a perfect vectorial vortex beam through a Sagnac-like interferometer is identical to the system shown in Fig. 7.3.8, except that a higher-order axicon grating is loaded on the modulator instead of a vortex grating [9]. However, this method can only generate a perfect vectorial vortex beam with a Pancharatnam topological charge of 0.

8.5 Perfect Optical Vortices Array

Perfect optical vortices arrays, where multiple perfect optical vortices of different orders are distributed at different spatial locations, are generally implemented by designing diffraction gratings. Perfect optical vortices arrays can also be regarded as a technique of simultaneous generating multiple perfect optical vortices, which is of great interest in many domains. In this section, two separate perspectives on perfect scalar vortex beam arrays [10, 11] and perfect vectorial vortex beam arrays [12] are discussed.

8.5.1 Perfect Scalar Vortices Array

A perfect scalar vortex beam, the perfect optical vortices mentioned earlier, does not have a spatially anisotropic polarization distribution. The difference between a perfect optical vortex and a common vortex beam is that, the transverse intensity distribution of perfect optical vortex is independent of the topological charge. Therefore, based on the method of generating vortex beam arrays described in Chap. 4, a perfect optical vortices array can be obtained if certain measures can be taken to make each vortex beam in the generated array have the same intensity distribution.

A perfect optical vortex is a far-field manifestation of the Bessel Gauss beam and can be obtained by Fourier transforming a Bessel Gauss beam. Thus, a Bessel Gauss beam array can be generated in the near field and then Fourier transformed to obtain a perfect optical vortices array. Recalling what was introduced in Sect. 3.7.1, an l-order Laguerre Gaussian vortex beam is diffracted by an axicon to produce an l-order Bessel Gauss beam in the region $z < z_{\max}$. This suggests that an axiconatic phase can be superimposed on the diffraction grating that generates the vortex beam array, which in turn gives a Bessel Gauss beam array in the near field and a perfect optical vortices array in the far field [10], as shown in Fig. 8.19.

In Fig. 8.19, (a) shows a diffraction grating that can produce +1 and +2 order vortex beams at diffraction orders (b_x, b_y) of (−1, 0) and (1, 0). The grating (c) is obtained by superimposing the one-axis prism phase (b), which generates perfect optical vortices of orders +1 and +2 at the far-field (−1, 0) and (1, 0) diffraction levels (e). This method of generating an array of perfect optical vortices essentially uses an axicon to generate a Bessel Gauss beam which is then Fourier transformed. The ring size of the resulting perfect optical vortices is only determined by the radial phase period d of the axicon, the wavelength λ and the focal length f of the Fourier transformation lens, which satisfies Eq. (8.2.4). For an optical system, both λ and f are constant, so the transverse dimensions of the individual beams in a perfect optical vortices array can be controlled by adjusting parameter d. Figure 8.20 gives a 1 × 2 perfect optical vortices array generated in the far field under the same λ and f but with various d. Obviously, the spot diameter increases significantly as d decreases.

The distance between two adjacent diffraction orders in a perfect optical vortices array is determined by the grating constants of the diffraction grating, as concluded in Chap. 4, which will not be repeated here.

The previous analysis showed that, superposing an axicon phase on a diffraction grating capable of generating a vortex beam array can producing a perfect optical vortices array. Thus, in combination with the various methods described in

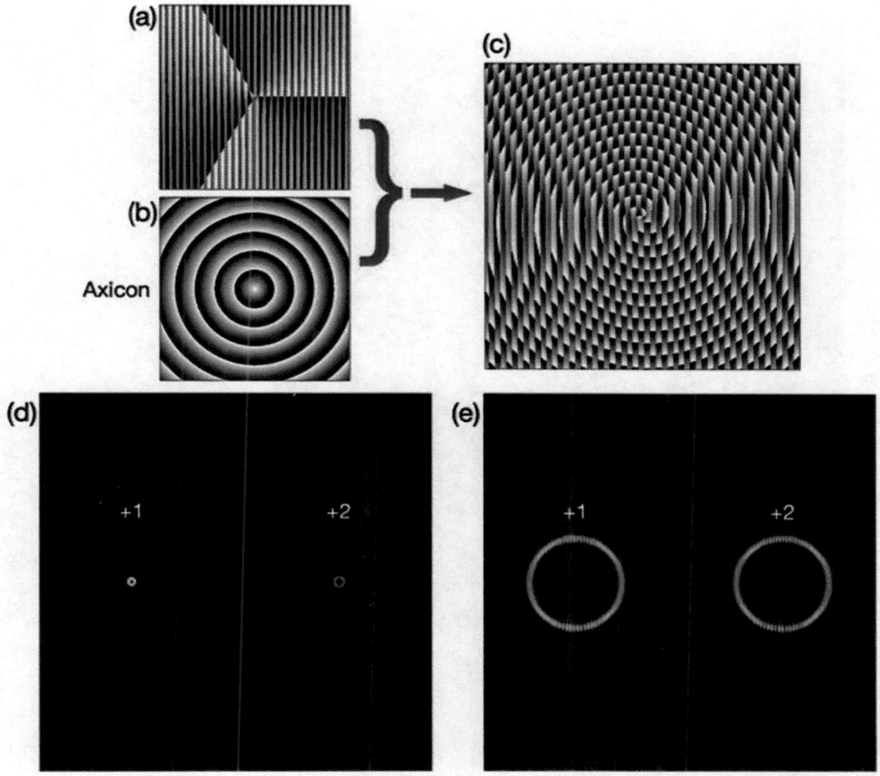

Fig. 8.19 Generating a perfect optical vortices array. **a** diffraction grating that can generate a 1 $\times$ 2 vortex beam array; **b** axicon phase; **c** diffraction grating that can generate a perfect optical vortices array from superimposing grating (**a**) with axicon phase (**b**); **d** far-field diffraction field of a Gaussian beam propagating through grating (**a**); **e** far-field diffraction field of a Gaussian beam propagating through grating (**c**)

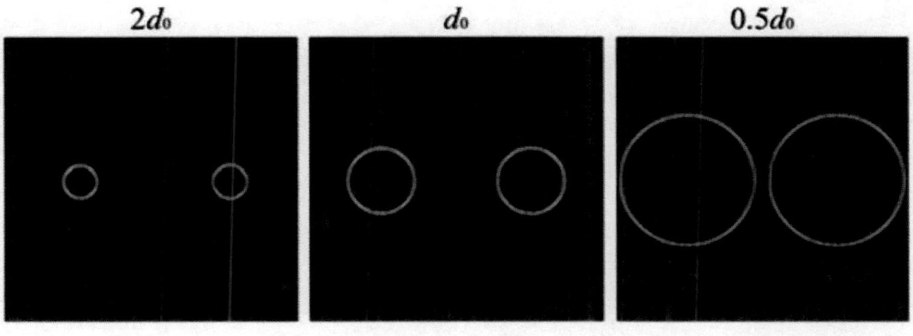

Fig. 8.20 A1 $\times$ 2 perfect optical vortices array generated under the same λ and f but with various d

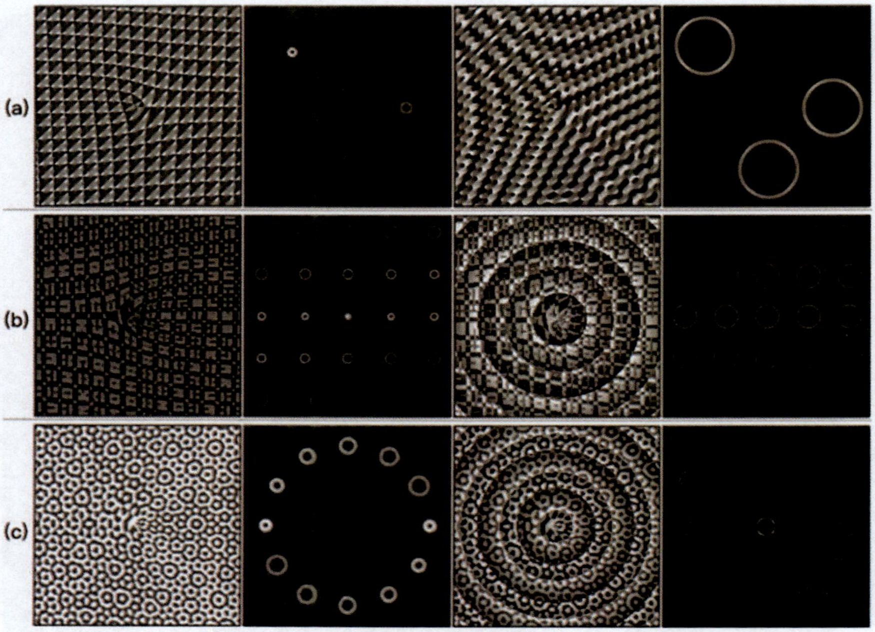

Fig. 8.21 Some examples of the generated perfect optical vortices arrays

Chap. 4, perfect optical vortices arrays with arbitrary position distribution and arbitrary topological charge distribution can be obtained. Some examples are given in Fig. 8.21. In this figure, the diffraction grating shown in the third column is obtained by superposing the grating in the first column with an axicon phase. The second and fourth columns are the far-field diffraction when Gaussian beams passing through the grating shown in the first and third columns, respectively.

8.5.2 Perfect Vectorial Vortices Beam Array

Similar to the vectorial vortex beam arrays described in Sect. 7.7, a perfect vectorial vortex beam array can be created by superposing two circularly polarized perfect scalar vortex beam arrays of the same dimension with opposite circular polarizations, where each diffraction orders of the two scalar arrays are required to correspond to each other.

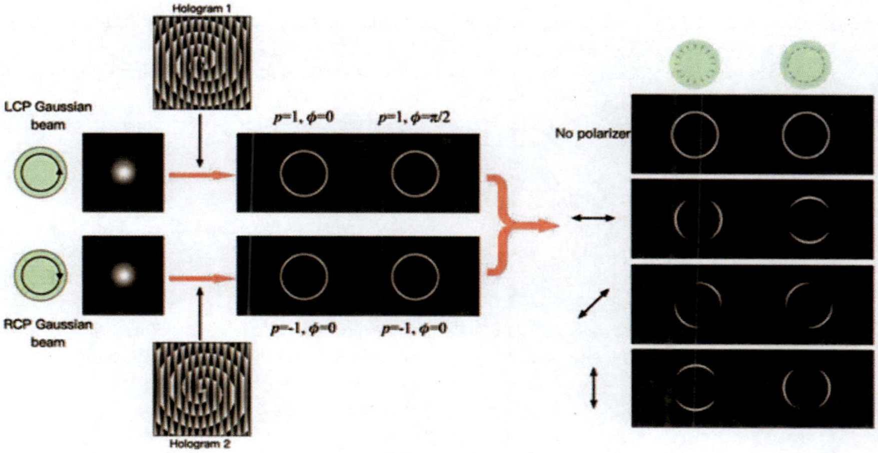

Fig. 8.22 Generating perfect vectorial vortex beam array, where p indicates the topological charge

$$\begin{bmatrix} \left|\psi_{(1,1)}^{\text{perfect}}\right\rangle & \cdots & \left|\psi_{(1,N)}^{\text{perfect}}\right\rangle \\ \vdots & & \vdots \\ \left|\psi_{(M,1)}^{\text{perfect}}\right\rangle & \cdots & \left|\psi_{(M,N)}^{\text{perfect}}\right\rangle \end{bmatrix} = \begin{bmatrix} \left(\psi_L^n|L_n\rangle\right)_{(1,1)} & \cdots & \left(\psi_L^n|L_n\rangle\right)_{(1,N)} \\ \vdots & & \vdots \\ \left(\psi_L^n|L_n\rangle\right)_{(M,1)} & \cdots & \left(\psi_L^n|L_n\rangle\right)_{(M,N)} \end{bmatrix} + \begin{bmatrix} \left(\psi_R^m|L_m\rangle\right)_{(1,1)} & \cdots & \left(\psi_R^m|L_m\rangle\right)_{(1,N)} \\ \vdots & & \vdots \\ \left(\psi_R^m|L_m\rangle\right)_{(M,1)} & \cdots & \left(\psi_R^m|L_m\rangle\right)_{(M,N)} \end{bmatrix} \tag{8.5.1}$$

This process can be illustrated as (Fig. 8.22).

The generation system and mode adjustement of the perfect vectorial vortex beam array is identical with that of the vectorial vortex beam array in Sect. 7.7 and will not be repeated here.

References

1. Ostrovsky AS, Rickenstorff-Parrao C, Arrizón V. Generation of the "perfect" optical vortex using a liquid-crystal spatial light modulator. Opt Lett. 2013;38(4):534–6.
2. Arfken GB, Weber HJ. Mathematical methods for physicists. Harcourt/Academic, 2001.
3. Vaity P, Rusch L. Perfect optical vortices: Fourier transformationation of a Bessel beam. Opt Lett 2015:40(4):597.
4. Gradshteyn IS, Ryzhik IM. Table of integrals, series and products. Elesvier Academic Press, 2001.
5. Arrizón V, Ruiz U, Sánchezdelallave D, et al. Optimum generation of annular vortices using phase diffractive optical elements. Opt Lett. 2015;40(7):1173–6.

6. Fu S, Gao C, Wang T, et al. Detection of topological charges for coaxial multiplexed perfect vortices. Singapore: CLEO-PR/OECC/PGC 2017.
7. Fu S, Wang T, Gao C. Generating perfect polarization vortices through encoding liquid-crystal display devices. Appl Opt. 2016;55(23):6501–5.
8. Li P, Zhang Y, Liu S, et al. Generation of perfect vectorial vortex beams. Opt Lett. 2016;41(10):2205–8.
9. Wang T, Fu S, Gao C, He F. Generation of perfect polarization vortices using combined gratings in single spatial light modulator. Appl Opt. 2017;56(27):7567–71.
10. Fu S, Wang T, Gao C. Perfect optical vortex array with controllable diffraction order and topological charge. J Opt Soc Am A. 2016;33(9):1836–42.
11. Li X, Ma H, Zhang H, et al. Close-packed optical vortex lattices with controllable structures. Opt Express. 2018;26(18):22965–75.
12. Fu S, Gao C, Wang T, et al. Simultaneous generation of multiple perfect polarization vortices with selective spatial states in various diffraction orders. Opt Lett. 2016;41(23):5454–7.